ARCHITECTURAL PRECAST CONCRETE

SECOND EDITION

PRECAST/ PRESTRESSED CONCRETE INSTITUTE
175 WEST JACKSON BOULEVARD
CHICAGO, ILLINOIS 60604

312 786-0300   FAX 312 786-0353

This Design Manual for the Architect represents years of intensive work and study within and outside the Precast/Prestressed Concrete Institute. The task of reflecting and refining these many viewpoints has been accomplished by the following Committee:

PCI ARCHITECTURAL PRECAST CONCRETE MANUAL COMMITTEE

This Design Manual is endorsed by the
ARCHITECTURAL PRECAST ASSOCIATION

TABLE OF CONTENTS

PREFACE

Architectural precast concrete is a child of the 20th Century and modern technology that can trace its lineage back into ancient history. As such, it is a building material almost without precedent. Concrete in its cruder forms was used by the Romans in the construction of their aqueducts. Europe refined the time tested formula in the 19th Century, developing a reinforced concrete that combined the compressive properties of concrete and the tensile strength of steel. The continuing growth and industrialization of the West created a genuine need for new techniques and materials that could be used in prefabricated construction. Architectural precast concrete was developed to fulfill this need.

The first documented modern use of precast concrete was in the Cathedral "Notre Dame Du Haut" in Raincy, France by Auguste Perret in 1923 though used just as screen walls and infill in an otherwise in-situ concrete solution. The depression years followed soon after, and then the cataclysm of World War II. Following the end of world conflict, when labor and material costs began to increase, the use of architectural precast concrete began to flourish.

Today, architectural precast concrete can be provided in almost any color, form or texture making it an eminently practical and esthetically pleasing building material. Architectural precast concrete encompasses all precast concrete units employed as elements of architectural design whether defined to stand alone as an architectural statement or to complement other building materials. It is difficult to imagine an architectural style that cannot be expressed with this material. Precast concrete is not only compatible with all structural systems, it can be designed to harmonize with, and complement, all other materials.

However, throughout the formative years, the architect, the engineer, and the builder, as well as the precaster, lacked any definitive reference volume that defined and illustrated this interesting material. This lack was both world- and language-wide.

The Precast/Prestressed Concrete Institute (PCI), a non-profit corporation founded in 1954 to advance the design, manufacture and use of precast and prestressed concrete, had long recognized the need for a manual to provide guidelines and recommendations pertinent to the design, detailing and specification of architectural precast concrete. In 1973, PCI published the First Edition of *Architec-* *tural Precast Concrete* and for the first time there was available a comprehensive design manual on the subject of architectural precast concrete. Compiled, edited and published by the Precast/Prestressed Concrete Institute, this Manual presented a single authoritative reference for the architectural decision maker.

New developments in materials, manufacturing and erection procedures have expanded the role of architectural precast concrete in the construction industry since the initial Manual was written. In keeping with its policy of being a leader in the concrete industry, PCI is publishing this Second Edition of *Architectural Precast Concrete* in order to make state-of-the-art technology available to the architects and engineers who design and build with this versatile material.

The second edition of the design manual is a major revision incorporating much of this new technology. The sections dealing with color, texture, and finishes; weathering; tolerances; and connections have been extensively revised. Detailed guide specifications have been modified to meet today's construction needs. In addition, the photographs used to illustrate pertinent points throughout the Manual have been selected to represent the potential design opportunities for architectural precast concrete.

Numerous manufacturing and erection techniques are included in the text to provide a better understanding of design concepts and elements requiring design decisions. Design, contract drawings and specifications are all vitally important, and should be combined with an assessment of the capability and experience of the precasters who bid on the project.

The guidelines and recommendations presented show current practices in the industry. These practices should not, however, act in any way as barriers to either architectural creativity or to potential innovations on the part of the precaster.

The practices described in this Manual may be used as a basis by both architect and precaster in the development of exciting new concepts utilizing advances in technology. This initiative will undoubtedly lead to deviations from some of the stated recommendations in the text.

The editors of both editions of the Manual worked closely with the PCI Architectural Precast Concrete Manual Committee and with the professional members of the Institute. Technical accuracy has been reviewed by architects, engineers, precast concrete producers, material and equipment suppliers, and affiliated industry organizations. This unique combination of various disciplines and viewpoints provides an interaction that ensures a knowledge of all aspects in design, engineering, production and erection of architectural precast concrete. Since conditions affecting the use of this material are beyond the control of the Precast/Prestressed Concrete Institute, the suggestions and recommendations presented are provided without guarantee or responsibility on the part of PCI. It is assumed that each project and architect is unique, and requires different solutions for different problems. For this reason, all examples shown must be considered as suggestions rather than definitive solutions.

Architectural precast concrete combines maximum freedom of architectural expression with the economies of mass production of repetitive precast concrete elements. For this concept to function most effectively, it is strongly recommended that the architect seek counsel from local PCI and CPCI* architectural precast concrete producers in the early design stages and throughout further development of the contract documents. Many consulting engineering firms specializing in the development and design of precast concrete are also available to the project architect.

With this precaster/consultant help, proper esthetic, functional, structural, and mechanical features can then be detailed. These will accurately reflect material characteristics, manufacturing and erection efficiencies, cost factors, quality control standards, and local trade practices. Properly implemented, a continuing dialogue between designer and precaster will ensure optimum product quality at a minimum installed construction cost.

This Manual should be used in conjunction with the *PCI Design Handbook—Precast and Prestressed Concrete (MNL 120);* the PCI *Manual for Quality Control for Plants and Production of Architectural Precast Concrete Products (MNL 117);* and PCI *Recommended Practice for Erection of Precast Concrete* (MNL 127).

This Manual is arranged in a sequence which corresponds to the steps that an architectural/engineering firm might employ in evaluating, selecting and incorporating materials into a construction project.

Chapter 1 provides a general background concerning the **State-of-the-Art of Architectural Precast Concrete** and covers the applications and advantages of architectural precast concrete.

Chapter 2 considers **Design Concepts Related to Usage and Economics** for the initial evaluation and selection of architectural precast concrete for a project. The conceptual

architect would primarily utilize this information during wall analysis. Repetition and the master mold concept would be of most interest to the production architect and detailer.

Chapter 3 contains **Surface Esthetics** design considerations and concerns the critical decisions which the designer must make among the many available options as to color, shape, and texture. This chapter covers everything from initial samples and mix design to acceptability of appearance and weathering. Weathering should be reviewed by both designers and detailers.

Chapter 4 presents **Design.** This chapter covers design responsibilities, structural design considerations, contract drawings, connections, tolerances and joints. These factors must be carefully considered by the job captain, draftsman and detailer. These group leaders should also be familiar with the design considerations included in Chapters 2 and 3, if a sound, economical finished product is to result.

Chapter 5 reviews **Other Architectural Design Considerations,** and covers interfacing with other materials including windows, energy conservation, acoustical properties, fire resistance, and roofing. Each requires careful consideration in developing the design criteria and working details for the architectural precast concrete panels or other elements.

Chapter 6 is intended as an aid to **Specification Writers.** The information contained in this chapter should be evaluated in close coordination with the project designer and detailer to avoid creating unnecessary pitfalls in the project by providing the best possible contract documents. Specifications should be neither open to interpretation nor unnecessarily restrictive.

An **Index** is provided at the end of the Manual for easy reference along with a photograph identification section with architectural credits.

A design manual by its very concept can only illustrate what has been accomplished, not what can be. Any attempt to categorize and define architectural precast concrete with its myriad expressions and possibilities is not fully possible. Precast concrete is a fluid material which offers the designer the opportunity to be innovative and obtain desired design objectives that cannot be accomplished with other materials. This Manual will help architects to define their own potential and will provide a basis for reaching it, not by giving design alternatives, but by pointing out all available options in using architectural precast concrete. The significance of precast concrete as a building material lies not in its ability to do new things, but in its inherent quality of being flexible enough to make a design concept become a reality.

_____

*Canadian Prestressed Concrete Institute

STATE-OF-THE-ART

1.1 MANUAL CONTENT AND CONCEPTS

In order for the reader to derive maximum benefit from this Manual, the concepts which guided its preparation are briefly outlined in this section.

Concrete is an old material but its present qualities and usages characterize it as modern and versatile. The infinite variety of forms and shapes possible with architectural precast concrete, as a result of its plasticity, allows the designer to overcome the monotony of many proprietary systems, the confining regularity of masonry units and the design limitations of materials subject to the shortcomings of dies, brakes and rollers. With precast concrete, the designer is in complete control of the ultimate form of the facade and is free of any compromise that might be traceable to a manufacturing process. Also, precast concrete has excellent inherent thermal, acoustical and fire resistive characteristics.

Research and quality control of the ingredients of concrete have led to a better understanding of the unique potentials of precast concrete. Improvements in proportioning, mixing, placing, finishing and curing techniques have established reliable concrete qualities. These improvements have allowed design strength to be substantially increased in the last few decades. Transportation, erection and other important qualities, such as durability and appearance, have kept pace with these developments.

By continuing to provide the designer with complete design freedom, the precast concrete industry has experienced steady growth since the explosive 1960's, particularly in the ever widening range of precast concrete applications. The widespread availability of architectural precast concrete, the nearly universal geographic distribution of the necessary raw materials, and the high construction efficiency of prefabricated components all add to the appeal of architectural precast concrete construction. Established precast concrete engineers and precasters have a high level of craftsmanship and ingenuity along with a thorough knowledge of the material and its potential in converting the designer's vision into a finished structure. Related to this knowledge must be an understanding by the designer of the important design considerations dictated by:

1. Material characteristics.

2. Wall analysis.

3. Weathering effects.

4. Color, finishes and textures.

5. Manufacturing and erection efficiencies including cost factors, quality control standards and local trade practices.

These design considerations are described in considerable detail throughout this Manual. It is important that the designer evaluate the applicable design considerations carefully in choosing the shape, color or texture that will be emphasized on a project. Design considerations based upon manufacturing and erection factors are complex and varied. Because of the limitless expressions of precast concrete, such considerations can rarely be stated in unqualified terms. It is important for the designer to know which are valid and to assess their influence on a specific application. The design considerations for architectural precast concrete are not any more numerous or difficult than the ones associated with other materials. Optimum utilization of precast concrete will result from a thorough understanding of these applicable design considerations. Consider, for example, the interplay between the configuration of a precast concrete unit and its structural capacity, and between its shape and realistic finishes. Add to this the available options in reinforcement, including the possibility of prestressing the unit to offset tensile stresses caused by handling or service conditions, and the importance of understanding these design considerations becomes clear.

The architect with or without practical experience with precast concrete will benefit from a detailed study of the entire Manual in order to obtain an understanding of inter-related design considerations. Subsequent to such a study, the Manual may be used as a reference for future design of architectural precast concrete.

"Industrialized" precast concrete products comprise an important segment of the precast concrete market. These units are normally produced in standard shapes such as double or single tees, channel, solid or hollow slabs with many different finishes often machine applied. Although many of the design considerations discussed in this Man-

Fig. 1.2.1

Fig. 1.2.2

ual apply to these products, additional design considerations may also govern. These design considerations are thoroughly covered in the companion *PCI Design Handbook—Precast and Prestressed Concrete.* Similarly, information concerning regional availability of products and product variation between individual plants is readily available from the producers. Thus, industrialized products are not specifically mentioned in this Manual. This Manual concentrates on the architectural precast concrete applications that are custom designed in shapes and finishes for each individual project.

The Manual contains some recommendations with respect to job requirements and contract conditions relating to precast concrete. These procedural recommendations may help to minimize complications or disappointments during the bidding and construction stages.

The design considerations together with the procedural recommendations should be geared to local practices. The importance of coordinating the development of architectural precast concrete projects with local precast concrete manufacturers cannot be over-emphasized. The ultimate aim of this Manual is to encourage this communication and interrelationship.

1.2 APPLICATIONS OF ARCHITECTURAL PRECAST CONCRETE

Architectural precast concrete has been used for many decades in the United States and Canada. Its full potential in terms of economy, versatility, appearance, structural strength, quality and permanence continues to expand as witnessed by the new projects changing the skylines of many cities throughout the North American continent.

Buildings with precast concrete cladding dating back to the 1920's and 1930's today attest to the fine craftmanship of that period and the permanence of the material. Two of the classic examples are the exact replica of the Greek Parthenon, (Fig. 1.2.1), cast in concrete with a precast concrete skin built in Nashville, Tennessee between 1920 and 1931, and the Baha'i Temple, Wilmette, Illinois (Fig. 1.2.2), designed by Canadian architect Louis Bourgeois. This structure, started in 1920 with final completion in 1953, is one of the most beautiful and delicately detailed ever conceived in the United States. It is built of white concrete panels with exposed quartz aggregate over a steel superstructure. Also, buildings shown in the first edition of this Manual continue in trouble free service today with little or no change in their functional or esthetic appearance.

Today architectural precast concrete demands equal craftmanship in the design and tooling aspects of the manufacturing process. Production has progressed from reliance on individual craftsmanship to a well controlled and coordinated production line method with corresponding economic and physical improvements. These state-of-the-art manufacturing techniques do not sacrifice the plastic qualities of concrete, nor do they limit the freedom of three-dimensional design. In knowledgeable, sympathetic hands, these techniques can be adapted to fit specific performance and esthetic requirements of the contemporary designer.

Treatment of texture and color can be rich, extracting the best qualities of the raw materials. Coarse aggregates are selected for their size, shape, and color. Fine aggregate and cement are selected according to the desired texture and color of the finished element. The mixing of aggregates and cement is similar to the artist's mixing of colors on a palette.

The photographs in this chapter afford a glimpse of the variety of expressions possible with precast concrete; bold yet simple to emphasize strength and durability; intricate as well as delicate to mirror elegance; three-dimensional sculpturing to display individuality. Other photographs or drawings throughout the Manual illustrate the above variety of expression along with specific concepts and details.

The exterior face of the precast concrete panels for the restaurant/boathouse, Fig. 1.2.3, were cast with a 60 ft radius to enhance the cylindrical shape of the building while the interior face is flat to aid in connection to the structural system.

Fig. 1.2.4 shows architectural precast concrete spandrels and column covers extensively articulated to reduce this medical office building's apparent mass. The material allows the surfaces to be strong in detail with numerous striations, notched backs, and rustications. These precast concrete panels are of a warm gray color with a medium sandblast finish at the base and a light sandblast finish at the upper levels.

An on-going program of restoration and historic preservation dictated the design approach for the development in Fig. 1.2.5, which encompasses office space, retail space, pedestrian plaza with fountain and 1077-car parking garage. Architectural precast concrete wall panels were chosen for the complex to produce the level of craftsmanship and detail displayed in the original building, and to match the existing texture and colors of the original building.

The magnificent fountain in the center of the pedestrian plaza utilizes the same architectural precast panels used in the new buildings, Fig. 1.2.6. The elaborate double curve shape of the panels demonstrates the graceful and intricate detail that can be expressed using precast concrete design.

A special design feature of the 19-story hotel, in Fig. 1.2.7, are the 45 deg. chamfers on the architectural precast concrete panels. The chamfers give the building a varying degree of relief throughout the day. Depending on the location of the sun, the building texture changes and gives an interesting constrast with its surroundings.

The gray-green precast concrete sheathed the 9-level vertical shopping mall, Fig. 1.2.8, in combination with polished and thermal finish granite and bronze-painted aluminum grilles set in front of back-painted spandrel glass. The precast concrete exterior relates to the rhythm, proportions and scale of the neighboring neo-classical buildings.

The project in Fig. 1.2.9 consists of an 18-story office and 4-story commercial building. The structural columns are clad with semi-circular precast concrete column covers, typi-

Fig. 1.2.3

Fig. 1.2.4

Fig. 1.2.5

Fig. 1.2.6

Fig. 1.2.7

Fig. 1.2.8

Fig. 1.2.9

Fig. 1.2.10

Fig. 1.2.11

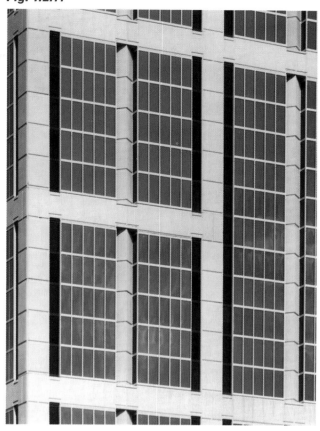

cally 23 ft 4 in. high with a wall thickness of 4 in. Radius of curve in plan of the column covers is 1 ft 8 in.

The 31-story, landmark office building, Fig. 1.2.10, is a tour de force in the use of precast concrete and precast detailing to articulate complex facades. The architects clearly aspired to make a compelling visual statement that could be appreciated at all scales, whether seen from close up or from many blocks away, Fig. 1.2.11. The result is a highly distinctive, artfully composed and crafted commercial high-rise unlike any other in the city.

Designed to reflect the architectural character and scale of a historic retail district, precast concrete provided the means to fulfill the contextual relationships in development of a highly sculptured and articulated facade, Fig. 1.2.12. The quality of finish, uniformity of color, textural aspects and distinctive characteristics of architectural precast concrete were the determinants in its selection for this mixed-use project. The thinness and delicacy reflected in the precast concrete provides a unique architectural expression.

Green marble and gray granite aggregates in the panels complement the green glass and aluminum window wall, Fig. 1.2.13. Narrow vertical rib spacing, horizontal reveals and deep recesses articulate the precast concrete wall panels.

The monumental strength of the international airline terminal is derived from sixteen structural towers clad with beige precast concrete panels, Fig. 1.2.14. The panels are hung on the steel frame in such a manner to allow a reveal be-

tween each panel, thus creating a surface which enhances the expression of the building's structure, Fig. 1.2.15.

Fig. 1.2.16 is a mixed-use development including office, hotel, condominium, retail and marina components. The details here were developed with the intention of establishing a connection between the new buildings and the traditional urban fabric. The acid-etched precast concrete window units and wall panels refer to limestone window frames, lintels, string courses, trim, and ornament on brick buildings, Fig. 1.2.17.

Over 250 pieces of architectural precast concrete panels were used to form the cladding and free-standing arcade for an upscale fashion store, Fig. 1.2.18. The owner wanted a soft stone look to create warmth. Therefore, the white limestone aggregate and white cement panels were sandblasted to achieve the desired look and feel. In addition, the use of precast concrete to form the curved elements and provide the mass at the columns accomplished a very pleasing structure.

A singular feature belonging to this 11-story corporate office building is the adaptability of a typical panel design to non-rectangular and atypical panel conditions, Fig 1.2.19. The 600 architectural precast panels easily accommodate obtuse, acute and curved corners, along with changes in spandrel dimensions. An image of traditional solidity is provided by panel depths of 1 ft 11 in. plus the added accentuation created by continuous 8 in. diameter half-rounds and reveals.

Fig. 1.2.12

Fig. 1.2.13

Fig. 1.2.14

Fig. 1.2.15

Architectural precast concrete was the clear choice for a facade concept centered on developing an expression sympathetic to the classical origins of the 1906 Medical School quadrangle, Fig. 1.2.20. Using a contemporary interpretation, forms and details including pilasters, column capitals, decorative fascia and cornices were created. Color compatibility was a second objective in capturing the unique character of the original facade's weathered marble. As a result of precast concrete's versatility, formulating special colors and textures was easily explored in arriving at a two-tone scheme that mirrored the patina and range of the marble, mimicking its light and dark highlights.

The corporate office highrise, Fig. 1.2.21, features 1400 architectural precast panels with embedded polished granite to achieve a level of detail that was not practical using handset granite. Acid etched precast concrete resembling limestone was used at exposed corners on the street level, and throughout the tower where more intricate detail was desired.

The flexibility of precast concrete allowed the creation of the attractive covers, textures and shapes which help the 16-story, multi-use, hospital addition integrate with the surrounding architecture, Fig. 1.2.22. Five different colors/hues are incorporated into the precast concrete facade: white concrete, beige concrete—both received a medium sandblast to expose the aggregate and, in some areas, both are combined in the same panels; and white, terra rosa, and walnut glazed tiles set in the precast concrete.

Fig. 1.2.16

Fig. 1.2.17

The Police Administration Building in Philadelphia, Figs. 1.2.23 and 1.2.24, made history as one of the first major buildings to utilize the inherent structural characteristics of architectural precast concrete. The 5 ft. wide, 35 ft. high (three-story) exterior panels carry two upper floors and the roof. The building is unusual in its plan configuration, consisting of two circles connected by a curving central section, demonstrating the adaptability of concrete to unusual plan forms. This structure was an early model for the blending of multiple systems into one building. Precast concrete and post-tensioning were the relatively new techniques which were then successfully combined.

The headquarters building for the Department of Housing and Urban Development in Washington, D.C., completed in 1968, Figs. 1.2.25 and 1.2.26, has loadbearing panels which house air-conditioning units below sloping sills and form vertical chases for mechanical services. Gasketed windows are recessed in the deeply sculptured panels for solar control. The panels lend remarkable plasticity to what might otherwise have been a rather "dull" facade.

From a design standpoint, the architectural precast concrete loadbearing spandrels, Fig. 1.2.27, allows the surfaces to be articulated with numerous reveals and staggered joints, thus reducing the visual scale of a five-story parking structure.

The 13 in. loadbearing window wall panels were produced as one unit from grade level to the roof, Fig. 1.2.28. Designed for a limestone appearance, by using brown

Fig. 1.2.18

Fig. 1.2.19

Fig. 1.2.20

Fig. 1.2.21

Fig. 1.2.22

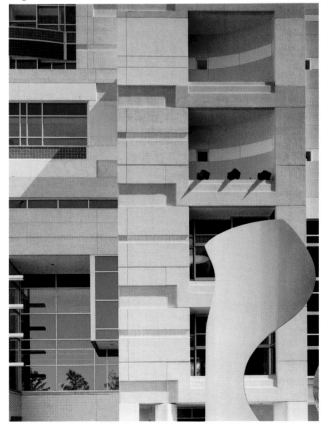

coarse aggregates and gray cement, and by casting rustications into the panels, the panels were given a medium sandblast finish.

The seventy-five loadbearing window wall panels on the three top floors were designed to describe horizontal bands separated by delicate fins, Fig. 1.2.29. To construct a building on such a small site, a typical floor has about 6300 sq ft, precast concrete offered no need to store materials on site. Once the foundation was completed, the erection continued until the shell was complete. The structural shell then offered five floors of storage for the related trades to stockpile their materials.

The five-story speculative office building, in Fig. 1.2.30, has white concrete, loadbearing wall panels, in combination with precast floor slabs, to produce a column-free interior. Panels are detailed to emphasize the window pattern and panel joints, Fig. 1.2.31.

Loadbearing sandwich window wall panels for the 20-story office building, Fig. 1.2.32, are 10 ft wide by 13 ft high. They have a 16 in. interior wythe, 2-1/2 in. of insulation and a 3 in. exterior skin. The corner columns are cladding at the base and then serve as insulated formliners for cast-in-place concrete for the rest of the height.

Three-story columns are notched to carry the L-shaped spandrels, 32 ft 6 in. long, which support double tee floor and roof members, Fig. 1.2.33. There are twin columns at each bay.

Fig. 1.2.23

Fig. 1.2.24

Three high-rise appartment towers have insulated, 10 x 30 ft, loadbearing panels that were precast to give the appearance of stonework, Fig. 1.2.34. Three dimensional ledges, integral columns and keystone joints at the window locations were formed in the panel faces. While the stone patterns are complex, there was a high degree of repeitive casting.

Precast concrete columns and loadbearing spandrels may be an integral part of the building's structural design, Fig. 1.2.35. Overhangs and sunshades protect windows from the sun while enhancing the architectural appeal of the structure.

With the growth of the architectural precast industry, the facilities for plants have become impressive as shown in Fig. 1.2.36. The plants are located near highways, railroads or water, enabling the industry to provide products throughout the United States and Canada.

The buildings shown in this section give an overview of the current state-of-the-art,'' but are in no way complete. Further photographs throughout the Manual will expand and amplify this view.

Fig. 1.2.25

Fig. 1.2.26   *Fig. 1.2.27*

Fig. 1.2.28

Fig. 1.2.29

Fig. 1.2.30

Fig. 1.2.31

Fig. 1.2.32

Fig. 1.2.33

Fig. 1.2.34

Fig. 1.2.35

Fig. 1.2.36

1.3 ADVANTAGES OF ARCHITECTURAL PRECAST CONCRETE

The advantages of using architectural precast concrete can be enhanced when the entire design group and general contractor have the opportunity to jointly develop a project from the initial design stage. Types of finishes, shapes, efficient and economical building modules, structural systems, site delivery, erection procedures and construction schedules, all become important considerations for the successful development of the project. The time spent in development always pays off in accelerated construction and significant cost savings. Therefore, it is recommended that the designer seek counsel from local architectural precast concrete producers in the early design stages and throughout the development of the contract documents. Through a coordinated team effort, maximum economy and optimum utility and quality can be achieved. The functional, structural, and mechanical applications warrant special attention.

For convenience, the principal advantages are summarized below:

DESIGN ADVANTAGES

DESIGN FREEDOM: A diversity in design expression is possible with custom-made solutions that utilize industrialized production techniques and result in unique buildings. Design flexibility is possible in color, texture, size and shape to obtain desired esthetic expressions, such as strength, massiveness, grace or openness. Moreover, precast concrete can be made to harmonize with other building materials.

PLASTICITY: Precast concrete is a quality material with initial plasticity that is incredibly responsive to the designer's creative needs. Repetitive precasting has made the use of complex shapes and configurations economically feasible.

QUALITY CONTROL: Precast concrete, manufactured in a plant, permits rigid factory-controlled conditions, thus ensuring a uniformly high quality facade in the desired shapes, colors and textures. Precasting makes it possible to inspect the surface finish prior to installation.

FUNCTIONAL ADVANTAGES

STRUCTURAL CAPABILITIES: Loadbearing wall panels serve as an important part of the structural framing. They form the supporting structure for floors and roof at the building perimeter. Depending on floor plan, this can lead to interior space free of columns or interior bearing walls providing flexibility of partition layout. Elimination of separate structural frames from exterior walls will result in savings exceeding the cost of increased reinforcement and connection materials required for loadbearing units. This savings is most apparent in buildings with a large ratio of wall to floor area.

Loadbearing wall panels, comprising structural-esthetic-functional features, provide the opportunity to construct an economical, attractive building. Such structures contribute significantly to the development of contemporary architectural philosophy, specifically, a system in which the walls

are actually doing the structural work they seem to be doing.

Precast concrete wall panels may be connected to form structural walls by casting in place spandrel beams and columns using the wall panels as forms. This transforms the usually inactive cladding (curtain wall) into lateral load (seismic and wind) resisting elements. Seismic loads are resisted primarily by the building's central core and partly by the ductile concrete exterior frame. The floor slabs act as diaphragms. Panels may also serve as shear walls or wall-supporting units.

EFFICIENT BUILDING ENVELOPE: The high density and good crack control possible with precast concrete panel construction make these elements air and watertight with fewer joints thus offering excellent protection from climatic conditions.

THERMAL PROPERTIES: Precast concrete panels can be designed to provide a high degree of energy efficiency for the buildings they enclose. Recessed window walls, vertical fins and various other sculptured shapes facilitate the design of many types of shading devices for window areas. Specific wall thermal characteristics can be designed for each face of the structure to suit its sun orientation and natural environment. To obtain a range of U values, precast concrete walls may either have insulation applied to the back, or the insulation may be fully incorporated into a sandwich wall panel. Also the thermal mass inertia of concrete reduces peak heating and cooling loads.

ACOUSTICAL INSULATION: Efficient and economical sound control is provided by the density of precast concrete.

FIRE RESISTANCE: Architectural precast concrete is noncombustible with an inherent fire resistant capability. Precast concrete can also be used as a fire protective covering for steel.

DURABILITY: Precast concrete offers superior weathering and corrosion resistance qualities.

LITTLE OR NO MAINTENANCE: Architectural precast concrete provides trouble-free service. The reduced maintenance characteristics of precast concrete carry over into interior applications. Precast concrete wall panels offer smooth, dense, clean surfaces that minimize the collection of dirt and bacteria.

CONSTRUCTION SITE ADVANTAGES

TIME SAVING: Prefabrication combined with speed of erection saves valuable overall construction time. Panels are manufactured while foundation and site work proceed at the same time, allowing delivery and erection from truck to structure on precise and predetermined construction schedules. On larger projects, this "telescoping" of critical path functions results in significantly reduced construction time, Fig. 1.3.1.

The time required for construction can be reduced by the use of precast concrete as integrated formwork. All the formwork required for a structure can be manufactured in advance of concreting. This permits greater flexibility and

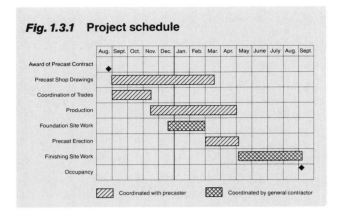

Fig. 1.3.1 Project schedule

	Aug.	Sept.	Oct.	Nov.	Dec.	Jan.	Feb.	Mar.	Apr.	May	June	July	Aug.	Sept.
Award of Precast Contract	◆													
Precast Shop Drawings														
Coordination of Trades														
Production														
Foundation Site Work														
Precast Erection														
Finishing Site Work														
Occupancy													◆	

▨ Coordinated with precaster   ▧ Coordinated by general contractor

continuity in concrete operations. Delays in concreting due to time required for initial curing of concrete, form removal, and re-erection of forms can be eliminated. The precast concrete units may be erected quickly and the structure is complete when the concrete is placed and joints caulked.

ECONOMICAL ERECTION: On-site labor time is minimized, and erection is possible in all kinds of weather.

RAPID ENCLOSURE: Use of precast concrete allows earlier access by finishing trades and minimizes delays caused by bad weather, permitting year-round construction. Faster completion reduces interim financing costs and results in earlier cash flows.

TRADE SCHEDULING: Trade overlap problems decrease to the extent that electrical, mechanical, plumbing and HVAC sub-systems can be integrated into the precast concrete components. This saves costly site work and speeds the installation of auxiliary services.

LOW NOISE LEVEL: The relatively low noise level involved in erection as opposed to full construction activity can be important when construction occurs near areas where excessive noise must be avoided.

ELIMINATION OF FORMWORK: Form removal with its subsequent patching of form anchor holes and surface imperfections, and repair of other flaws, is eliminated.

Architectural precast concrete units used as exterior forms eliminate the need for temporary outside forms. Exterior scaffolding is not generally required, as work is carried out from inside the building, resulting in increased safety on the job. The problem of matching of finishes on cast- in-place concrete with precast concrete is eliminated.

Bottom horizontal faces of spandrel beams can be formed with precast concrete units designed as beams supporting the weight of wet concrete and construction loads above. Scaffolding or shoring normally required to support overhead removable forms, can be partly or completely eliminated, thus saving the time and cost involved in its erection.

ECONOMIC ADVANTAGES

The economic advantages of precast concrete are inherent in most of the above listed advantages. They become even more apparent as design and tooling innovations increase

productivity and pre-assembling of total walls helps to reduce on-site labor. Decreasing site operations will assist in stabilizing the overall cost of the finished building. Financing costs will be reduced by the shorter overall construction time.

The significance of these advantages will increase as architectural precast concrete is utilized beyond purely decorative or cladding (curtain wall) applications.

1.4 DEFINITIONS

Fig. 1.4.1 illustrates some of the terms used in this Manual. Others are defined below.

ADMIXTURE is a material other than water, aggregates and cement used as an ingredient of concrete or grout to impart special characteristics. These are usually employed in very small amounts.

AIR-ENTRAINING ADMIXTURE is a chemical added to the concrete for the purpose of providing minute bubbles of air in the concrete during mixing to improve the durability of concrete exposed to cyclical freezing and thawing in the presence of moisture.

APPROVAL (SHOP DRAWINGS OR SUBMITTALS) is an action with respect to shop drawings, samples and other data which the general contractor is required to submit. Approval when used in this context is only for conformance with the design requirements and compliance with the information given in the contract documents. Such action does not extend to means, methods, techniques, sequences or procedures of construction, or to safety precautions and programs incident thereto, unless specifically required in the contract documents.

ARCHITECTURAL PRECAST CONCRETE refers to any precast concrete unit of special or occasionally standard shape that through application or finish, shape, color or texture contributes to the architectural form and finished effect of the structure; units may be structural and/or decorative, and may be conventionally reinforced or prestressed.

BACKUP MIX is the concrete mix cast into the mold after the face mix has been placed and consolidated.

BOND BREAKER is a substance placed on a material to prevent it from bonding to the concrete, or between a face material such as natural stone and the concrete backup.

CLADDING (NON-LOADBEARING PANEL) is a wall unit that resists only wind or seismic loads and its own weight.

CLEARANCE is the interface space between two items. Normally it is specified to allow for product and erection tolerances and for anticipated movement.

CONNECTIONS are a structural assembly or component that transfers forces from one precast concrete member to another, or from one precast concrete member to another type of structural member.

CONTRACT DOCUMENTS are the design drawings and specifications, as well as general and supplementary conditions and addenda, that define the construction and the terms and conditions for performing the work. These documents are incorporated by reference into the contract.

Fig. 1.4.1 Terminology for precast concrete units

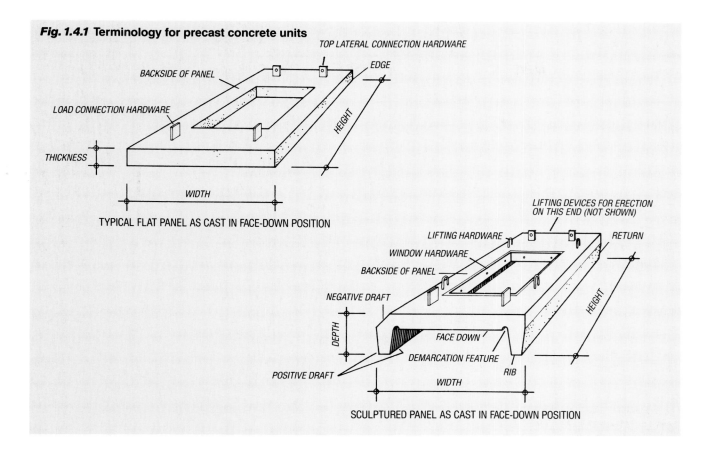

TYPICAL FLAT PANEL AS CAST IN FACE-DOWN POSITION

SCULPTURED PANEL AS CAST IN FACE-DOWN POSITION

CREEP (OR PLASTIC FLOW) is the time-dependent deformation of steel or concrete due to sustained load.

COLD JOINTS are joints necessitated by several casting stages but designed and executed to allow the separate castings to appear and perform as one homogeneous unit.

CURTAIN WALL UNITS see Cladding.

DESIGNER (PRIME CONSULTANT) is the architect, engineer or other professional responsible for the design of the building or structure of which the precast concrete forms a part.

DRAFT is the slope of concrete surface in relation to the direction in which the precast concrete element is withdrawn from the mold; it is provided to facilitate stripping with a minimum of mold breakdown.

EXPOSED AGGREGATE CONCRETE is concrete with the aggregates exposed by surface treatment. Different degrees of exposure are defined as follows:

Light Exposure — where only the surface skin of cement and sand is removed, just sufficiently to expose the edges of the closest coarse aggregate.

Medium Exposure — where a further removal of cement and sand has caused the coarse aggregate to visually appear approximately equal in area to the matrix.

Deep Exposure — where cement and sand have been removed from the surface so that the coarse aggregate becomes the major surface feature.

FACE MIX is the concrete at the exposed face of a concrete unit, used for specific appearance reasons.

FALSE JOINT is scoring on the face of a precast concrete unit, used for esthetic or weathering purposes and normally made to simulate an actual joint.

FORM —see Mold

GAP-GRADED CONCRETE is a mix with one or more normal aggregate sizes eliminated and/or with a heavier concentration of certain aggregate sizes over and above standard gradation limits. It is used to obtain a specific, more uniform exposed aggregate finish.

HARDWARE is a collective term applied to items used in connecting precast concrete units or attaching or accommodating adjacent materials or equipment. Hardware is normally divided into three categories:

Contractor's Hardware — Items to be placed on or in the structure in order to receive the precast concrete units, e.g., anchor bolts, angles or plates with suitable anchors.

Plant Hardware — Items to be embedded in the concrete units themselves, either for connections and precast concrete erector's work, or for other trades, such as mechanical, plumbing, glazing, miscellaneous steel, masonry, or roofing trades.

Erection Hardware — All loose hardware necessary for the installation of the precast concrete units.

HOMOGENEOUS MIX is a uniform concrete mix used throughout a precast concrete element.

LOADBEARING PRECAST CONCRETE UNITS are those precast concrete units which form an integral part of the building structure and which are essential to its stability. They resist and transfer loads applied from other elements. Therefore, a loadbearing member cannot be removed without affecting the strength or stability of the building.

MASTER MOLD is a mold which allows a maximum number of casts per project. Units cast in such molds need not be identical provided the changes in the units can be simply accomplished as pre-engineered mold modifications.

MATRIX is the portion of the concrete mix containing only the cement and fine aggregates (sand).

MOLD is the container or surface against which fresh concrete is cast to give it a desired shape; sometimes used interchangeably with form. (The term is used in this Manual for custom made forms for specific projects while the term form is associated with standard forms or forms of standard cross section.)

OPTIMUM QUALITY is the level of quality, in terms of appearance, strength and durability, which is appropriate for the specific product, its particular application and its expected performance requirements. Realistic cost estimates for producing it within stated tolerances are factors which must be considered in determining this level.

NON-LOADBEARING PRECAST CONCRETE UNITS see Cladding.

PRESTRESSED CONCRETE is concrete in which permanent internal stresses have been induced by forces caused by tensioned steel. This may be accomplished by:
   Pretensioning — the method of prestressing in which the tendons (prestressing steel) are tensioned before the concrete has been deposited.
   Post-Tensioning — the method of prestressing in which the tendons (prestressing steel) are tensioned after the concrete has hardened.

RETURN is a projection of like cross section which is 90 deg. to or splayed from the main face or plane of view.

REVEAL is a groove (rustication) in a panel face generally used to create a desired architectural effect. Also, the projection of the coarse aggregate in an exposed aggregate finish from the matrix after exposure.

RUSTICATION see Reveal

SAMPLES

   DESIGN REFERENCE SAMPLES are usually small, 12 in. x 12 in., specimens made by the precast concrete plant laboratory to provide the designer with early conceptual ideas of color and texture.

   BID REFERENCE SAMPLES are usually small, 12 in. x 12 in., specimens made by a producer to show the designer what can be made locally and is used as a basis for the producer's bid.

   SELECTION SAMPLES are usually 3 ft x 3 ft in size and are made by the successful producer before casting any units; they become the basis for accepting the appearance of finishes. (Full-size production run elements are then approved and become the final standard for acceptance.)

SANDWICH WALL PANEL is a wall panel consisting of two layers (wythes) of concrete fully or partly separated by insulation.

SEALANTS are a group of flexible materials used to seal joints between precast concrete units and between such units and adjacent materials.

SEALERS OR PROTECTIVE COATINGS are clear chemical compounds applied to the surface of precast concrete units for the purpose of improving weathering qualities or reducing water absorption.

SET-UP is the process of preparing molds for casting including locating materials (reinforcement and hardware) prior to the actual placing of concrete.

SHOP DRAWINGS are graphic diagrams of precast concrete units and their connecting hardware, developed from information in the contract documents. They show information needed for both field assembly (erection) and manufacture (production) of the precast concrete. They are normally divided into:

   Erection Drawings — all drawings used to define the location, connections, joint treatment, and interfacing with other materials for all precast concrete units within a given project. Special handling instructions and information for other trades and the general contractor are also shown.

   Anchor Setting or Contractor's Setting Drawings — giving location of all anchoring hardware cast into or fastened to the building or structure.

   Production Drawings — the actual detail drawings necessary for production of the precast concrete units. Such drawings may be shape drawings, set-up drawings, reinforcement and hardware drawings and should include details of all materials used in the finished precast concrete units.

SHRINKAGE is the volume change in precast concrete units caused by drying normally occurring during the hardening process of concrete.

SYSTEMS BUILDING is essentially the orderly combination of "parts" into an "entity" such as sub-systems or the entire building. Systems building makes full use of industrialized production, transportation and assembly.

THERMAL MOVEMENTS are the volume changes in precast concrete units caused by temperature variations.

TOLERANCES are (a) the permitted variation from a basic dimension or quantity, as in the length or width of a member; (b) the range of variation permitted in maintaining a basic dimension, as in an alignment tolerance; and (c) a permitted variation from location or alignment.

TOOLING refers to most of the manufacturing and service processes preceding the actual set-up and casting operations.

WEATHERPROOFING is the process of protecting all joints and openings from the penetration of moisture and wind.

WEATHER SEALING is the process of treating wall areas for improved weathering properties; see Sealers.

DESIGN CONCEPTS RELATED TO USAGE AND ECONOMICS

2.1 GENERAL COST FACTORS

The buildings reviewed in Section 1.2 illustrate the wide range of projects utilizing architectural precast concrete. The variations in scope, complexity, and detailing make it difficult to provide accurate cost information for such work in terms of price per square foot of wall area prior to completion of the design concept stage. During the conceptual stage the designer must consider the cost implications of material selection, textures, surface geometry, cross section, unit repetition, and erection methods.

After a design has advanced to the sketch stage, with determination of size, shape and finish, more accurate cost estimates can be provided if the total wall areas for the different units are known. Until this stage is reached, architects must be guided by their own expertise, coupled with the advice of precasters or consultants. Selected guidelines regarding cost of architectural precast concrete are included in this chapter for further assistance.

The architect who desires a more detailed understanding of the cost factors involved in precast concrete construction is advised to study both **Surface Esthetics** in Chapter 3 and **Design** in Chapter 4. Many of the recommendations in these chapters are, in the final analysis, based upon economy.

The nature of precast concrete is such that nearly anything which can be conceived, structurally designed, and readily transported, may be constructed. But to do this within reasonable and stated cost limits requires careful consideration of design and detailing. Fortunately, several cost factors influencing architectural precast concrete are independent of each other. For example, a sculptured or intricate design may be achieved within a limited budget by selecting economical aggregates and textures combined with repetitive units and effective production and erection details.

Fig. 2.1.1 shows a small office building completed well within a low budget. Only 32 panels were involved, each having 4 windows. With this small utilization of molds, the architect managed to obtain a sculptured panel of intricate and well defined form by using a standard local aggregate and exposing it only in the flat areas. In addition, the archi-

tect chose a simple window detail consisting of a neoprene glazing gasket secured to a reveal in the concrete by an aluminum angle.

Fig. 2.1.2 shows a small college library clad in exposed aggregate architectural precast concrete. The smooth horizontal banding and window trim detailing contrasts with the rough texture of the aggregate facing. The effect achieved is similar to the molded stone banding contrasting with the rough cut stone on many of the surrounding older campus buildings. Use of repetition and "opposite-hand" matching panels made the precast concrete panels more cost effective than other acceptable stone finishes.

Repetition is also the key to achieving quality and economy

Fig. 2.1.1

Fig. 2.1.2

Fig. 2.1.3

in the design for walls. During the design stage, the design of the exterior walls of a typical office building can be analyzed at three basic locations:

1. Below typical floors, normally the ground floor and mezzanine.

2. At the top floor, mechanical floors, penthouse, etc.

3. At all typical floors.

A cost breakdown of these three parts will usually illustrate that the lowest square foot cost is for the walls on the typical floors because of the volume and repetition. If the unit cost of either Part 1 or 2 appears expensive in relation to Part 3, a review may be useful, unless the areas involved are insignificant in relation to the overall area of the building. The elements which may be considered in such a review of Parts 1 and 2 follows:

1. Walls for lower floor(s)

    The optimum solution for a ground floor would be a design that allows the precast concrete units to be cast in the same master mold as contemplated for the typical floors. This holds true even with the larger openings, doors or entrance areas normally required for such a floor. The illustration in Fig. 2.1.3 is a good example of this.

    If the ground floor design does not lend itself to the master mold concept in relation to typical floors, it is often the simplest solution to use precast concrete

solely as cladding consisting of flat sections with returns only as necessary. One flat mold can then be the master mold and modifications obtained by relatively simple adjustment of bulkheads. Flat panels may also be economical because the layout of a ground floor often precludes much repetition.

A cladding solution with flat panels and minimum returns may lead to numerous and fairly small panels in contradiction to later recommendations made in this Manual. Smaller flat units in large numbers may still be more economical than larger units with little repetition and high tooling costs. Erection of the larger number of units may be partially offset by the use of small cranes or forklifts which are often adequate for this low elevation.

A cladding solution for a ground floor design often allows the choice of more expensive finishes for this area only. Consideration should be given at the ground level to the proximity of the units to the eyes and hands of the public. In some cases accumulation of snow or dirt against ground floor panels may influence choice of finish, since ease of cleaning becomes an important factor. The finish requirements for the typical floors may be less demanding, suggesting a different finish, provided the combination is esthetically pleasing or has a logical separation.

2. Walls for top floor(s)

The architect should balance the shape and size requirements for the top floor panels with a reasonable utilization of molds. Top floor panels are sometimes difficult to erect, and therefore expensive. The top floor is not the place to use excessively large units unless the design and the budget warrant the additional crane cost.

Size limitations for top floor panels are cost considerations for projects lending themselves to erection by mono-rails which cannot be used for top floors, or climbing cranes because of the limited capacity of such equipment. The architect has added freedom of design within a given budget where erection of top floor panels is feasible with mobile cranes. Because top floors often have additional height, the architect should think in terms of units which are narrower in width than those of lower floors. Fig. 2.1.4 illustrates this point and the different emphasis of the joints. Top floor panels can also benefit from flat cladding designs.

The project in Fig. 2.1.5 shows an application of some of the major points made in this chapter. The center bay of the building (see Fig. 2.1.6) contains stairs, special service areas and storage rooms with heavy floor loads. It was designed to transmit all horizontal loads to the foundations. Construction of this cast-in-place concrete bay provided reasonable lead-time for the precaster, enabling him to work with a minimum number of molds. The actual erection of the precast concrete walls and floors for the two side bays required less than three weeks and afforded complete enclosure of the building.

Fig. 2.1.4

A center mullion served as the loadbearing part for the two-window panels on typical floors. The balance of the panel was cantilevered from this mullion with the vertical edges providing cover for air-conditioning ducts. The wall panel for the typical floors was modified for the first floor to accommodate larger windows and/or doors by bulkheading the master mold as indicated in Fig. 2.1.7.

The master mold on this project was the four-window panel at the end of the bays, Fig. 2.1.8. This panel had to carry its own weight between the center bay and the precast concrete corner support incorporated within the panels. It was reinforced accordingly, and cast in one piece. The four-window master mold was later modified with a center divider to provide a mold for casting two typical window panels simultaneously. This was followed by the bulkhead modifications for the ground floor panels as previously described.

The architect may be highly creative in some aspects of the design that might normally increase costs, yet accomplish the effect desired within a given budget by paying close attention to other remaining cost factors. Many advances in precast concrete have been achieved in this manner.

Sound design, careful detailing, and clear specifications are also important factors in obtaining optimum quality and economy using architectural precast concrete. Realistic bidding is aided by avoiding ambiguous terms and uncertain details in the preparation of the contract documents.

Fig. 2.1.5

Fig. 2.1.6 Structural layout of building with rigid core and loadbearing panels

Fig. 2.1.7 Typical application of master mold concept

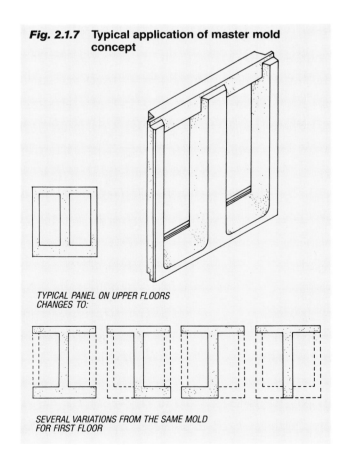

TYPICAL PANEL ON UPPER FLOORS
CHANGES TO:

SEVERAL VARIATIONS FROM THE SAME MOLD
FOR FIRST FLOOR

Fig. 2.1.8

2.2 REPETITION AND THE MASTER MOLD CONCEPT

Maximum economy combined with optimum quality are fostered by designing and detailing precast concrete with an understanding of the importance of repetition for such work. To accomplish repetition without sacrificing design freedom, the architect should have an appreciation of the effects of both tooling and mold requirements on economy and quality in precast concrete.

2.2.1 PLANNING AND TOOLING CONCEPTS

The success of a good precast concrete project can usually be traced to the fact that someone took the time and effort to do a thorough job of planning the shapes and forms for that particular project. This is essentially the task of the precaster. However, this work is made easier if the design and detailing of the precast concrete units reflect good tooling concepts. The Architect must have a thorough understanding of the requirements of tooling to produce a truly economical design.

Tooling, as used throughout this Manual, is the combination of:

1. Planning and fabricating of all molds and devices to produce concrete in desired shapes.

2. Provision for proper fabrication and placing of reinforcement and hardware.

3. Provision for safe stripping and handling of the units.

4. Provision for uniform and efficient finishing of the units.

The choice of material for the molds is usually determined by the number of uses and is essentially a matter of economics. If the molds are properly designed and used, the quality of product obtained from any mold material is equally good; the choice of mold material should therefore be left to the precaster.

2.2.2 REPETITION

Repetition is essential to maximize economy for precast concrete components. Careful planning is necessary to achieve good repetition in the design without sacrificing design freedom. For example, many design variations may be developed by incorporating two basic architectural panel types (spandrel panels and floor-to-floor panels with openings) on the same structure. These panel types may also be varied with different architectural finishes and textures. Attention should be focused on the overall geometry of the structure, not on the shape of the panel, although the cost of complex shapes becomes economical through repetitive precasting. The hotel, Fig. 2.2.1, had to meet the owner's requirements of economy and rapid construction, while remaining compatible with the historic masonry buildings in the immediate vicinity. Cladding the 36-story luxury hotel with highly repetitive precast concrete panels was the solution to both challenges. The architects used rustication joints to permit banding of the panels. The fiberglass forms were coated with a chemical retardant to form bands on the panels. The rest of the forms were coated with a normal form release agent. After the panels cured, light sandblasting exposed the pink granite aggregate on the retarded bands and created a smoother, paler finish where the form surface was coated with a release agent. The final result was a single precast panel with two surface textures and two colors, permitting the architects to break up the mass of the tower with 14.6 ft bay modules that relate to the scale of the 19th Century brick rowhouses nearby.

Mold costs are usually an expensive item. Fig. 2.2.2 shows the panel reinforcement and back forming for the production of six Tuscan arches for a project, see Fig. 3.5.39. The pieces are so complex it took five weeks and 600 man-hours for the carpenters to build the wooden mold. Thus, the cost of tooling should be spread over as many units as possible to obtain the optimum economy. The production of each element becomes more economical, the more elements that can be cast with a given mold. It is simple to demonstrate the difference in cost between 25 mold uses, 100 uses, or even 250 uses which is quite feasible with good molds. If the panel measures 20 ft 0 in. by 12 ft 6 in., this tooling cost represents 15, 5 and 2 percent per square foot of wall, respectively, provided the casting schedule allows the use of one mold. Approached in this manner, it is easy to see why special elements should be designed so they can be made by modifying the mold of a typical panel.

Present bidding practices often result in late award of subcontracts. In practice, this means that additional molds

Fig. 2.2.1

Fig. 2.2.2

must be built in order to meet the project deadline. The necessity for extra molds increases costs and partially offsets the intent of designing for high repetition.

It is often the case that, in the initial design stage, a high degree of repetition appears possible. However, as the details are finalized, considerable discipline is required on the part of the designer if the creation of a large number of non-repetitive units is to be avoided. Any budget costs given at the initial design stage should take into account the possibility that the number of different units will increase as the design progresses. If non-repetitive units are unavoidable, costs can be minimized if the units can be cast from a "master mold" with simple modifications without the need for completely special molds. However, even relatively minor variations, e.g., a dimensional change of a rail, blockout location, connection hardware position or blockouts of any kind, are mold changes that cause increased costs.

2.2.3 MASTER MOLD

The architect can make a significant contribution to economic production by designing precast concrete panels with a knowledge of the "master mold" concept, mold types and by providing the precaster with sufficient lead time to make duplication of tooling unnecessary. The master mold concept is based on the fabrication of one master mold (with its appropriate additional tooling) which allows a maximum number of reuses per project. Units cast in this mold need not be identical provided the changes in the units can be accomplished through pre-engineered mold modifications. These modifications should be achieved with a minimum of idle production time and without jeopardizing the usefulness or quality of the original mold. Typical applications are shown in Fig. 2.1.7. Different panels from this type of mold are shown being erected, Fig. 2.1.8.

It is relatively easy to alter a mold if the variations can be contained within the total mold envelope by use of bulkheads or blockouts, rather than by cutting into the mold surface. When a large number of precast concrete units can be produced in each mold, then the cost per square foot will be both reasonable and affordable. The master mold concept is illustrated in Figs. 2.2.3 and 2.2.4. Here, all or a large number of panels can be produced from a single mold, built to accommodate the largest piece, and variously subdivided to produce the other sizes required. Whenever possible, the largest pieces should be produced first to avoid casting on areas worn and damaged by placing and fastening the side form bulkheads. Although every project will have atypical conditions (which are unavoidable), the most successful and cost-effective projects will maximize the repetition of elements.

An example of pre-engineered mold changes is shown in the models for a loadbearing panel in Fig. 2.2.5. The forms were made in two pieces that bolt together. The head section was removable and could be replaced with a modified upper fascia to achieve design variations. The process is illustrated starting with the two basic panels joined at the jobsite with the addition of a corner unit (Fig. 2.2.5a). By redesigning the upper fascia only, the panel on the left takes on a new look, while keeping costs substantially lower than if a complete new mold were needed. All of the engineering and load factors of the loadbearing section remain unchanged (Fig. 2.2.5b).

Even though the panel on the left is essentially the same as the one on the right, the slight change in height and the addition of fins to the right panel creates a different appearance (Fig. 2.2.5c).

The wide flat center section of the panel can be blocked out for full windows, window frames or doorways. Also, different types of concrete finishes and textures are possible in the center area. A demarcation groove makes casting of the different finishes a simple operation. By having the

Fig. 2.2.4 Master mold concept

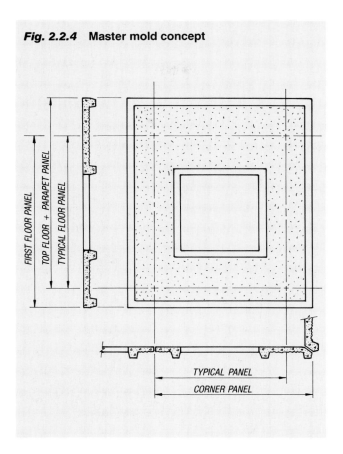

TYPICAL PANEL
CORNER PANEL

Fig. 2.2.5

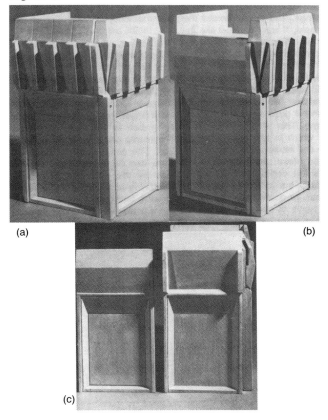

(a)
(b)
(c)

Fig. 2.2.3 Mold concepts

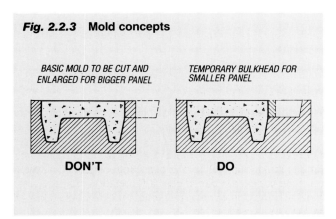

BASIC MOLD TO BE CUT AND ENLARGED FOR BIGGER PANEL

TEMPORARY BULKHEAD FOR SMALLER PANEL

DON'T
DO

Fig. 2.2.6

Fig. 2.2.7

flexibility of varying expressed material usage, it is possible to relate the basic tone of the design to the existing surrounding structures.

The number of forms required for a given job is determined by the time allowed to complete the job and the facilities available. Casting can usually proceed during most of the erection process, if the panels have been manufactured in the correct sequence to accommodate this operation. The number of forms is also affected by the original planning of the master mold. If most conditions can be covered by modifying the master mold with bulkheads, or blockouts, fewer forms will be required. Such modifications may be made to clear columns in specific locations, or for precast concrete units which are identical except for window openings, Figs. 2.2.6. A scale model for a pattern is shown in Fig. 2.2.7. The form was designed for three different conditions. These included the typical precast concrete panel, an extra long unit at the top three floors, and special conditions at the third floor. All were produced on the same mold by changing bulkheads.

Simple modifications are also accomplished by the use of bulkheads for units which comprise only a part of the master mold.

Fig. 2.2.8 shows an office building where all panels were cast from one master mold. The mold contained the corner projections (left and right), the spandrel on the lower floor, and the parapet on the top floor. The master mold was never used to its potential volume, Fig. 2.2.9; nevertheless

Fig. 2.2.8

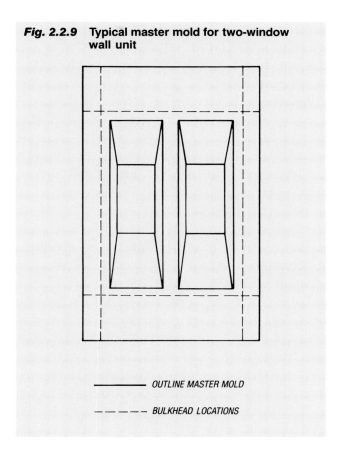

Fig. 2.2.9 Typical master mold for two-window wall unit

——————— *OUTLINE MASTER MOLD*

— — — — — *BULKHEAD LOCATIONS*

the one-type mold resulted in an extremely economical wall.

The details for casting individual panels should always be left to the precaster. Elevations, wall sections, and details of each different type of wall panel should be drawn by the architect. When using large elements, if the appearance of smaller panels is desired for esthetic reasons, false joints (rustications) can be used to achieve this effect. Compromise may be required between the finish and the shape of a precast concrete panel. Wherever possible, the designer should avoid fragile edge details which increase handling costs. Chamfered edges reduce edge damage and mask irregularities in alignment. A gradual transition from one mass of concrete to another within a precast concrete unit is also important. A large radius or fillet is preferred for the transition from a mullion, sill or eyebrow to a flat portion of the unit. Too sudden a change in mass increases the danger of drying shrinkage cracks at the point where two concrete sections of distinctly different mass meet. A large radius reduces cracking and enhances the finish of the precast concrete unit, Fig. 3.3.11. In addition, sufficient thickness of concrete is needed to develop insert capacity for handling and for connections. The architect must visualize simultaneously both the mold and the method of stripping. If this aspect is ignored, it may be clearly reflected later in either the cost of the product or its quality.

The optimum economy in production is attained if the panel can be separated from the mold without disassembling the mold. This is accomplished by providing draft on

the sides of all openings and edges. Drafts are a function both of shape and production techniques. The designer is urged to consult the local precasters for specific recommendations. Generally, the minimum draft for ease of stripping is 1 in. in 12 in. (1:12). This draft should be increased for narrow sections or delicate units where the suction between unit and mold becomes a major factor, Fig. 3.3.8. The draft should be increased to 1:6 for screen units pierced with many openings, for ribbed panels, or for smooth concrete. Vertical sides or reverse (negative) draft will create entrapped air voids which, if exposed, may be objectionable. If negative draft is required, mold and production cost will increase, since extra effort is required to minimize surface blemishes. It may be necessary to make mold modifications such as incorporating slip (removable) blocks in the rail to aid in stripping the precast concrete panel. In general, the greater the draft, the more economical and uniform the finish.

2.2.4 ENVELOPE MOLD

The complete envelope mold is a box mold where all sides remain in place during the entire casting and stripping cycle. Such molds have good economy and quality potentials because they are usually simple to build and maintain with enough strength to withstand concrete pressure during and after consolidation.

An envelope mold is shown in Fig. 2.2.10 and the corresponding panel in Fig. 2.2.11. Fig. 2.2.12 shows a complete envelope mold, while Fig. 2.2.13 illustrates the more conventional mold with removable side and end bulkheads.

The initial cost of envelope molds is high, but the daily costs for set-up in preparation for casting are low compared to those for molds with loose bulkheads. The daily preparation for casting is very labor-intensive when loose bulkheads are involved, since the bulkheads must be measured, aligned, and fastened for each casting operation to ensure proper placement. The reduced labor cost with envelope molds becomes an important factor when the repetition is high.

Envelope molds will produce superior products because all exposed corners are permanently formed, eliminating any danger of concrete leakage. In addition, the permanent side forms produce very consistent, accurate panel dimensions. The quality of joints produced from such molds is shown in Fig. 2.2.14.

This type of mold does have some drawbacks. The configuration of the form requires that panels be stripped flat from the mold and rotated to a vertical position later. Also, side forms cannot be penetrated by lifting loops. The slightly wider joint between panels caused by the draft required on the side forms is another disadvantage of envelope forms.

Proper draft is an extremely important factor in the ease or difficulty of stripping the mold. The architect should endeavor to design units with the drafts required to accommodate an envelope mold. Fig. 2.2.15 shows a generic envelope mold designed with the minimum workable drafts.

Several modified versions of the complete envelope mold

Fig. 2.2.10

Fig. 2.2.11

Fig. 2.2.12   Total envelope molds

LIFT PANEL   LIFT PANEL

FLAT PANEL   SCULPTURED PANEL

Fig. 2.2.13   Conventional molds

LIFT PANEL   REMOVE BULKHEAD   LIFT PANEL   REMOVE BULKHEAD

FLAT PANEL   SCULPTURED PANEL

which will accommodate precast concrete units without drafts along one or more edges are illustrated in Fig. 2.2.16. Since the loose side rails or back forming are stripped with the unit, the mold allows 90 deg. returns or returns with negative drafts to be cast using an envelope mold. The modified envelope molds are much easier to reassemble than loose bulkhead forms, because daily measuring and aligning is not necessary with side rails. When properly designed, a modified envelope mold will provide the same good corner details and high quality finish found on units cast from a complete envelope mold. Thought must be given to preventing leakage especially where removable side or end rails attach to forms. This point of leakage can mar the finish. A return as indicated in Figs. 2.2.16c and d will cause the leakage to take place where it will not be seen.

Such modified envelope molds, however desirable for quality and daily cost savings, cannot justify their initial cost unless a reasonable repetition exists. Many precasters instinctively avoid the higher cost of envelope molds unless the master mold usage is above 60 to 75 castings, or unless the specified quality justifies the increased tooling cost.

Chapter 3 further describes shapes and forms to illustrate how an architect might promote repetition and the use of these mold types, and stresses detailing which bears on the same concepts.

Fig. 2.2.14

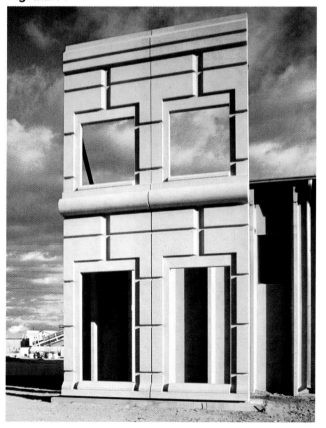

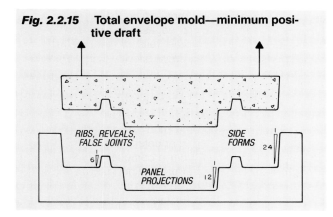

Fig. 2.2.15 Total envelope mold—minimum positive draft

RIBS, REVEALS, FALSE JOINTS

SIDE FORMS

PANEL PROJECTIONS

2.2.5 LEAD TIME

Provision for the longest possible lead time is an important factor in the optimum usage of tooling. To fully plan the tooling (including modifications), shop drawings or at least shape drawings must be substantially complete. The development of proper tooling requires considerable effort and time.

If the precast concrete manufacturer is not provided with the full lead time given to the general contractor, the cost of the precast concrete job invariably rises, or the project schedule must often be lengthened. It is desirable to specify that major subcontracts, including precast concrete, be let within a short, defined time period after award of the general contract.

Pre-bidding and awarding contracts authorizing engineering and drafting costs (subject to stated project cancellation fees) can result in substantial savings in production costs and are recommended. Part of this savings can be passed on to the owner directly. Pre-awarding precast concrete subcontracts also benefits the owner by decreasing the start-up time required after award of the general contract. Such pre-awarded subcontracts may then be assigned to the successful general contractor.

The total number of molds of a given type required on a project (apart from different types of molds) is solely a question of economical production within the time available. In many cases, this time factor, rather than tooling economy, demands duplication of molds. Other factors can also affect the decision to duplicate molds. The precaster's other production commitments and plant and site storage limitations affect the optimum number of molds. Fortunately, duplicating a mold is less expensive than building a new type of mold. Duplicate molds do not require additional engineering, and the materials for identical molds can be prepared simultaneously, shortening the time required. When several identical molds are required, the precaster can make a positive, or pattern, over which several identical molds may be easily formed.

The importance of sufficient lead time should not be overlooked. With enough time to fully plan the tooling, a better quality end product can be achieved. In addition, sufficient lead time offers cost savings in the precast concrete portion of a project and important savings in the construction time for the overall project.

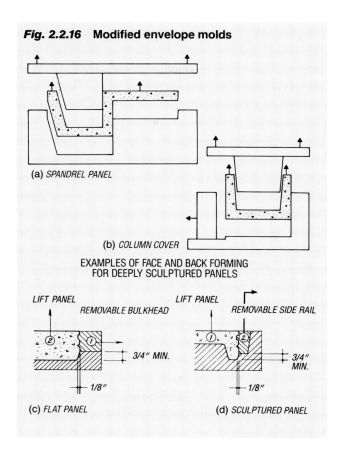

Fig. 2.2.16 Modified envelope molds

(a) SPANDREL PANEL

(b) COLUMN COVER

EXAMPLES OF FACE AND BACK FORMING FOR DEEPLY SCULPTURED PANELS

LIFT PANEL
REMOVABLE BULKHEAD
3/4″ MIN.
1/8″
(c) FLAT PANEL

LIFT PANEL
REMOVABLE SIDE RAIL
3/4″ MIN.
1/8″
(d) SCULPTURED PANEL

2.3 OPTIMUM QUALITY AND THE ECONOMICS OF MATERIALS

2.3.1 OPTIMUM QUALITY

The selection and description of materials and performance qualities for architectural precast concrete should be clearly stated. They should not be left open to variable interpretations nor should they be overly restrictive.

There should be a reasonable balance in quality between the different components going into a project.

Industrialization of the building industry has led to new concepts of quality. One of these is optimum quality which may be defined as follows:

Optimum quality is the level of quality, in terms of appearance, strength and durability, which is appropriate for the specific product, its particular application and its expected performance requirements. Realistic cost estimates for producing it within stated tolerances are factors which must be considered in determining this level.

The contract documents should make reference to the PCI *Manual for Quality Control for Plants and Production of Architectural Precast Concrete Products* (MNL 117) as the industry guideline for production of architectural precast concrete elements. Exceptions or specific requirements should be clearly set forth in the contract documents.

Chapters 3 and 4 covering SURFACE ESTHETICS and DESIGN CONSIDERATIONS will emphasize this optimum quality concept, while Chapter 6, dealing with SPECIFICA-TIONS, will highlight the items to be covered in the specifications in order to define the optimum quality for a specific project. This in turn should help to obtain reasonably identical quality from potential bidders.

2.3.2 ECONOMICS OF MATERIALS

A short discussion of the economics of materials may be useful to ensure that the optimum quality concept is not neglected in the selection of such materials.

The cost of cement in the finished, erected product will normally vary between 4 and 8 percent of the total cost, depending upon the amount of concrete volume per sq ft of wall, and whether gray or white cement is required. The premium for white cement is not a great percentage of the overall cost, although white cement is from 2.0 to 2.5 times more expensive than gray. In addition to specifying white cement for color effect, the architect will also obtain better uniformity in color than is possible with gray. See Section 3.2.1.

The aggregate cost may vary from approximately 5 to 20 percent of the total erected cost of the product, depending on the cost of the aggregate and the amount of face mix needed. For reasons of appearance and cost, the choice of aggregates will be an important factor. However, even the most expensive aggregates are often practical in exposed aggregate concrete especially when they are used only in the thin face mix.

Since the selection of aggregates (and to a smaller degree that of cement) have a substantial influence on final cost, the architect may obtain estimates or bids with a base price corresponding to the lowest cost combination suitable for the project, and with quoted premiums for upgrading of materials. This should not, however, include more than one or two alternatives, as the interaction between the cost of the materials, finishes, shape, and production techniques may complicate evaluation of such premiums.

The cost of reinforcement may fluctuate less as the result of the architect's decision than due to load requirements, such as handling, loadbearing or other structural functions. An exception is the architect's choice with respect to the shape and depth relationship covered in Section 4.2.8. Another exception is choice of protection of the reinforcement discussed in Section 4.4.7. The architect should realize that the cost of galvanizing or epoxy coating reinforcement is substantial and is not a substitute for adequate concrete cover or concrete quality.

The cost of hardware is mainly governed by load requirements including special structural functions and possible earthquake conditions. Hardware cost may be minimized by making the precast concrete units as large as is consistent with the size limitations given in Sections 3.3.9 and 4.2.9. Four connections are the minimum required for most precast concrete units. The labor cost of producing and handling small individual pieces of hardware normally exceeds the material costs making the relative cost of hardware high for small units.

The architect should evaluate all the factors influencing

the economics of a particular precast concrete project in order to achieve the full potential of the project. A major cost indicator is the degree of repetition, as already covered in Section 2.2.2. Another important economic consideration is the choice of shapes and textures as discussed in Chapter 3.

2.4 TOTAL WALL ANALYSIS

An architect analyzing the cost of architectural precast concrete should estimate the cost of the total wall as is normally done with other facing materials. However, the total cost may be lowered by taking full advantage of the precast concrete portion.

In addition to acting as exterior walls, the precast concrete panels may perform other functions: they may be load-bearing, wall-supporting, or serve as formwork or shear walls; they may be used to incorporate insulation and vapor barriers; they may be the interior finish; they may serve partly or fully as containers of mechanical/electrical services; or they may combine several of these functions to become a wall sub-system and be partially pre-assembled prior to reaching the site. Precast concrete panels may also be cast compositely with other materials to provide an entirely different finished surface. Clay products (brick, tile, terra cotta) and natural stone (granite, marble and limestone) have all been used as veneer facing.

Pre-assembling of wall units should be geared to local practices. This development requires coordination between local sub-contractors to solve problems concerning contract and trade practices. The present acceptance of this approach varies considerably from one region to another.

Where window openings are fully contained within a precast concrete panel, the structural capacity of the concrete should be utilized to reduce requirements for window frames. For example, by having the precast concrete panels absorb the pin loads from pivoted windows in one office building, the savings in sash requirements amounted to 5 percent of the total wall cost. For glazing of windows in concrete panels, refer to Section 5.2.

2.5 PRECAST CONCRETE PANELS USED AS CLADDING OR CURTAIN WALLS

The use of non-loadbearing precast concrete cladding or curtain walls has been the most common application of architectural precast concrete. Cladding panels are those precast elements which resist and transfer negligible load from other elements of the structure. Generally, they are normally used only to enclose space, and are designed to resist wind, seismic forces generated from their self weight, and forces required to transfer the weight of the panel to the support. Cladding units include wall panels, window wall units, spandrels, mullions and column covers. Their largest dimension may be vertical or horizontal. These units may be removed from the wall individually without affecting the stability of other units or the structure itself. For the purpose of discussion, cladding or curtain wall units do not extend in height beyond a typical floor-to-floor dimension and are normally limited in width to less than the bay width of the structure.

The use of precast concrete cladding is a practical and economical way to provide the desired architectural expression, special shapes and uniform finishes. When used over steel columns and beams, cladding can provide the required fire-resistance rating without resorting to further protection of the steel. When used over cast-in-place concrete columns and beams, it will often permit achieving a uniformity of finish in combination with a special architectural shape, all in the most economical manner. Cladding can be multi-functional, for example, by providing space behind for services and exterior grooves for vertical window-washing machinery.

The amount of repetition, and the choice of size, shape and finishes are the major design and cost considerations for cladding units. The economy of precast concrete cladding (curtain wall) units is achieved by paying close attention to the design and detailing of the precast concrete units. This is a basic requirement for all precast concrete, but particularly so for units which function only as cladding.

Typical wall panel system cross sections are shown in Fig. 2.5.1. These walls may be solid wall panels, window wall panels or spandrels. In addition, column covers and mullions are a common application of cladding units.

In high-rise buildings three characteristic facade patterns can be identified that impact considerably on the panel design. The first is that of cladding that plates the structural

framing, vertically and horizontally, the large opening then being infilled with glass (Fig. 2.5.2).

The second pattern eliminates the column covers, and the facade then becomes alternating horizontal bands of spandrel panels and glazing (Fig. 2.5.3). In this pattern the panels and glazing are placed in front of the columns, which are then individually suppressed.

The third pattern is a return to the traditional facade design of rectangular window openings "punched" into a plane surface (Fig. 2.5.4). This pattern originated from the requirement of loadbearing walls, that wall area must be provided between glazing to carry vertical loads, and so windows were relatively small. The re-appearance of this pattern derives some rationale from the needs of energy conservation which mitigates against large areas of poorly insulated glazing. A much stronger impetus comes from the dictates of architectural fashion and the desire to return to modeled facades and the visual interest that can be obtained by the traditional manipulation of voids and solids. This trend has resulted in some ingenious precast concrete configurations with the use of L- and T- shaped panels to reduce the number of costly joints. These panel shapes are derived from the requirements of erectors and their efforts to reduce installation cost. Some typical panel arrangements are shown in Fig. 2.5.5.

Solid Wall Panels use panel finish, shape, size and repetition as the major design and cost considerations. The high level of design flexibility possible with custom wall panels

Fig. 2.5.1 Typical wall systems

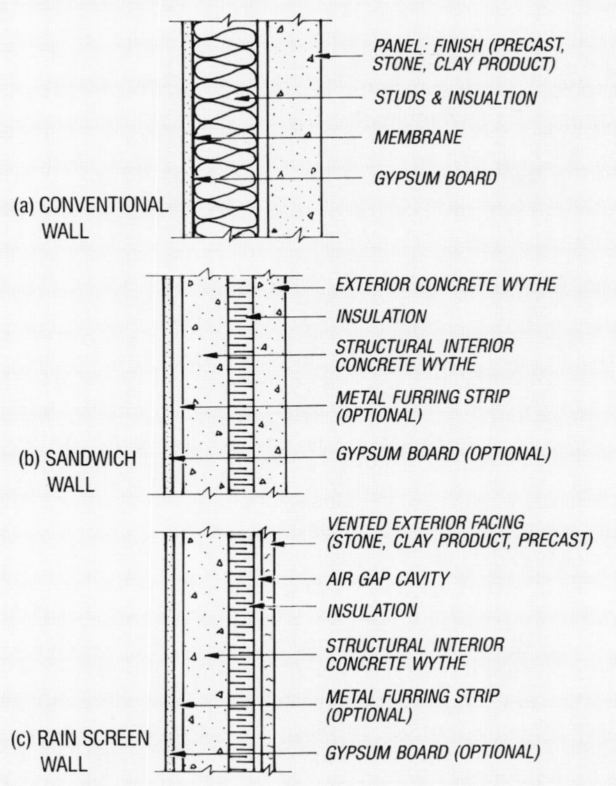

(a) CONVENTIONAL WALL
- PANEL: FINISH (PRECAST, STONE, CLAY PRODUCT)
- STUDS & INSUALTION
- MEMBRANE
- GYPSUM BOARD

(b) SANDWICH WALL
- EXTERIOR CONCRETE WYTHE
- INSULATION
- STRUCTURAL INTERIOR CONCRETE WYTHE
- METAL FURRING STRIP (OPTIONAL)
- GYPSUM BOARD (OPTIONAL)

(c) RAIN SCREEN WALL
- VENTED EXTERIOR FACING (STONE, CLAY PRODUCT, PRECAST)
- AIR GAP CAVITY
- INSULATION
- STRUCTURAL INTERIOR CONCRETE WYTHE
- METAL FURRING STRIP (OPTIONAL)
- GYPSUM BOARD (OPTIONAL)

Fig. 2.5.2

Fig. 2.5.4

allows for a wide variety of architectural appearances. The precast concrete skin was articulated by the use of a textured pattern to reduce the scale of the massive windowless walls of a computer center, Fig. 2.5.6. Within each panel, the light-colored, six-foot square was lightly sandblasted. The surrounding two-foot-wide grid, set off from the squares with a 2 in. wide by 1/2 in. deep reveal, was given a darker, rougher finish through the use of retarders. The deep set reveals on the facade are incorporated to give visual relief.

Window Wall Panels may be flat or be heavily sculptured. They may contain a single opening or a series of windows. They are either one-story in height and made as wide as possible or cast narrower to span vertically for two to four floors. The project in Fig. 2.5.7 uses mostly window box units that span 30 ft column to column. The large size is disguised by the use of both horizontal and vertical reveals as well as the strong display of the continuous horizontal projections.

Spandrel Panels are horizontal units which separate adjacent strips of glass. They may be cast flat, have returns at the top and/or bottom or be heavily sculptured. A designer will sometimes require the structural frame of a building to express itself in the building's facade. In such cases, the use of precast concrete spandrel elements made up either as a series of individual units or as one unit extending between columns with support located on the floor or on the column is an esthetically appropriate solution. The typical 27 ft long precast concrete spandrels in Fig. 2.5.8 have

Fig. 2.5.3

Fig. 2.5.5 Typical arrangements of precast concrete panels

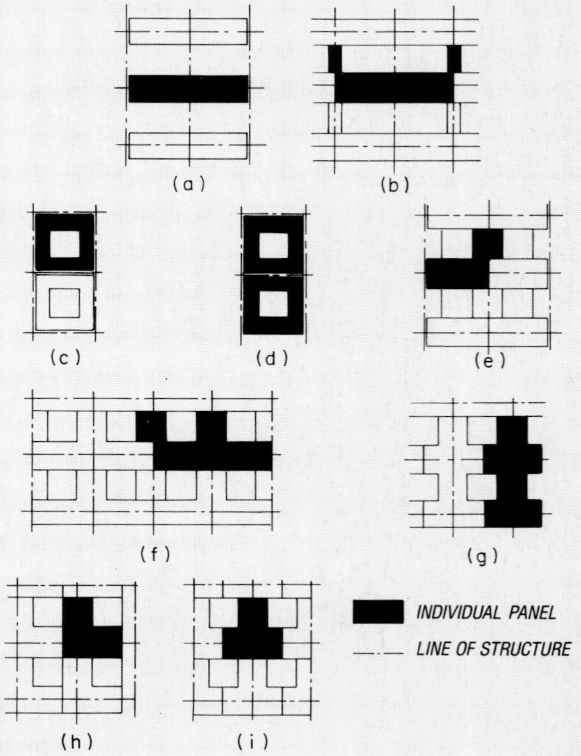

(a)

(b)

(c)

(d)

(e)

(f)

(g)

(h)

(i)

⬛ INDIVIDUAL PANEL

—·— LINE OF STRUCTURE

Fig. 2.5.6

Fig. 2.5.7

Fig. 2.5.8

a 4 ft splayed sill, 4 ft fascia and a 4 ft soffit. At the upper two floors, the spandrels receive the raker beams which provide solar protection for the terraces.

Column Covers and Mullions are usually a major focal point in a structure. These units may be broad or barely wider than the column itself and run vertically up a structure. They are often used to conceal the structural columns and may completely surround them at the ground level. Column covers are usually manufactured in single story units and extend either from floor to floor or between spandrels. However, units two or more stories in height may be used. In order to minimize erection costs and horizontal joints, it is desirable to make mullions as long as possible, subject to limitations imposed by weight and handling. Also, in many cases it may be desirable to combine the column cover or mullion with adjacent spandrel to minimize joints.

The 5 in. thick column covers, Fig. 2.5.9, are 3 ft x 9 ft 3 in. and are the major focal area of the building. The precast concrete units were given a medium sandblast finish. In Fig. 2.5.10, the precast concrete mullions at 5 ft centers lend a verticality to the 20-story tower surfaces that emphasize the desired delicacy of scale and details.

Wall-Supporting Units are precast concrete cladding units which support a portion of the wall, but carry no loads from floors or roof slabs. These units cannot be removed from a wall without affecting the stability of other units and are normally designed so that their largest dimension is vertical.

Such units may be "stacked" to support the wall above them to the roof level or any portion of this height, Fig. 2.5.11. They may be quite slender and support considerable wall height, if they can be tied into the structure as required for lateral stability. Engineering considerations include thermal movements, building drift and structure deflections, and must be accounted for in proper detailing.

Wall-supporting units can be made in one piece through several stories, Fig. 2.5.12. The weight of these units should not be carried by more than one floor. Erection techniques to accommodate pre-determined partial load distribution between floors are not economically feasible. It has proven impractical to support adjacent units on alternate floors. The recommended practice is that a specific floor be designated and designed to take the load of all precast concrete units passing this floor. These considerations are illustrated in Fig. 2.5.13.

Wall-supporting units may answer a particular design consideration for structures where the exterior columns are set back from the edge of the floor slab. This is illustrated in Fig. 2.5.14. A cantilevered floor will deflect to a certain degree over a period of time due to the weight of the wall units. By using stacked units for this condition, the designer has responded to this consideration.

Precast concrete units designed to carry their own weight over a considerable width are also considered wall-supporting units. This width may be equal to the column spacing for the exterior wall or a multiple thereof. Where such

Fig. 2.5.9

Fig. 2.5.10

Fig. 2.5.11 Wall-supporting panels

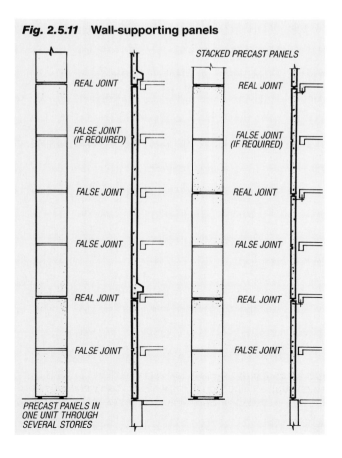

Fig. 2.5.12

Fig. 2.5.14 Precast units for buildings with cantilevered floor slabs

PANEL SUPPORTED ON
CANTILEVERED FLOOR SLAB

DON'T

LATERAL TIE CONNECTION ONLY

PANEL SELF-SUPPORTING VERTICALLY

DO

SUPPORT AT BEAMS ONLY

DO PANEL SELF-SUPPORTING
HORIZONTALLY BETWEEN BEAMS

Fig. 2.5.13 Connections for wall-supporting precast concrete wall panels

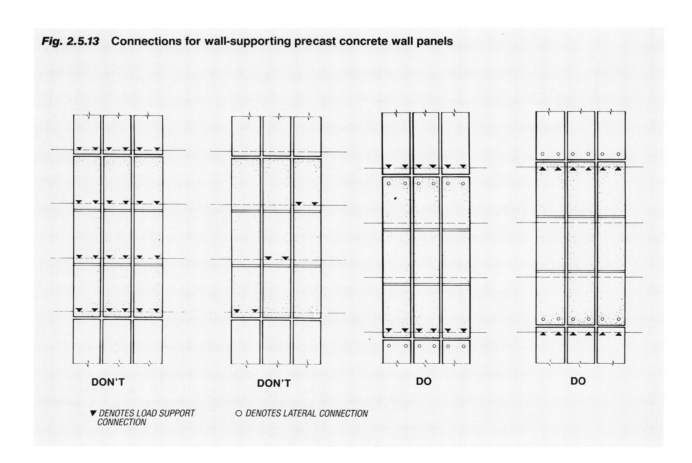

DON'T   **DON'T**   **DO**   **DO**

▼ DENOTES LOAD SUPPORT   ○ DENOTES LATERAL CONNECTION
   CONNECTION

Fig. 2.5.15

Fig. 2.5.16

units are bearing at, or in close proximity to, the columns, edge beams may be eliminated, or if needed for other structural reasons may afford savings in size and/or reinforcement. Because of size or weight limitations such units are normally made only one story high so that the width becomes the largest dimension. When the panel spans across several columns, the potential for deflection (or rotation) of the edge beams caused by the weight of wall units is reduced. Where several units are carried by one beam, extreme deflection may create tapered joints and the possible touching of units at their tops. Spanning the distance between the columns with one unit provides a deep beam and consequently much less deflection. By storing and supporting such units in a way similar to their ultimate position in the wall, any deflection in the unit will normally have taken place by the time of erection.

Panels, 30 ft long by one story high, serve as bearing panels for those above, with lateral restraint provided by the support structure, Fig. 2.5.15. This reduced the load the support structure is required to carry.

In Fig. 2.5.16, the hotel and low-rise support structure are clad with white precast concrete. By stacking each panel on the panel below, the gravity loads from the panels are carried directly to the foundation and do not introduce additional load into the superstructure. This approach resulted in significant savings in perimeter structure weight and simplified precast concrete connections. The low-rise structure is stacked up to 32 ft in height and the high-rise to 44 ft above its base.

To take maximum advantage of loadbearing and wall-supporting units, it is apparent that a decision as to their functions should be made before structural design has progressed to a stage where revisions become costly.

2.6 PRECAST CONCRETE PANELS USED AS LOADBEARING WALL UNITS

Often the most economical application of architectural precast concrete is as a loadbearing element, which resists and transfers loads applied from other elements. Therefore, a loadbearing member cannot be removed without affecting the strength or stability of the building.

Concrete elements normally used for cladding or curtain wall applications, such as solid wall panels, window wall or spandrel panels, have an impressive structural capability. With very few modifications, many cladding panels may function as loadbearing members. The steel reinforcement required to physically handle and erect a unit is often more than necessary for in-place loads. The slight increase in loadbearing wall panel cost (due to erection and connection requirements) can be offset by the elimination of separate structural frames (beams and columns) from exterior walls or reduction of interior shear walls. This savings is most apparent in buildings with a large ratio of wall to floor area. Also, the increase in interior floor space gained by eliminating columns can be substantial and, depending on floor plan, flexibility of partition layout can be obtained.

Precast concrete loadbearing wall panels, comprising structural-esthetic-functional features, provide the opportunity to construct an attractive building at an economical cost. Such structures contribute significantly to the development of contemporary architectural philosophy creating a system in which the walls are actually doing the structural work they seem to be doing. To realize the full potential of such usage with no sacrifice in esthetic advantages, the structural engineer should be involved from the initial concept stage of the project. Considerations should include the load effects on member dimensions, coordination of temporary bracing, connections, and erection sequencing.

Loadbearing panels and shear walls may be supported by continuous footings, isolated piers, grade beams or transfer girders. The bearing wall units can start at an upper floor level with the lower floors framed of beams and columns.

Loadbearing wall units or shear wall units should be the first design consideration if one or more of the following three conditions exist:

1. There is inherent structural capability of the units due to their configuration or to sufficient thickness of flat panels. The sculptural configuration of units often enable them to carry vertical loads with only a slight increase in reinforcement. For example, the precast concrete units may already have ribs or projections which could serve as column elements for the wall. In other cases, the thickness of flat panels (including sandwich wall panels) may be sufficient for loadbearing use.

Buildings with loadbearing panels in one- and four-story units are shown in Figs. 2.6.1 to 2.6.2. These buildings utilize standard prestressed floor and roof slabs with the loadbearing panels designed in widths to suit the standard module of the slabs. The *PCI Design Handbook—Precast and Prestressed Concrete* lists standard widths and other information regarding such slabs.

The structural exterior frame may also have architectural precast concrete units easily recognizable as columns and beams. Fig. 2.6.3 illustrates how elegantly this may be executed. Cruciform precast units were produced to tolerances of plus or minus 1/16 in. and resulted in the unique architectural feature at the column connection, Fig. 2.6.4.

Loadbearing panels without a recognizable column section, may have irregular rib projections sufficiently strong to allow the panels to carry loads from floor and roof.

Window wall panels can also be loadbearing. Since window wall panels are usually custom made for specific projects, the designer can take advantage of the self contained columns and girders inherent in the cross section of these panels by designing haunches to provide bearing for floors, Fig. 2.6.5

Spandrels may also be loadbearing. The total precast, 16 story office building in Fig. 2.6.6 has 8 in. thick, 7 ft high and 30 ft long curved, Fig. 2.6.7, and straight spandrel beams supporting the floor and roof slabs. The exposed structural system in Fig. 2.6.8 was emphasized by insets at the connections of spandrel beams and columns. An inset in the 14 in. thick, 45 ft long spandrel itself further defines the structural system by expressing the line of the unseen precast concrete floor.

The loadbearing flat walls, Fig. 2.6.9, are typically 8 to 10 ft wide and 8 in. thick. The panels are two-story (25 ft 4 in.) and are thickened at the vertical edges to form integral columns. This concentration of reinforcement carries the vertical loads and serves as lateral bracing. Some infill between the precast concrete wall units is the only cladding part of the facade.

2. An efficient structural layout of the building facilitates distribution of lateral forces from wind or earthquake loads and the design lends itself to repetitive use of panels. Floors, precast concrete (with a cast-in-place topping) or otherwise, act as diaphragms, distributing lateral forces to reduce the load on individual wall units and connections.

3. The building has a central core or bay designed to absorb lateral forces and transfer them to the foundation. The core or bay provides rigidity, enabling connections between precast concrete panels and floor slabs to be simple. While the core is being erected or cast, the precaster can proceed with the fabrication of the wall units and install them after the shear wall or core has been completed, resulting in a savings in construction time.

Fig. 2.6.1

Fig. 2.6.2

Fig. 2.6.3

Fig. 2.6.4

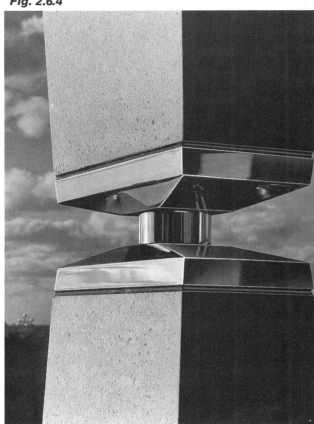

Fig. 2.6.5

Fig. 2.6.6

Fig. 2.6.7

Fig. 2.6.8

These conditions do not preclude other situations where loadbearing panels or shear walls may be used. When any of these conditions are met, however, utilization of precast concrete panels as structural components should be a primary design consideration. The importance of conditions 2 and 3 is diminished with decreasing intensity of lateral forces. Many buildings of other shapes in earthquake zones 0 and 1, and outside extreme wind load zones, may utilize loadbearing panels or shear walls.

In multi-story buildings, the loadbearing wall panels can be several stories in height up to the maximum transportable length, or one-story high and connected at every floor level. The architectural requirements generally govern. The variety of shapes and surface finishes commonly associ-

Fig. 2.6.9

Fig. 2.6.10 Vierendeel truss

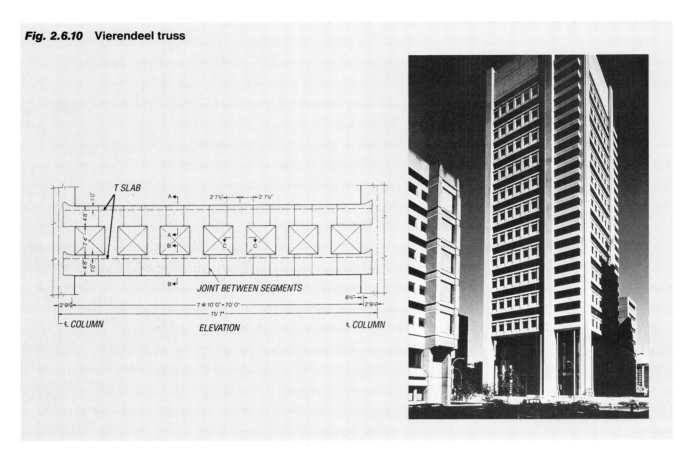

Fig. 2.6.11 Beam assembly

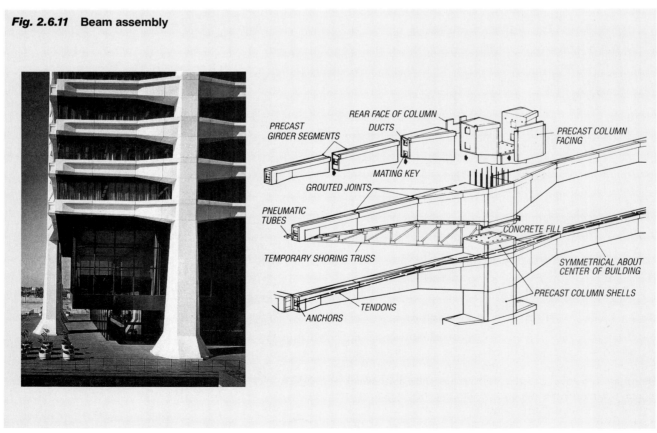

ated with cladding are possible, provided the structural and other technical requirements can be satisfied at the same time. By extending loadbearing panels vertically through several stories, complex connection details are minimized, and consequently the economic advantages of loadbearing wall panels are increased.

Architectural requirements normally dictate that building elevations have wall panels of the same appearance. Therefore, the wall panels receiving the greatest gravity loads may be determined and panel elements designed interchangeably with the same reinforcing steel in all panels. This permits any panel to be installed at any point on the exterior of the structure, since the floor plan of a load-bearing panel building is usually the same on all stories, producing uniform loads on the building perimeter. In most cases, there is little need to be concerned with differential foundation settlement. This is one of the important advantages for high-rise, loadbearing panel structures where the bearing walls also serve as shear walls.

Another application of loadbearing panels is segmental construction of frame and wall units (primary load carrying members), assembled by post-tensioning to create large structural elements. This allows precast concrete to be used for long horizontal or vertical spans within size limitations imposed by manufacturing, transportation, and handling equipment.

Segmental construction requires that the designer give special consideration to:

1. Size and weight of the precast concrete elements.

2. Configuration and behavior of the joints between elements.

3. Construction sequence, and the loads and deflections imposed at various stages.

4. The effect of normal tolerances and deviations upon the joints.

Although this use is still rather unique, it will undoubtedly be further explored in the future as a means of providing large column-free interiors combined with only a few exterior columns. Fig. 2.6.10 illustrates the design of a 70 ft long Vierendeel truss used as the wall of an office building in an earthquake zone. Each truss is composed of six I sections and ten short link sections, with four special segments forming the ends. The trusses are supported on cast-in-place hollow columns (column voids provide space for mechanical ducts), and the truss-column juncture is accomplished by reinforcing steel extending from the truss and splicing with the column steel. The large, 75 ft 7 in. by 16 ft 8 in. Vierendeel truss segments were post-tensioned together on-site. The building has a striking architectural appearance with the structure clearly expressed.

Fig. 2.6.11 illustrates the use of cantilever beams forming a frame constructed by use of segmental elements and shows the construction method and joint details. This structure employs precast concrete shells, into which column reinforcing steel and concrete is field placed, as well as precast concrete facade beams, thus providing a homogeneous exposed aggregate finish. The girders,

which are symmetrical about their centerline, are continuous through two columns, with a 49 ft center span and two 42 ft cantilevers. Beams are post-tensioned after assembly, and after field grouting of joints.

2.7 PRECAST CONCRETE PANELS USED AS SHEAR WALLS

In many structures, it is economical to take advantage of the inherent strength and rigidity of exterior precast concrete wall panels and design them to serve as the lateral load resisting system when combined with diaphragm action of the floor construction. Walls which resist and transfer horizontal loads (lateral forces) in or parallel to the plane of the wall from superstructure to foundation are referred to as shear walls. The walls thus act as vertical cantilever beams. Usually, a structure will contain a number of walls which will resist lateral forces in two orthogonal directions, Fig. 2.7.1. The effectiveness of such a system is largely dependent on the panel-to-panel connections.

The lateral forces may be wind forces or earthquake loads. The importance of the latter varies according to the location of the project; many areas now require analysis of structures for earthquake forces in varying degrees of intensity. Concrete panels have the inherent strength required for handling, and perform as shear walls with little or no additional reinforcement. It is important, however, that the connections be designed to transfer lateral forces, and still accommodate thermal movements and differential deflections (or camber) as covered in Section 4.5.2. The

Fig. 2.7.1 Exterior shear walls

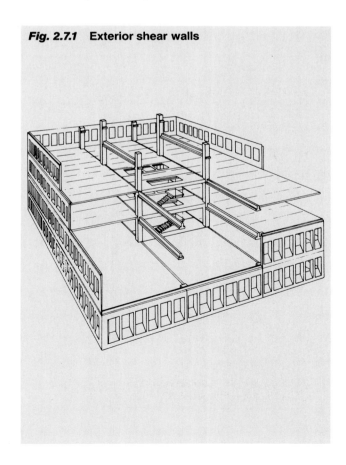

Fig. 2.7.2

ability to transfer lateral forces may be the only structural utilization of the panels, but is more often combined with the loadbearing or wall-supporting capabilities already described.

In seismic areas, the exterior shear wall system permits design flexibility because it eliminates the need for a structural core. The exterior walls also provide the vertical strength and horizontal connections necessary to allow the entire wall to function as a single unit, eliminating the need for exterior columns and beams, Fig. 2.7.2.

The 12 ft high by 29 ft long window panels, Fig. 2.7.3, act as a truss in transferring loads to the foundation, while the 12 x 21 ft solid end wall panels act as traditional shear walls.

In an interior shear wall system, cladding panels distribute the lateral forces to floor diaphragms, which transfer them to the interior shear walls. These precast concrete wall panels are frequently tied together at the corners to form a structural tube that cantilevers from the foundation.

2.8 PRECAST CONCRETE PANELS USED AS FORMWORK FOR CAST-IN-PLACE CONCRETE

There is a trend toward the use of architectural precast concrete units as forms for cast-in-place concrete. This system is especially suitable for combining architectural (surface esthetics) and structural functions in loadbearing

Fig. 2.7.3

Fig. 2.8.1

facades, or for improving ductility in locations of high seismic risk.

Houston's 20-story Control Data Building, which was completed in 1969, was the first composite frame building. On the exterior, slender, 8 in. wide flange steel sections reinforce the concrete columns, which were cast using precast concrete units as forms, Fig. 2.8.1. Composite building design significantly reduced steel tonnage and represented a major change in building design.

Because the cost of formwork is a significant part of the overall concrete cost in a structure, substantial savings can usually be achieved by designing architectural precast concrete units so they also function as formwork.

The advantages of using architectural precast concrete units as formwork will be achieved only when the architect, engineer of record, precaster and general contractor have the opportunity to develop a project jointly from the design stage. Types of finishes, shapes, repetitive use of efficient and economical building modules, structural systems, site delivery, erection procedures and construction schedules, all become important considerations for the successful completion of the project. The time spent in development pays off in accelerated construction and significant cost savings.

Precast concrete panels as formwork for columns and bond beams reduced field labor and the construction time schedule considerably for the five story building in Fig. 2.8.2. Only 210 working days elapsed from the start of construction until occupancy by the first tenants.

Panel size varies depending on the window fenestrations and exterior column spacing. The panel may involve several bays and may be more than one story. Transportation and erection requirements usually govern the size of the panels.

There are benefits for all members of the construction team:

1. The architect is free to shape the form and appearance of the structure, with virtually no restriction. Precast concrete, manufactured in a plant, permits rigid factory-controlled conditions, thus ensuring a uniformly high quality facade in the desired shapes, colors

Fig. 2.8.2

and textures. Precasting makes it possible to inspect the surface finish before the cast-in-place concrete is placed. Form removal with its subsequent patching of form anchor holes and surface imperfections, and repair of other flaws, is eliminated. The architect may choose to avoid complicated column connections by casting the exterior columns in place using precast concrete loadbearing panels as exterior formwork. The use of precast column covers as formwork can also be advantageous when it would be difficult to erect conventional covers after construction of the upper floors.

2. The engineer of record may, with proper details, employ the form as a contributing portion of the structural support system. Effective composite behavior will significantly increase the strength of members and reduce deflections. Continuity and ductility may be achieved by casting in place spandrel beams and columns using the wall panels as forms. The ductility of walls partially depends on the locations of reinforcement. Ductile behavior is significantly improved if the reinforcement is located at the ends of the walls. Thus, the usually inactive cladding can become a major lateral load (seismic) resisting element.

3. The general contractor will benefit from a reduction in the time required for construction. All the formwork required for a structure can be manufactured in advance of concreting. This permits greater flexibility and continuity in concrete operations. Delays in concreting due to time required for preliminary curing of concrete preceding form removal and re-erection of forms can be eliminated. The precast concrete units may be erected quickly and the structure is complete when the concrete is placed. The need for temporary outside forms is eliminated.

Exterior scaffolding is not generally required, as work is carried out from inside the building, resulting in increased safety on the job. Also, scaffolding or shoring normally required to support overhead removable forms, can be partly or completely eliminated, thus saving the time and cost involved in its erections. Precasting also permits the use of fewer forms for complex shapes such as tapered and curved columns.

For forms that are intended principally to achieve a desired architectural effect, the architect/engineer should specify surface finish and desired minimum thickness of architectural material, but design and layout of the forms and supporting systems will normally be the responsibility of the general contractor. The architect/engineer may require drawings of form panels, design details, and layout to be submitted for review before concreting.

Fine materials from the cast-in-place concrete must be prevented from escaping through the joints and running out onto the finished facing. If mortar is allowed to harden on the surface, particularly on an exposed aggregate finish, it will be almost impossible to remove. For this reason, precast concrete panels used as formwork should have a finish which is relatively easy to clean. Spilled or leaked concrete should be thoroughly removed before it hardens. Alternatively, the precast concrete elements may be wrapped or covered with non-staining materials to protect them from concrete spillage and other construction materials, debris, and mechanical damage.

Realistic assumptions with regard to construction techniques are required. It must be determined (or specified) how the precast concrete panels will be supported during concreting in order to design the proper reinforcement within the panels.

Concrete form panels should be erected and temporarily braced to proper elevation and alignment in such a way that the tolerances specified for the finished structure can be met. Temporary bracing for the panels generally consist of adjustable pipe bracing from panel to floor slab. Supports, braces, and form ties must be stiff enough so that their elastic deformation will not significantly affect the assumed load distribution. Form ties may be attached as shown in Fig. 2.8.3, to embedded strap anchors or threaded inserts provided in the panels for that purpose, or welded to plates cast in the panels, Fig. 2.8.4. Ties are then fastened in the conventional manner with hardware on the other side of the interior wood or steel forms. Column forms may use column clamps or be wrapped with steel bands to aid in resisting hydrostatic pressure. Care must be taken to protect corners of precast concrete units when wrapping forms.

Attachments between the precast concrete form and other elements, such as steel columns, must be detailed to provide the necessary field adjustments. Contact area

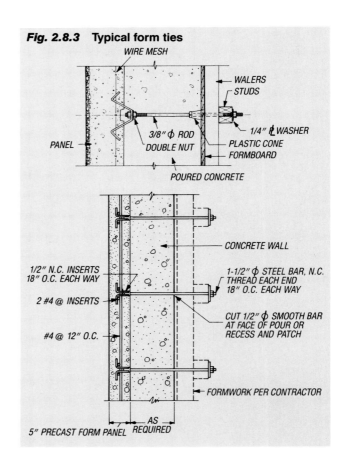

Fig. 2.8.3 Typical form ties

Fig. 2.8.4

between precast concrete and external braces, clamps or bands should be protected from staining and chipping.

Where effective bond between precast concrete form units and the cast-in-place concrete is essential to the design, it may be achieved by grooving or roughening the form face in contact with the cast-in-place concrete and by the use of mechanical anchoring devices such as ties or stirrups extending across the interface between form panel and cast-in-place concrete, Fig. 2.8.5.

Following installation of form and shear ties, the reinforcing steel and interior formwork are placed, and concreting carried out. Thorough consolidation of concrete behind the precast concrete forms is required to prevent voids which would interrupt the bond of the form to the concrete. Voids might also trap water which could show through as staining on the panel face and cause damage in freeze-thaw cycles. Care must be exercised to prevent damage to concrete panels by contact with vibrators.

Special consideration should be given to counteracting or controlling bowing of panels. Cambering of architectural precast concrete forms to compensate for deflections is expensive and should be avoided. Since the forms are generally supported to permit them to act as continuous beams while the concrete is being placed, the panels must be designed to hold deflections to an acceptable limit consistent with minimizing cracking. In order to control cracking, the modulus of rupture of the concrete should not be exceeded during placement of the concrete or under ser-

Fig. 2.8.5

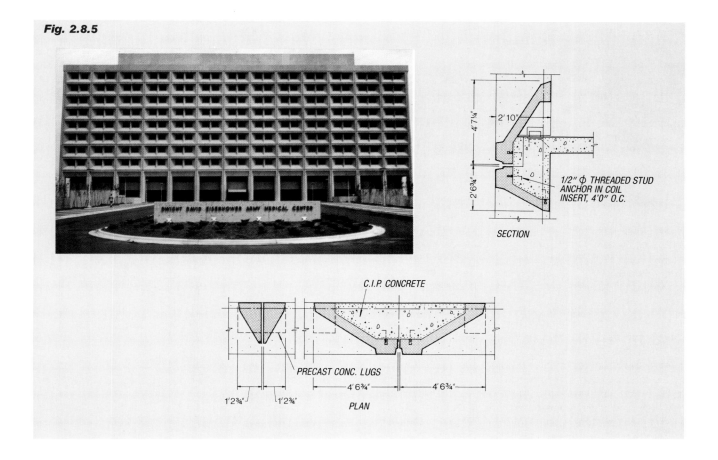

SECTION

1/2" φ THREADED STUD ANCHOR IN COIL INSERT, 4'0" O.C.

C.I.P. CONCRETE

PRECAST CONC. LUGS

PLAN

vice loads. Where the member is long enough to develop bonded strand, pretensioning may be used in the precast concrete form units.

Architectural precast concrete surfaces require tight joints so that concrete does not leak and mar decorative facings. Methods used to close joints include buttering on the inside with mortar (units many also be bedded on mortar when designed to accept load transfer through the joint), and gasketing with low density closed cell neoprene rubber or other durable, permanent, resilient materials to reduce cracking due to thermal movements. A shiplap joint may also be used to aid in preventing the concrete from leaking and marring surface finishes, Fig. 2.8.6.

Where form panels are non-composite, the joint material should prevent load transfer. However, since these joints usually occur only once in each floor and once in a bay, the composite properties of the compression member can be included in member stiffness calculations. This can be of particular importance for long slender columns. Where form panels are intended to act compositely with the cast-in-place concrete, the joint material must be mortar or other non-staining material of sufficient strength to transfer the intended loads. In this case, the composite section can be used for strength as well as stiffness. For joints exposed to the environment, mortar is usually raked back from the face, and the joint caulked.

Some precautions and special details may be required to accommodate differential shrinkage or creep between cast-in-place concrete and precast concrete used as forms for columns or loadbearing walls. Stress relief may be simply handled in the design of the joints in the precast concrete forms at the top and bottom of vertical structural components carrying axial loads.

Horizontal construction joints in the cast-in-place concrete should be made 3 in. below the top edge of the panels used as permanent forms rather than in line with horizontal form joints. This reduces the possibility of water leakage through the construction joints.

Design of composite flexural members using precast concrete as forms requires location of form joints in areas remote from points of high moment, since any reinforcement in the precast concrete must be discontinued at that location. With the joint so located, the shear at that section can usually be adequately resisted by the reinforcement contained in the cast-in-place concrete.

2.9 MISCELLANEOUS USES OF PRECAST CONCRETE

In addition to functioning as exterior and interior wall units, precast concrete finds expression in a wide variety of esthetic and functional uses, including:

1. Art and sculpture

2. Lighting standards and fountains

3. Planters, curbs, and paving slabs

4. Balconies

5. Screen sound barriers, and retaining walls

6. Screens, fences and handrails

7. Street furniture

8. Ornamental work

Artists have found precast concrete well suited to expressions of boldness and strength in a variety of shapes, forms, and textures as illustrated in Figs. 2.9.1 and 2.9.2. Figurines for the murals, Fig. 2.9.3, were cast in five separate molds, 4 ft x 7 ft 6 in. and then set into an 8 ft x 15 ft wood mold. Figurine panels are 6 in. thick and the figures project up to 1-1/2 in. from the flat areas. Highly figured panel design, Fig. 2.9.4, using extensive detailing including rosettes, flutes, bullnoses, glyphs, and rustication offers visual interest.

Precast concrete can be used to produce indentifying signs, Fig. 2.9.5. Rubber mats were used to form the vertical ribs on these 30 ft tall, very striking smooth white concrete entryway signs for an office park.

Bollards, Fig. 2.9.6, and lighting standards, Fig. 2.9.7, and other fixtures in precast concrete are being used in an increasing number of design applications that combine sturdiness with varied forms and shapes.

The esthetic and functional use of precast concrete has greatly expanded with the increased development of pedestrian malls and plazas. The artwork and functional sculpture, lighting standards, and fountains already discussed are frequently seen in these areas. Planters and

Fig. 2.8.6 Shiplap joint

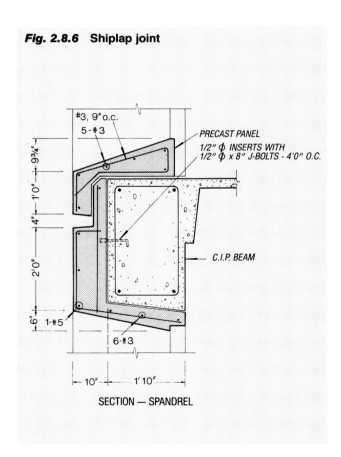

#3, 9" o.c.
5-#3
PRECAST PANEL
1/2" φ INSERTS WITH
1/2" φ x 8" J-BOLTS - 4'0" O.C.
C.I.P. BEAM
1-#5
6-#3
9¾"
1'0"
4"
2'0"
6"
10"
1' 10"

SECTION — SPANDREL

Fig. 2.9.1

street furniture are another application of precast concrete which has gained increased importance with the proliferation of malls and plazas. Multifaceted precast concrete wall caps and sandblast finished planter walls help to soften the building mass in Fig. 2.9.8. The design of the precast caps along the wall is reminiscent of the natural stone caps used on the original hospital structure replaced by this medical office building (see Fig. 1.2.2). Several other types of plaza planters are shown in Figs. 2.9.9 and 2.9.10.

Balconies are becoming more decorative as architects seek to explore the potential of their highly visible location. The attractive architectural precast concrete balcony panels, in Fig. 2.9.11, follow traditional Chinese architecture in this high rise apartment building. The balcony railings reflect a wall pattern found in the design of the ancient Peking Palace in China. Note the attention to detail, with crisp vertical reveals and small scuppers that extend well past the face of the panel to avoid staining due to water runoff.

Screens may be loadbearing, wall-supporting or part of curtain walls. They may also be free standing when used as dividers or fencing. Precast concrete sunscreens allowed an unbroken horizontal expanse of glass across both upper levels of the earthsheltered office building in Fig. 2.9.12. Architectural sunscreens and large solid wall panels reinforce a strong design statement. Screens are often used to decoratively shield the space from sunlight or to block specific areas from public view. They may also serve to renovate older buildings.

Fig. 2.9.2

Fig. 2.9.3

Fig. 2.9.4

Fig. 2.9.5

Fig. 2.9.6

Fig. 2.9.9

Fig. 2.9.7

Fig. 2.9.8

Fig. 2.9.11

Fig. 2.9.12

Fig. 2.9.13

Street furniture, can also benefit from the clean lines and variety of textures possible with precast concrete.

Barrier walls have become popular in recent years because of the growing need to control excessive noise pollution generated by busy airports and highways. Using exposed aggregate outer surface, the barriers may be designed to blend in with the adjacent neighborhood.

In Fig. 2.9.13, a 500 ft long and 20 ft high precast concrete waterwall conceals the two-level parking deck below the building. This feature wall consists of 32 panels of gray precast concrete between vertical columns of cream precast concrete. Each gray panel has a decorative relief design cast in the abstract form of a tree with the stylized "limbs" on each side of the "trunk" continuously sprayed by individual horizontal water jets concealed in the cream columns. Architectural precast concrete was the primary construction material for the water park in Fig. 2.9.14. It has as its focal point a serpentine waterfall, 12 ft high, bordered by watersteps and waterspouts.

As a replacement for terra cotta or stone ornamentation, precast concrete is an ideal material, Fig. 2.9.15. Impressions can be taken of the ornaments and molds made to cast the replacements in concrete with highly refined, intricately detailed relief patterns. With sufficient repetition to amortize initial mold-making costs, ornamental precast concrete can be economically feasible. For example, see the decorative freize of precast concrete inlaid with bright blue tile in Fig. 2.9.16.

Fig. 2.9.14

Fig. 2.9.15

Fig. 2.9.16

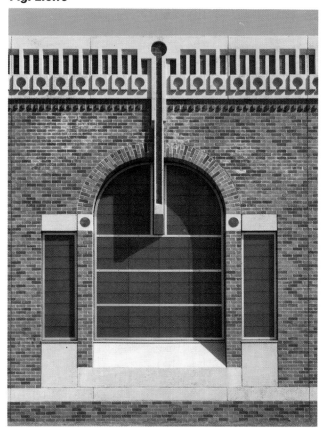

SURFACE ESTHETICS

3.1 GENERAL

Many facets in the design of architectural precast concrete are of vital importance to the architect. Two significant design considerations described in Chapter 2 were: total wall analysis; and repetition and the master mold concept. Chapter 3 discusses the surface esthetics of precast concrete panels which demand decisions by the architect, such as color, form and texture, and weathering. Because of the versatility of the material, the architectural focus can vary greatly from project to project, changing the relative importance of each facet of the design.

The design process is the creation of a multitude of expressions which relate to the owner and the general public. Architects are often bound by budget limitations that require the development of skills different from, and often in conflict with their creativity. The degree of design success will depend upon the architect's ability to extract from each component the ultimate in performance, utility, and appearance.

Each step between design concept and the final contract documents is important. At an early stage, consideration should be given to an orderly development of samples and recognition of uniformity requirements, and selection of shapes, sizes, colors, textures and finishes.

PRE-BID CONFERENCE should be held at least 3 weeks prior to bid. Prior to this meeting, the precast concrete manufacturers should have submitted their samples, technical literature, company brochures, and proposed materials. During the pre-bid conference, each precaster should outline his plans for the work to be accomplished in accordance with the requirements of the contract documents. The samples submitted should be discussed, material sources and adequacy of reserves reviewed, and production schedules evaluated. The precast concrete manufacturer's plant facilities and personnel must be sufficient to produce the work required for the project within the specified time. Key schedule items such as mockup panels, shop drawing/design submittals, mold production, and scheduled start of manufacture should all be covered. Generally, one pre-bid conference is held for all anticipated precasters and general contractors. At this meeting, potential problems are discussed, and the final architectural precast concrete panel finishes are clearly spelled out.

3.2 UNIFORMITY AND DEVELOPMENT OF SAMPLES

Development of samples should be treated as a means of translating the architect's design concepts into a standard for realistic production requirements. The major factors affecting uniformity of architectural precast concrete units are described in Section 3.2.1. These should be recognized through all stages necessary to prepare, assess and approve samples.

Samples of architectural precast concrete are intended to give information on the type of materials used and the type of finish. Samples for architectural precast concrete should only be regarded as a standard for performance within the variations of workmanship and materials to be expected. If the color or appearance of the concrete is likely to vary significantly, samples showing the expected range of variation should be supplied. The concrete placement and consolidation method used for samples should be representative of the actual procedures used in the production of the element.

3.2.1 UNIFORMITY

Concrete is a manufactured product, but contains natural materials. Actually, it is the inherent natural beauty of materials, such as sand and stone, which is most often expressed in architectural concrete. The limitations of these natural materials with respect to uniformity must be recognized, and the requirements for uniformity of the precast concrete product must be set within these limitations.

The architect should also recognize production limitations dictated by the rules of the market place. Difficulties in uniformity related to the quarrying, crushing, screening and transporting of aggregates are fairly easy to understand. Not quite so obvious are potential variations in uniformity caused by varying climatic conditions affecting final curing, or by planned production breaks for weekends and holidays or unplanned breaks. Another factor influencing uniformity is the time period required for storage of precast concrete units as dictated by contractual conditions or by operations at the construction site beyond the control of the precaster.

These material and production factors may cause differences in appearance either in color or texture and occasionally in both. Such variations in colors or textures are considered acceptable in other materials such as masonry and natural stone, but strangely are considered less acceptable in precast concrete.

Uniformity of texture and color within an individual panel should not be a problem if the recommendations of this section are followed, but the architect must determine to what degree uniformity is required for a complete facade. Where uniformity is essential, shapes, colors and textures must be chosen to assist in achieving this uniformity, and sometimes one aspect sacrificed for the other. Controlled variations in color and/or texture may, however, create design possibilities which should be explored.

3.2.2 DEVELOPMENT OF SAMPLES

Numerous choices in textures and colors due to the great range of various coarse aggregates, sands, cements, and coloring agents, combined with a variety of finishing processes, make architectural precast concrete one of the most versatile products in the construction industry.

For the architect to develop and select the color and texture for architectural precast concrete requires a combination of art and skill. The same is equally true of the precaster who must translate these requirements into workable concrete mixes and the proper finishing techniques.

Achieving the desired textures and colors together with feasible production techniques, is a process which re-

quires the precaster to produce samples that satisfy the architect's design concepts. This may be accomplished by producing a few samples, or it may be a slower process requiring numerous series of samples and considerable investigation of corresponding production and finishing techniques. Fig. 3.2.1 shows various samples to assist in selecting architectural finishes for shadowing and color.

The importance of this process is not always recognized by the architects or the precasters. To assure success, however, all research and development should be completed prior to formal bidding. It is also recommended that all precasters approved for a particular project should develop samples for approval as a prerequisite for bidding.

At this stage of the procedure, sample development may involve considerable expense in research and investigation by the precaster. The architect can be of considerable help by visiting precasting plants which have a selection of samples at hand to assist in selecting limits for the desired finish. By watching plant operations and talking with personnel, the architect can also obtain a good understanding of production considerations.

Since the architect is responsible for the final decision, design judgment must be supplemented with an assessment of the operating procedures and technical personnel from all plants likely to bid on the project.

Some precasters have small samples in stock showing color, finish, and textures used on previous projects, Fig. 3.2.2. Previous work of a similar nature can serve as a use-

Fig. 3.2.1

Fig. 3.2.2

ful visual standard and highlight potential problems. Even though an architect has seen the selected aggregates used with a similar finish in existing precast concrete units, it is important to develop specific project samples. These samples must reflect the relationship between materials, finishes, shapes and casting techniques, such as mold types, orientation of exposed surfaces during casting, and consolidation procedures.

3.2.3 PRE-BID SAMPLES

Due to individual preferences, differences in sources of supply, or different techniques developed in various plants serving the same area, the architect should not expect to select one sample and obtain exact matching by all precast concrete producers. If done, this may preclude the optimum quality from plants other than the one submitting the specified sample. Pre-bid samples can provide a solution to this problem.

Many architects have developed a practice of making sample selection and approval just prior to bid closing. Thus, for a specific project, the approved precasters' names and corresponding sample code numbers, may be published in an addendum or approval list given in writing to the general contractor.

The architect, when making pre-bid approval of samples part of the specifications, should adhere to the following requirements:

1. Sufficient time should be allowed for the bidder to submit samples or information for approval. Time should also be provided to enable such approvals to be conveyed to the precaster in writing so that the precaster can estimate and submit a bid.

2. Any pre-bid submittal should be treated in confidence, and the individual producer's solutions and/or techniques protected both before and after bidding.

If samples from individual plants are approved, the result may be a slight variation in color, aggregate or texture, but not necessarily quality. The individual precaster, within specification limits, selects the material and employs the placing and finishing techniques best suited to their plant operation. By making approval of pre-bid samples a prerequisite for bidding, the architect and client are protected by requiring equivalent optimum quality from all pre-

casters. All involved then know the result to be achieved in color and texture of the finish.

The size of a sample should relate to the maximum size of aggregate to be used and allow for realistic placement of the concrete and accurate expression of detail. Samples should be at least 12 x 12 in. Larger samples are recommended but handling difficulties should be recognized.

All submitted samples should be clearly identified by precaster's name, date produced, identifying code number, and name of project for which it was submitted. If the precast concrete units are to have an exposed interior finish, samples should also be provided for this purpose.

Information should be provided with the submission of samples regarding the type of materials, quality of concrete and type of finish proposed. This information should be submitted both for face mixes and backup mixes, if different. The precaster should also certify that the samples are indicative of the production of the precast concrete elements for the specified project.

CONCRETE MIX. The submission should include the following information relative to the concrete mix:

1. Uniformly graded, or gap-graded.
2. Cement content in pounds per cubic yard of concrete.
3. Water-cement ratio.
4. One-day and twenty-eight day compressive strength.
5. Unit weight.
6. Slump range.
7. Percent air entrainment, if required.
8. Admixtures (types, source and amount).
9. Water absorption at 28 days. (This is an indication of the unit's ability to remain clean under normal weathering conditions.)

AGGREGATES. The submission should include the following information relating to aggregates:

1. Type of material.
2. Inertness.
3. Hardness or abrasiveness.
4. Maximum size.
5. Cleanliness. Silt and organic impurities.
6. Soundness.
7. Gradation.

EXPOSED FACES. The submission should include the following information regarding exposed faces, whether exterior or interior:

1. Type of finish.
2. Data on weather sealing, if applicable.
3. Gradation, all standard sieves. Where percentage passing No. 50 and 100 sieves are high (Section 3.4.1) anticipated fluctuations in these percentages should be given.

As the precaster must work with these criteria throughout production, latitude should be allowed for some of the data as long as this range is reasonable and acceptable.

DEVIATIONS FROM SPECIFICATIONS. If the characteristics of submitted pre-bid samples in any way deviate from the specifications, the precaster should make this clear to the architect. This is done when the samples and other required information are submitted. For proper evaluation and approval of the samples, the precaster should state the reasons for the deviations. These reasons might be the precaster's concern over controlling variation in either color or texture within specified limits. In regard to adequacy of specified materials, concerns about satisfying all conditions of the specifications must be based upon practical plant production requirements, and the performance or weathering of the product in its final location.

If such deviations are approved with samples, the original project specifications and contract drawings should be changed accordingly by the architect.

3.2.4 PRODUCTION APPROVAL SAMPLES

Immediately after award of the contract, the precast concrete manufacturer should prepare and submit for approval, before casting any units, a representative sample or samples of the required color and texture, providing pre-bid samples prepared by their plant were not the basis for the specifications, or providing the pre-bid approval method was not used. When approved, the samples should form the basis of judgment for the purpose of accepting the appearance of architectural finishes. These samples also should establish the range of acceptability with respect to color and texture variations, surface defects and overall appearance.

Although 12 in. square samples provide valuable information on texture and color tone for the architect's initial esthetic evaluation, small samples are unlikely to give a true picture of the possible variations of finish over a large area, demonstrate normal surface blemishes, or show the effects of the natural day-to-day variations of aggregates and cement. Fig. 3.2.3 shows that a 12 in. x 12 in. x 1 in. sample bears little relationship to the appearance and physical characteristics of a production panel. Differences in mass, density, and curing rate between the sample and the production panel may make direct comparison difficult. For instance, a different chemical retarder may be needed because of the higher heat of hydration generated by the production panel, and the matrix of the panel may show more resistance to sandblasting because it has cured more at the time of finishing. In addition, a flat sample cannot show the relationships between materials, shapes, details, and casting techniques. Thus, 1/4 or 1/2 scale samples and full-scale mockups are essential to a realistic evaluation of the production methods and the finished product. The production of uniform, blemish-free samples, which demonstrate the abilities of a single master craftsman, will be completely misleading and could cause endless difficulties when the production personnel using actual manufacturing facilities have to match "the sample." Both large

Fig. 3.2.3

mockups and small samples should be made as nearly as possible in the same manner intended for the actual units.

Two of the samples used to evaluate the requirements for a particular window panel are shown in Figs. 3.2.4 and 3.2.5. In this case, the precaster produced a quarter scale sample of the window corner. The architect increased the depth of the groove when this sample indicated how little it would "read" as originally dimensioned. It was then the decision of the precaster to go beyond the quarter scale sample to build a portion of the full scale panel, Fig 3.2.5, to verify production requirements.

Architects and owners will frequently authorize the expenditure for mockups, either of a full scale portion of a panel or the entire typical unit. A mockup may be several modules wide by one or two stories high. Investing in such mockups will remove uncertainties in the minds of architect and owner alike or lead to modifications which may improve the appearance or reduce the overall cost of the project, Fig. 3.2.6. Occasionally, an enterprising owner will incorporate a full mockup panel, including a mockup of all interior finishes, into a rental office at the actual jobsite. These larger samples require considerable time in their manufacture, and should not be specified unless considerable lead time is allowed for the entire project. Also, it may be desirable not to include the mockup costs in the base bid so that this cost can be evaluated separately.

Mockups are prepared to represent the approximate final effect of materials, color, texture, scale, patterns, shadow

Fig. 3.2.4

Fig. 3.2.5

Fig. 3.2.6

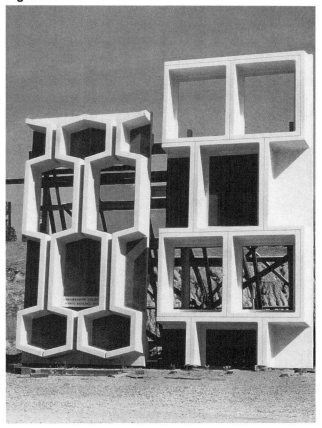

Fig. 3.2.7

and nuances of form desired by the architect. Changes in aggregate orientation, color tone, and texture can easily be noted on full scale mockup panels. Mockups also offer the opportunity for validation of design and details at an early stage in the construction cycle, when modifications can still be made to details, molds, window gaskets, and connections. Mockup panels should contain typical cast-in-inserts, reinforcement, and plates as required for the project. Handling the mockup panels serves as a check that the stripping methods and lifting hardware will be suitable.

The objective of the mockup sample can also be to demonstrate the more detailed conditions that may be encountered in the project (recesses, reveals, outside/inside corners, multiple finishes, textures, veneers, etc.). This sample may not be fully representative of the exact finishes that can be reasonably achieved during mass production.

Mockups are particularly important on veneer-faced panels for evaluation of stone finishes and acceptable color variations. The mockup in Fig. 3.2.7 was used to select the color of stone, windows, and caulking as well as to judge overall building appearance. Acceptance criteria for both the stone and the anchorage should be established in the project specifications, and tests should be made on the sample panels to confirm the suitability of the stone anchors and the effects of bowing on the panel's performance. Tests may also be required to evaluate the behavior of the unit during anticipated temperature changes.

Esthetic mockups can offer the opportunity to evaluate the following factors:

1. Range of acceptable appearance in regard to color, texture, details on the exposed face, and uniformity of returns.

2. Sequence of erection.

3. Available methods of bracing units prior to final structural connections being made.

4. Desirability of the method of connection in light of handling equipment and erection procedures.

5. Colors and finishes of adjacent materials (window frames, glass, sealants, etc.).

6. Dimensional accuracy of the precast concrete work

and the constructibility of the specified tolerances.

7. The acceptability of the precast concrete panel inside surface finish (where exposed).

8. Available methods for the repair of chips, spalls or other surface blemishes. The mockup will also establish the extent and acceptability of defects and repair work.

9. Suitability of the selected sealers.

10. The weathering patterns or rain run-off on a typical section of precast concrete panel facade.

It is desirable at the time of manufacture and erection of the mockup that all interested parties be present and able to discuss the various problems as they arise. All information should be recorded and data fed back to the designer, precaster, or erector if changes are desired. Only after such changes have been implemented and the mockups approved should production begin.

Where mockup units are not used, the architect should insist on examination and approval (sign and date) of early production units, although it may be too late to make any significant design changes. To avoid possible later controversies, this approval should precede a release for production. The architect should realize, however, that delays in visiting plants for such approvals may upset normal plant operations and the job schedule. It should be clearly stated in the contract documents how long the production units or the mockup structure should be kept in the plant or jobsite for comparison purposes. It is recommended that the contract documents permit this approved full-sized unit to be used in the job installation.

Selection of at least three typical panels from the initial production will permit establishment of a range of acceptability with respect to color and texture variations, surface blemishes, and overall appearance. These panels should not be scheduled for erection until the late stages of construction. The panels should be stored outdoors and adjacent to each other to allow proper lighting (sun and shade) for comparisons. At the plant, the panels are generally stored on adjacent racks. At the jobsite, an independent steel frame or cast-in-place support structure may be required. When erected, the panels should be mounted adjacent to each other on the building to allow continued comparison, if necessary. Completing the sample panel process is extremely important and helps to develop communications among all parties.

Plant inspection by the architect during panel production is encouraged. This helps assure both the architect and the precaster that the desired end results can and are being obtained.

3.2.5 ASSESSMENT OF SAMPLES

In assessing samples nearly everyone has a tendency to apply a little wishful thinking when trying to visualize how the entire unit—or even the entire facade—might look from viewing and touching a small 12 in. square sample. This may lead to disappointments later when production is in progress and remedies are expensive or impossible.

Fig. 3.2.8

If small samples are used to select the aggregate color, the architect should be aware that the general appearance of large areas of a building wall will tend to appear lighter than the samples.

For example, exposed gray granite (salt and pepper) may look good on a small sample, but frequently comes out "mottled" in an actual panel, if the aggregate pieces are small. If the predominant part of the granite is white, the mottling will be made worse with gray matrix, and vice versa. This finish may be made more acceptable if the face is sandblasted because of the resulting dulling of the colors, but it is still better to increase the size of the aggregate to eliminate visual merging of the colors.

Mockups can be effectively assessed only if mounted in their final orientation; horizontal, vertical, or sloping. Samples viewed from a distance of a few feet will reveal details that are lost on a building which cannot be viewed from closer than 50 to 100 ft. Details should be appraised from a distance typical for viewing the installed panel. Overlooking this fact may lead to demands for shapes, textures and drafts which not only are expensive, but may not even be identifiable in the finished building.

Another good example is the fluted panel. When viewed from a distance, the ribs should be reasonably deep to read. They should also have a draft (see Section 3.3.2) related to the depth and spacing of ribs in order to facilitate stripping without damage. Some precasters advocating increased draft have shown that draft does not detract from appearance by making panel samples with different drafts, and by having them appraised by the architect from a distance, typical for viewing the installed panel.

Hand-tooled or hammered ribs are expensive but may be justified if the panels will be viewed at close hand such as stairway panels and interior walls or on ground floors. On upper floors a similar effect may be achieved at much less cost by retarding or sandblasting the ribs. Hammered ribs are shown in Fig. 3.2.8. Well proportioned ribs in exposed aggregate finish are pictured in Fig. 3.2.9. The ribs in Fig. 3.2.10 also in exposed aggregates are rather delicately dimensioned. The panels have a flat border area to accommodate variations in panel sizes, thus eliminating the need for any bulkheading in the ribbed area.

The architect should let the samples "weather" and observe them under the climatic conditions to which the

Fig. 3.2.9

Fig. 3.2.10

precast concrete units will be exposed, such as sunshine, rain and shadows. The samples should also be judged in relation to adjacent buildings.

Time rarely allows weathering of samples over an adequately long period, but it is particularly important where a project with precast concrete is contemplated for production in stages. The architect is advised to limit the choice of aggregates and finishes in such projects to those which are in common use and which are easy to duplicate in later stages. To counter weathering effects, cleaning of an earlier stage upon completion of the next one will often provide a reasonable match.

To obtain a reasonable uniformity of appearance, a balance may have to be struck between configuration of the precast concrete unit and the choice of a concrete mix. Returns in some finishes will not appear exactly like the front face (down-face) due to casting techniques and aggregate shapes. See Section 3.3.7. This should be recognized and accepted within certain limits, because it may well influence the architect's choice of shape, materials and finishes.

Difficult mixes and finishes with respect to uniformity may be appropriate and economical in flat panels cast face-down and without any appreciable return, but not in highly sculptured panels.

Lightweight aggregates of known and acceptable quality (meeting ASTM C330) may be used for architectural precast concrete if locally available. Lightweight aggregates

should meet reasonable absorption requirements as given in Section 3.6.12. Precasters should demonstrate that they have the required experience to work with lightweight materials. Special requirements for mixing of lightweight and normal weight concrete within one unit are covered in Section 3.2.6.

Special aggregates can lead to unusual and dramatic finishes. The cost, while higher, will not be prohibitive. A two inch face mix will use approximately 15 lbs of coarse aggregate per sq ft. Cost of aggregate will be approximately $0.20 per sq ft for each $25 per ton of delivered aggregate cost to the precast concrete plant. When small aggregate sizes allow the use of a thinner face mix, the cost will be lower. The cost of special aggregates is not much of a premium compared to the cost of natural stone cladding, such as granite, marble or limestone. Local aggregates, e.g., gravels, however, should never be overlooked. They will be more economical and may look very attractive with the proper matrix and finish. The precaster is a good judge of locally available aggregates and will often have different samples of such material on display, see Fig. 3.2.2, along with those using special aggregates.

The architect should look at the many existing precast concrete applications and also recognize that added variations and new design concepts are possible.

3.2.6 ASSESSMENT OF CONCRETE MIXES

The architect should specify the parameters of concrete

performance requirements, but the actual design of the concrete mix should be left to the precaster.

So that the architect may appraise the appearance and the expected performance of a precast concrete unit using a specific mix, information should be obtained about the mix to assess its anticipated performance and appearance. Such an assessment should be part of the pre-bid sample procedure described in Section 3.2.3.

This section discusses concrete characteristics to help the designer specify the proper concrete requirements and evaluate the mixes proposed for the project.

A design strength for concrete should be determined by the architect, based upon in-service requirements, not forgetting production and erection considerations.

Since precasting involves stripping of units at an early age, rapid strength development is of prime importance. Transportation and erection involve the next strength requirement to which precast concrete units are exposed. The precaster should establish minimum stripping and transportation strength requirements. These strength levels will depend on the shape of the unit, handling, shipping and erection techniques, and will normally result in a high 28-day strength.

A 5000 psi compressive strength at 28 days normally satisfies production requirements and also assures proper durability.

In cases where the 5000 psi strength is not structurally necessary, or may be difficult to attain due to special cements or aggregates, the architect may still achieve sufficient durability and weathering qualities by stating proper air-entraining and absorption limits at a strength level as low as 4000 psi.

The strength of facing concrete is usually determined by using 6x12 in. standard cylinders. If fabrication of cylinders is impractical, 4 in. cubes may be used. The measured cube strength should be reduced 20% (unless strength correlation tests to 6x12 in. cylinders have been made) to obtain an estimate of cylinder strength. It may be impractical to prepare a standard test cylinder, for example, in the case of a facing mix containing a high percentage of coarse aggregate. The 4 in. cube will provide an adequate size for practically all facing mixes. Such cubes may be prepared as individual specimens or they may be sawed from 4 in. thick slabs. The slabs may be more convenient and are probably more representative of the final product.

In assessing the strength of concrete, statistical probabilities should be considered. Many variables can influence the strength of concrete even under close control. The strength level of the concrete should be considered satisfactory if the average of each set of any three consecutive cylinder strength tests equals or exceeds the specified strength and no individual test falls below the specified value by more than 500 psi. Alternatively, compressive results from a predetermined number of consecutive tests may be processed statistically, and the standard deviation established. This approach will measure the overall uniformity in performance. See also ACI 214, *Recommended*

Practice for Evaluation of Compression Test Results of Field Concrete.

It is advisable to specify air-entraining requirements for face mixes in precast concrete units exposed to freeze and thaw cycles in the presence of moisture.

An air-entrainment of 4% is normally desirable. Taking into consideration the many special consolidation techniques used for placing face mixes, a fairly liberal variation of this percentage, such as +2% and −1% should be allowed. An amount of air specified as 4% is thus acceptable if actually measured between 3% and 6%.

For most architectural facing mixes it may be impractical to entrain a specific air content due to gradation, high cement contents, and low slumps. Instead, the use of an air-entraining agent to provide a "normal" dosage is recommended. Normal dosage of air-entraining agent should produce an 8 to 10 percent air content when tested in accordance with ASTM C185, but using only the mortar (material passing the No. 4 sieve—less than 1/4 in.) portion of the proposed mix. The addition of normal amounts of an air-entraining agent to harsh gap-graded facing mixes will improve the workability and increase resistance to freezing and thawing even though only a small amount of air is usually entrained.

Since precast concrete units are generally erected in an above-grade vertical position, which is a moderate environment, air contents as low as 3 to 5% appear to provide the required durability. Low levels of air-entrainment are preferred, since the compressive strength of cast-in-place concrete is reduced by approximately 5% for each 1% of entrained air (when the water-cement ratio is held constant). Strength reductions tend to be greater in mixes containing more than 550 lb. of cement per cubic yard. Since most architectural precast concrete mixes contain a high cement factor, relatively high reductions in strength may be anticipated with high levels of air entrainment.

Apart from stating air-entraining requirements when necessary, the choice of admixtures should be left entirely to the precaster. However, it is the architect's prerogative to demand information about admixtures in the concrete mixes proposed for the project.

As a control measure for staining of concrete due to weathering, it is recommended that maximum water absorption limits be established. This subject is covered in greater detail in Section 3.6.12.

A concrete mix designed for purely structural reasons or for acid etched finishes (light exposure of aggregate) is normally fully (continuously) graded, which means that it contains all the sizes of aggregates (below a given maximum) in amounts which ensure an optimum density of the mix.

Concrete finishes exposing the aggregates are obtained by removing the cement/sand matrix from the exposed surfaces. During the removal process, coarse aggregate in the middle size range may not be able to adhere to the remaining concrete surface. If these sizes are not eliminated from the mix, the percentage of the surface covered by the matrix (sand and cement) may be too large and aggregate distribution too uneven to provide a good surface appearance.

Consequently, face mixes are usually gap-graded, which means that one or more of the intermediate sizes of coarse aggregate is left out of the mix. This leads to a concentration of certain aggregate sizes in excess of standard gradation limits, which are normally waived for architectural concrete face mixes.

While gap-grading is an established and well-proven practice, it should not be carried to extremes. This may cause separation of the paste and aggregates and thus create uniformity problems, especially where the mix is not deposited in its final location as illustrated in Figure 3.2.11. The amount of fines, cement and water should be minimized to ensure that shrinkage remains within acceptable limits and that surface absorption will be low enough to maintain good weathering qualities. The durability of the concrete would normally not be affected by any degree of gap-grading as long as proper concrete cover is maintained over the reinforcement.

The degree of gap-grading should be based on appearance, but related to production considerations and the weathering qualities desired for the specific exposure of the concrete.

Where a precast concrete unit is manufactured with an architectural concrete face mix and a structural concrete backup mix, these mixes should have reasonably similar shrinkage and thermal coefficient of expansion characteristics in order to avoid possible undue bowing or warping. Consequently, these two mixes should have similar water-cement and cement-aggregate ratios.

The combination of normal weight face mix and a backup mix with lightweight aggregates may increase the possibility of bowing or warping. Before accepting such a combination of mixes, sample units produced, cured and stored under anticipated production conditions, are often desirable to verify satisfactory performance.

The use of a face mix and a subsequent backup mix, or the use of a uniform mix throughout a unit, depends on the practice of the particular plant, the size and shape of the unit, and the type and extent of finish being produced. Units with complicated shapes and deep, narrow sections may require a uniform mix throughout the unit, in which case it should be called for in the specifications. Otherwise the choice should be left to the precaster. If a uniform mix is required, costs will be higher unless economical aggregates are used. However, a uniform mix may also be used when the material savings do not warrant the added costs of working with two mixes.

If a separate face mix is to be used, a minimum thickness should be determined. The thickness of a face mix after consolidation should be at least one inch or a minimum of 1-1/2 times the maximum size of aggregates used; whichever is the larger. If larger aggregates are hand-laid in the mold, these dimensions should apply to the concrete mix used as the matrix.

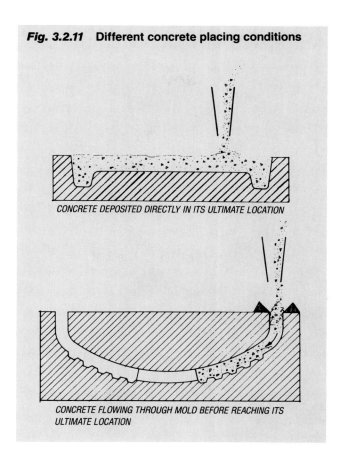

Fig. 3.2.11 Different concrete placing conditions

CONCRETE DEPOSITED DIRECTLY IN ITS ULTIMATE LOCATION

CONCRETE FLOWING THROUGH MOLD BEFORE REACHING ITS ULTIMATE LOCATION

The one inch dimension is chosen because the consolidated face mix is often used to support the reinforcing steel cage and thus provide the proper concrete cover over the reinforcement. For units not exposed to weather or for face mixes applied face-up, this dimension may be reduced provided the backup mix does not bleed through the face mix.

In addition to sample approval and assessment of the concrete mix for expected performance, the architect should check the following requirements:

1. Documentation from the precaster that the concrete mix is properly designed for appearance, strength, durability and weathering. Also that it is suitable for the particular panel configuration and the anticipated production techniques.

2. Materials, particularly aggregates, are suitable and available in sufficient quantities.

3. The precaster has facilities and procedures for uniform batching and proper mixing.

4. The precaster has the facilities, experienced personnel, and established quality control and recordkeeping procedures.

3.3 SHAPE, FORM AND SIZE
3.3.1 OPEN OR CLOSED UNITS

The shape of a precast concrete unit can be an important cost consideration. A major factor is whether the units'

Fig. 3.3.1 Panel shapes

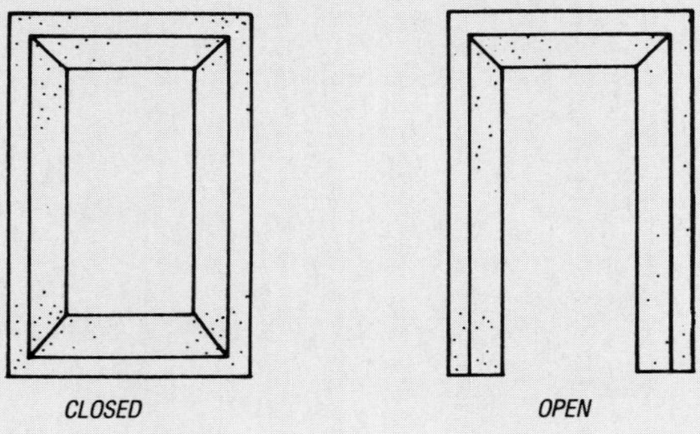

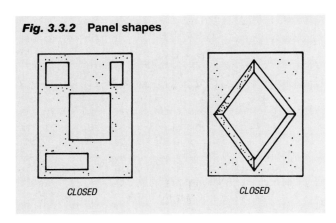

CLOSED   OPEN

Fig. 3.3.2 Panel shapes

CLOSED   CLOSED

Fig. 3.3.3

Fig. 3.3.4 Panel shapes

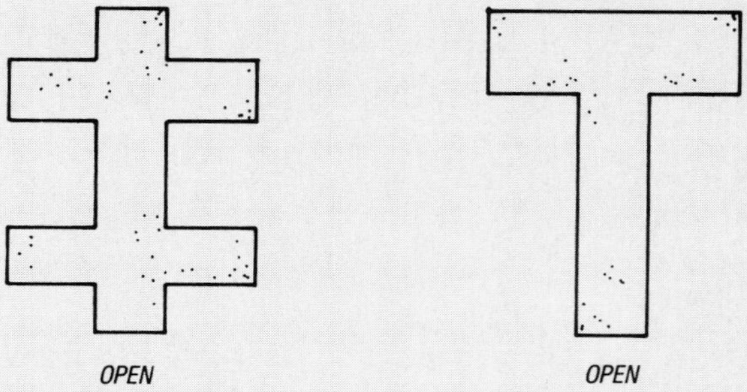

OPEN   OPEN

Fig. 3.3.5 Proportioning of open units

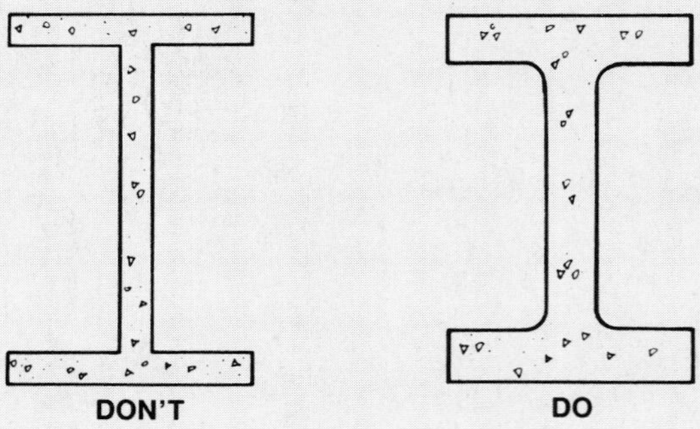

DON'T   DO

shape can be characterized as open or closed. Precast concrete units should be rigid, to allow for easy handling, and closed units afford this rigidity because of their shape. Spandrel panels are normally easy to handle as closed shapes. They may occasionally have large returns which require special attention.

A window unit is a typical example of a closed shape; the same window unit without the sill portion is an open shape, Fig. 3.3.1. Other examples of closed and open unit shapes are shown in Figs. 3.3.2, 3.3.3, and 3.3.4.

Open units are normally more delicate and may require temporary stiffeners or strongbacks for safe handling, thus adding to the cost. Also, some open panels may be difficult to store without the risk of developing excessive bowing or warping. This does not mean that open shapes should not be used, for their basic weaknesses can be overcome by proper proportioning, Fig. 3.3.5.

Combinations of closed and open shapes, as shown in Fig. 3.3.6, have better rigidity, but the cantilevered sections should be proportioned to minimize deflection and tolerance problems. Close tolerances must be maintained during production and curing to properly match units with open shapes during installation. The architect may help by choosing details which will minimize deviations, see Fig. 3.3.7.

The interfacing of windows and precast concrete panels is fairly simple in the case of closed shapes since connections and joint details are independent of site conditions

and tolerances and governed only by tolerances which relate to the manufacturing for the two products. In the case of open units and spandrel panels, the interfacing of windows in the facade will have to allow for slightly larger and more uncertain site construction tolerances. Where window openings occur between such units, glazing can only be accommodated by a window frame, which considers the appropriate tolerances of the opening.

As described in Section 4.5.2, all panel connections must allow for minor movements of the panels in relation to the supporting structure. In the case of open units and spandrel panels, it is important that similar allowance for movements be designed into the panel-to-window connections and the joints between the concrete and window sash. In

Fig. 3.3.6 Combinations of open and closed units

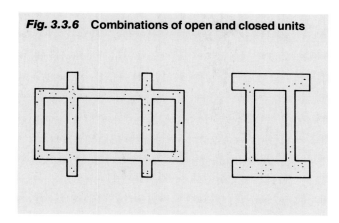

Fig. 3.3.7 Joining of open shaped units

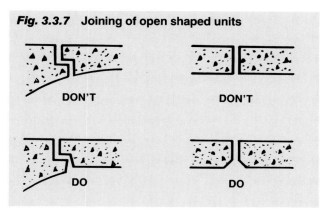

Fig. 3.3.8 Draft concepts

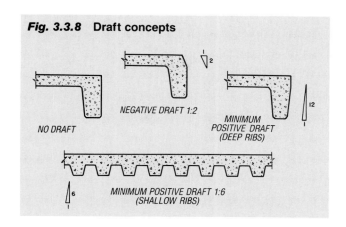

the case of closed panels, these movements are accommodated only in the joints between the concrete units. Furthermore, closed panels may, in some cases, allow direct glazing into the concrete, eliminating window sashes, see Section 5.2.5. Also, windows in closed shapes can be installed and glazed on the ground (most often in the precast concrete manufacturer's plant) which may result in overall cost savings for the facade.

3.3.2 DRAFTS

In establishing the shape of a panel, the designer should take into consideration the draft required to strip the precast concrete unit from the mold, as well as the draft required to facilitate a specific finish. Drafts on the sides of all openings and along the edges of the panel allow the panel to be separated from the mold without disassembling the mold. Generally, the minimum positive draft for ease of stripping the unit from a mold is 1 in. in 1 ft (1:12), with 1:8 preferred. This draft should be increased for narrow sections or delicate units where the suction between the unit and the mold becomes a major factor in both strength requirements and reinforcement of unit. The draft should be increased to 1:6 for units pierced with many openings, for narrow ribbed panels, smooth concrete, and for very delicate units, Fig. 3.3.8. Drafts for ribbed panels should be related to the depth, width, and spacing of the ribs. In deep sections, a 1:6 draft may result in an unattractive, bulky design. In order to minimize mold changes, discussion on draft should be initiated as early as

possible after award of contract.

The drafts required for finish consideration are a function of the shape of the panel, the specified stripping strength of the concrete, the mold release agent selected, the production techniques and the desire for long term durability. The architect is urged to consult the local precasters for specific recommendations. Vertical sides or reverse (negative) drafts will create entrapped air voids which may be objectionable. Without repetition, mold and production costs increase with negative draft since a slip block would have to be incorporated with the side rail and removed with each panel during stripping or the side rail removed in order to strip the panel. When the side rail must be removed, dimensional tolerance becomes a daily variable. Before requiring a negative draft on the top of a parapet panel, consideration needs to be given to the roofing or flashing details required for the parapet and the finish. The greater the draft the architect can allow, the better will be the uniformity and economy of the finish. A compromise may have to be established between the finishing requirements and the shape of the precast concrete unit. A precast concrete unit exposed all the way around but with good draft for the use of a complete envelope mold is shown in Fig. 3.3.9.

3.3.3 TRANSITIONS

The transition from one mass of concrete to another within a precast concrete unit is a prime consideration in finalizing shape. Wherever possible, this transition should be gradual. As an example, a large radius or fillet is preferred for the transition from a mullion, sill, or eyebrow to a flat portion within a precast concrete element, Fig. 3.3.10. This is necessary to minimize the possibility of shrinkage stresses causing crack formation at the point where two concrete sections of distinctly different mass meet. A gradual flowing transition of shape along with well distributed reinforcement will reduce the risk of this type of cracking, Fig. 3.3.11.

Where a reasonable transition is not possible, a concrete mix with low shrinkage characteristics (low water and cement content) becomes increasingly important. Alternatively, the transition may be accomplished by two stage (sequential) casting as illustrated by the example in Fig. 3.3.12. In this instance, the slender horizontal mullions

Fig. 3.3.9

Fig. 3.3.10

Fig. 3.3.11 Transition concepts

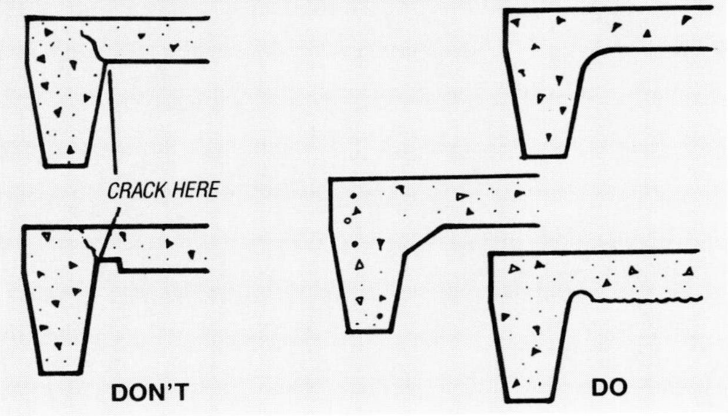

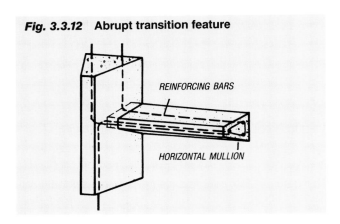

CRACK HERE

DON'T   DO

Fig. 3.3.12 Abrupt transition feature

REINFORCING BARS

HORIZONTAL MULLION

were precast separately and, after curing, placed in the mold for a four story window unit. Two stage precasting involves an extra operation that will add to the cost of the panel.

3.3.4 DEMARCATION FEATURES

When the surface of a precast concrete panel has two or more different mixes or finishes, a demarcation (reveal) feature is a necessary part of the design, Fig. 3.3.13. A deep demarcation separates the lightly sandblasted concrete from the exposed aggregate center section of the panel in Fig. 3.3.14. The depth of the groove should be at least 1-1/2 times the aggregate size and the width should be in dimensional lumber increments such as 3/4 or 1-1/2 in. The groove should generally be wider than it is deep in order to strip the panel without damaging the mold.

The importance of the separation provided by a demarcation feature depends on the configuration of the unit on which the finishes are combined. For example, a groove or offset is necessary when an exposed aggregate flat surface is located between widely spaced ribs with a different surface finish, but not necessary when a similar flat surface lies between closely spaced ribs. Proper samples should be used to assess the problem. The importance of the separation also depends on the specific types of finishes involved. See Section 3.5.13 for a discussion of finish combinations or variations on the same panel.

If a demarcation groove occurs near a change of section, it

Fig. 3.3.13 Demarcation features

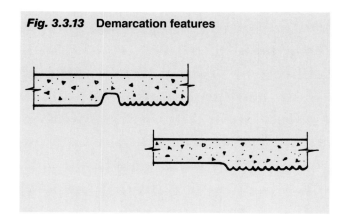

may create a plane of weakness (potential crack) and counter any attempt to provide a gradual transition from one mass to another. It may be necessary to thicken the section to compensate for the groove or provide a more rounded groove than would normally be used, see Fig. 3.3.11. Reglets, window grooves and false joints (rustications) will similarly reduce the effective section of the unit. In some cases, these features may determine the minimum section thickness required for the unit.

3.3.5 SCULPTURING

Sculptured panels can produce building facades with distinctive, strongly modeled elevations having flat interior wall surfaces. The light and shadow effect achieved by

Fig. 3.3.14

Fig. 3.3.15

sculpturing the exterior surface produces the major visual overall effect of precast concrete units; textures and colors are only of secondary importance when a building is viewed in its entirety, or from a distance, see Fig. 3.3.15. The intricacy and depth of articulation of the facade in Fig. 3.3.16 provides a feel similar to the terra cotta buildings built at the turn of the century.

Contrary to common belief, sculpturing of a wall unit will not constitute a cost premium where sufficient repetition of the unit will keep mold costs within reason and where the sculpturing will aid the unit's structural capacity.

Most precasters, for reason of economy, prefer to make the units as large as consistent with normal handling and shipping limitations. This will be discussed in Section 3.3.9. In order to handle large units and maintain structural integrity, they should have a certain depth-to-span ratio.

Sculpturing may increase the structural strength of the units and thus simplify handling. The panels should be shaped for sufficient stiffness in the direction of handling-induced stresses. Precast concrete panels molded around windows are often set forward of the glazing, adding stiffness and giving sculptural form, Fig. 3.3.17. Sculpturing may increase the depth-to-span ratio by providing ribs or projections in either direction of a unit, Fig. 3.3.18 and Fig. 3.3.19. The depth of panels is defined in Fig. 1.4.1, page 21, and the span would be either the height or width of a unit. With sufficient panel repetition, and where the depth and volume of the projections do not greatly exceed the opti-

Fig. 3.3.16

Fig. 3.3.18

Fig. 3.3.17

Fig. 3.3.19

Fig. 3.3.21

mum required for handling, there should be no cost premium beyond the cost of the added volume of materials. One guideline is to limit the depth of a sculptured unit in inches to less than the panel length in feet. For example, if an 8 X 13 ft panel has a configuration creating two-way ribs, or projections, of approximately 8 in. in the short direction and 13 in. in the long direction, the reinforcement required for safe handling is not likely to exceed its in-service requirements.

The projections do not have to be continuous or straight to serve this purpose, but may be overlapping or curved as long as no weak section is created within the units, Fig. 3.3.20. An example where the projections do not add to structural capacity because they are interspersed between

weaker sections is illustrated in Fig. 3.3.21.

Ribs may be part of the architectural expression or, where flat exterior surfaces are required, ribs may be added to the back of panels for additional stiffness. Although back-forming for the rib on the back of panels is an added expense, it may also be necessary to use to reduce the weight of the panels.

In units with ribs in only one direction, the dimension in the other direction might either be shortened or strengthened by using ribs on the back. A panel which has reasonable stiffness in the vertical direction but is weak horizontally is shown in Fig. 3.3.22. Because panels were to be transported several hundred miles (after pre-glazing in the plant), the precaster chose to improve the structural

Fig. 3.3.20

Fig. 3.3.22

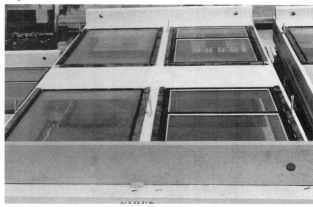

strength by incorporating a concrete rib and a steel beam on the back. These were subsequently used as connections to the structure.

Dimensions of ribs will in most cases, be determined as part of the architectural features of the units. Minimum dimensions as determined by design and practical considerations are treated in detail in Section 4.2.8.

3.3.6 EDGES AND CORNERS

It is strongly recommended that all edges of precast concrete units be designed with a reasonable radius or chamfer, rather than leaving them as sharp corners. This is particularly important where the panels may be close to traffic, whether pedestrian or vehicular. When the edge is sharp, only fine aggregate collects there and this weakens the edge. Also, voids occur due to the interference of larger aggregate. Sharp corners chip easily, both during handling and during service in the finished building. Chamfered or radius edges also partially mask irregularities in the alignment of the precast concrete panels.

The size of the radius for edges should be discussed with the local precaster, because the optimum size of the edges will depend on the aggregate size selected, mold materials, and production techniques. Fig. 3.3.23 shows an example of edges with a 5/16 in. radius, but still appearing distinct because its shape is uniform and straight, thus creating sharp shadow lines. This 5/16 in. minimum radius should be satisfactory for smooth concrete, Fig. 3.3.24(a)

or a light texture surface. A 1/4 or 5/16 in. rounded edge may also be satisfactory for an exposed aggregate finish with a maximum aggregate size of 1/2 in., Fig. 3.3.24(b). Above that size, the edge must be progressively more rounded, Fig. 3.3.24(c). When 1 in. diameter and larger aggregates are used, it may be advisable to have the exposed aggregate stop at a rustication, and have the edge detail be smooth concrete with a 5/16 in. radius, Fig. 3.3.24(d).

Design of corners in buildings clad with precast concrete units demands special attention by the architect. Economy results if the building elevations are designed from the corners inward using typical panels, as this avoids specially sized end or corner pieces.

Mitered corners are difficult to manufacture and erect within tolerances that are acceptable from either an appearance or a jointing standpoint. Concrete cannot be cast to a sharp 45 deg. point because of the size of the aggregates. Therefore, this edge must have a cut-off or quirk, Fig. 3.3.25. The size of the quirk return should never be less than 3/4 in., nor less than 1.5 times the maximum size of the aggregate used in the concrete mix. Normally a 3/4 to 1-1/2 in. quirk will read as a well-defined vertical edge on the corner of the building. A table showing the recommended size of the quirk return for different panel joint sizes is included in Fig. 3.3.25. A well detailed and fabricated miter and quirk miter is shown in Fig. 3.3.26.

Even with good size quirk returns, a mitered corner may tend to resemble the sketches in Fig. 3.3.27 with panels

Fig. 3.3.23

Fig. 3.3.24 Effect of aggregate size on corner details

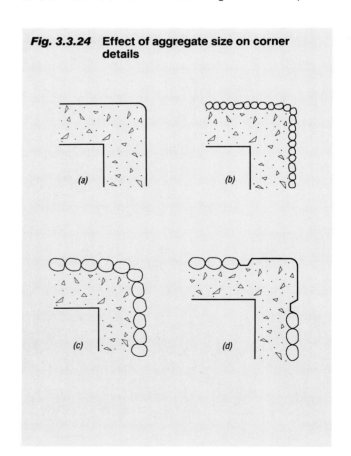

Fig. 3.3.25 Quirk miter dimensions

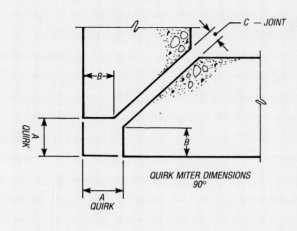

QUIRK MITER DIMENSIONS
90°

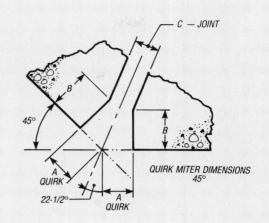

QUIRK MITER DIMENSIONS
45°

A	B	C
1 1/8	3/4	1/2
1 1/4	7/8	1/2
1 1/2	1 1/8	1/2
1 3/4	1 3/8	1/2
2	1 5/8	1/2

A	B	C
1 1/4	3/4	3/4
1 1/2	1	3/4
1 3/4	1 1/4	3/4
2	1 1/2	3/4
1 1/2	13/16	1
1 3/4	1 1/16	1
2	1 5/16	1

A	B	C
5/8	13/16	1/2
3/4	1 1/8	1/2
7/8	1 7/16	1/2

A	B	C
3/4	13/16	3/4
7/8	1 1/16	3/4
7/8	13/16	1
1	1 1/16	1

Fig. 3.3.26

Fig. 3.3.27 Unfortunate mitered corner details

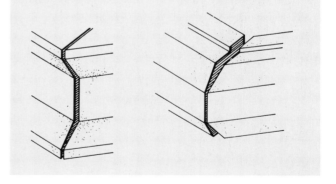

converging at top and bottom or at the center, depending on the vertical configuration of the panels. If the building design demands corners with mitered edges, the architect is urged to specify a mockup of the two initial corner panels at the precast concrete plant before approving the panels and releasing the balance for production.

Some precasters occasionally choose to pre-assemble corner pieces at the plant. This procedure can be both efficient and economical. Fig. 3.3.28 shows an example of such an assembly done in a jig in the precaster's yard and Fig. 3.3.29 pictures another pre-assembled corner piece in place.

The variations in the overall length of a building elevation,

Fig. 3.3.28

assuming that these stay within stated tolerances, may either be accommodated in the joints or in the design of the corner pieces, see Section 4.6 on tolerances. Fig. 3.3.30 shows reasonable corner details which rely on these variations being absorbed by the tolerances in the vertical joints. Due to the reduced size of the corner panels, they will normally undergo less thermal movement and can therefore tolerate greater joint width variation. In such cases the corner pieces may economically be designed and produced as part of one of the adjacent typical panels. Fig. 3.3.31 shows details where considerable variations can be accommodated at the corners. It should be noted that such corner pieces may present special connection problems.

Special corner pieces can be cast by using modified standard unit molds, part of the master mold concept discussed in Section 2.3.3. If the size of the project or the available time constraints warrant multiple molds, a separate corner mold is recommended. The expense of using a special corner mold to construct the elegant corner panels in Fig. 3.3.32 on a stone veneer-faced precast concrete panel building was justified based on a value engineering study. A more intricate detail was desired and flat areas of granite were able to be panelized. Corner molds are generally small and simple, and they can add interest to the facade. On a high-rise building, the cost of a small corner mold and the handling of an extra piece may still offset the modification cost of the master mold and/or be justified by the additional flexibility in erection tolerances. Separate

Fig. 3.3.29

Fig. 3.3.30 Corner details with tolerances accommodated in joints

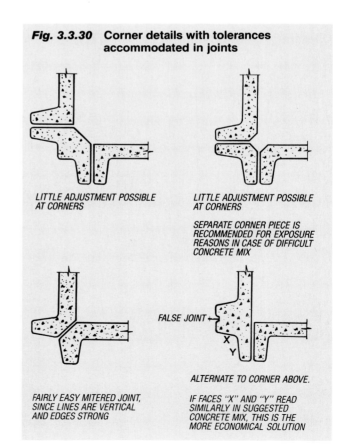

LITTLE ADJUSTMENT POSSIBLE AT CORNERS

LITTLE ADJUSTMENT POSSIBLE AT CORNERS

SEPARATE CORNER PIECE IS RECOMMENDED FOR EXPOSURE REASONS IN CASE OF DIFFICULT CONCRETE MIX

FALSE JOINT

X
Y

ALTERNATE TO CORNER ABOVE.

FAIRLY EASY MITERED JOINT, SINCE LINES ARE VERTICAL AND EDGES STRONG

IF FACES "X" AND "Y" READ SIMILARLY IN SUGGESTED CONCRETE MIX, THIS IS THE MORE ECONOMICAL SOLUTION

corners may also be advantageous in providing similar orientation of corner surfaces for matching finishes. Matching finishes can be very difficult when one is on a down-face and the other is on a return (vertical) face, see Fig. 3.3.33.

Each individual project requires special attention to the design and detailing of its corners in order to determine the optimum solution with respect to appearance, jointing, and economy.

3.3.7 SHAPES IN RELATION TO FINISHES

Before the shape of a precast concrete unit is finalized by the architect, its finish should be considered. Many finishes cannot be achieved with equal visual quality on all faces of the unit.

The reason for this encompasses a number of factors such as mix proportions, variable depths (and pressures) of concrete, and small differences in consolidation techniques, particularly in the case of intricate shapes with complex flow of concrete. The effect of gravity during consolidation forces the large aggregates to the bottom and the smaller aggregates, plus the sand and cement content, upwards. Consequently, the down-face in the mold will nearly always be the more uniform and dense than the returns or upper radius.

Consolidation of the concrete results in a more or less uni-

Fig. 3.3.31   Corner details with flexibility for erection tolerances

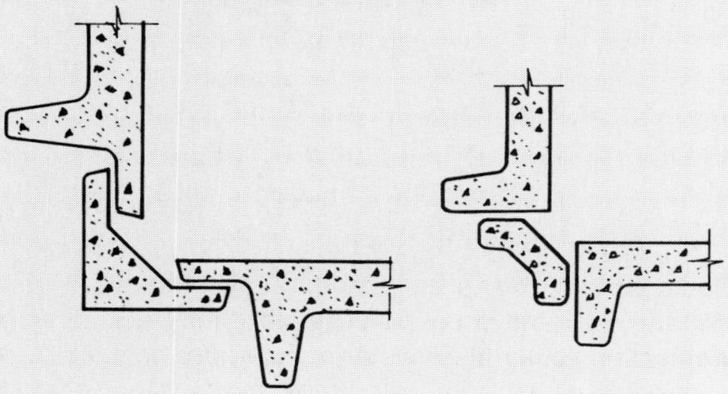

Fig. 3.3.33   Molds for corner units for even exposure

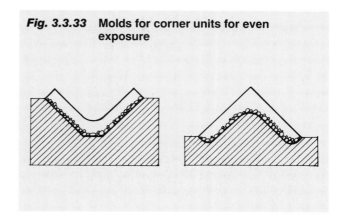

Fig. 3.3.32

form orientation of the aggregate, with the flat, long portion horizontal to the bottom of the mold. On returns and on the upper radius of curved panels the sharp angular points of the aggregate will show upon exposure. This can give the finished texture of returns greater than 12 in. a distinctly different appearance from that of the down-face, Fig. 3.3.34. With deep returns, a more uniform finish is obtained with a retarded exposed aggregate finish. When an exposed aggregate finish is specified, emphasis should be placed on choosing suitable concrete mixes with aggregates which are reasonably spherical or cubical to minimize differences between down-faces and returns. For panels with large returns, or other situations where variations in appearance must be minimized, the two-stage or sequential production technique, as described in Section 3.3.8, should be used, if feasible, or concrete mixes should have a continuous graded coarse aggregate and an ASTM C33 sand. Exposure of aggregates should be medium to deep with minimal color differences between mix ingredients.

If the units are cast so that surfaces with identical orientation on the building are cast in similar positions, small differences in finish between areas cast face-down and as a return should be acceptable, Fig. 3.3.35. When this mold and casting approach is taken, the choice of finish in relation to the configuration of the unit becomes one of determining the acceptable differences in textures between down-face and varying return surfaces. This is best judged by observing realistic sample panels from distances and positions simulating the viewing possibilities of the finished building.

Sculptured panels, channel panels, and panels with deep returns may have visible air voids on the returns. These air voids or "bug/blow holes," become accentuated when the surface is smooth, acid-etched or lightly sandblasted. If the air holes are of a reasonable size, 1/8 to 1/4 in., it is recommended that they be accepted as part of the texture. Filling and sack-rubbing could be used to eliminate the voids. However, this procedure is expensive and may cause color differences. Samples or the mockup panel should be used to establish acceptable air void frequency, size, and distribution.

The architect should accept a small difference between molded and non-molded surfaces, but the orientation of these surfaces should be determined on the drawings or

Fig. 3.3.34 Exposures variances caused by aggregate

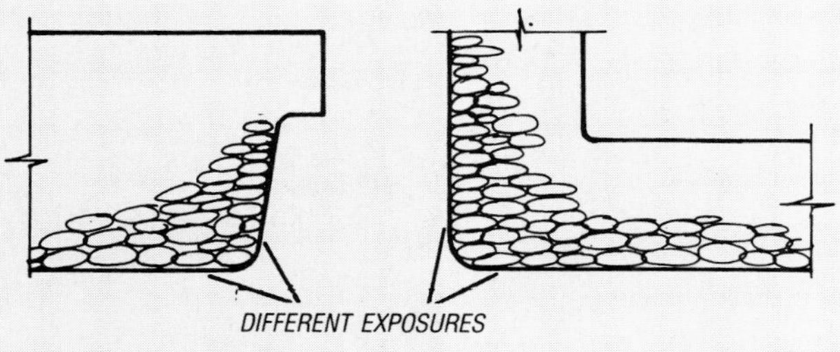

DIFFERENT EXPOSURES

specifications and be consistent throughout the project. The architect should also limit the choice of aggregates and finishes to those which lend themselves to a reasonable matching of the molded surfaces with a hand-finished open surface. Therefore, a smooth finish or one with a light exposure is not appropriate. The open surface may have to be seeded with the larger aggregates as part of the finishing process. When the shape of the units precludes any logical demarcation of formed and unformed surfaces, closed molds cast in a vertical position may be the only answer. This will normally require complicated and expensive molds and/or deep castings.

3.3.8 TWO-STAGE OR SEQUENTIAL PRECASTING

Panels with large or steep returns (such as channel column covers and some spandrels) may be cast in separate pieces in order to achieve matching high quality finishes on all exposed faces and then joined with dry joints as illustrated in Figs. 3.3.36 and 3.3.37. This method of casting enables all faces to be cast face-down with the same aggregate orientation and concrete density using conventional precast concrete forming methods; backforming is not required. Also, a combination of face mix and backup mix can be used, rather than 100% face mix. If this is the indicated production method, attention must be paid to suitable corner details and reinforcement at the dry joints. Although the dry joint may not show with certain mixes and textures, a groove or quirk will help to

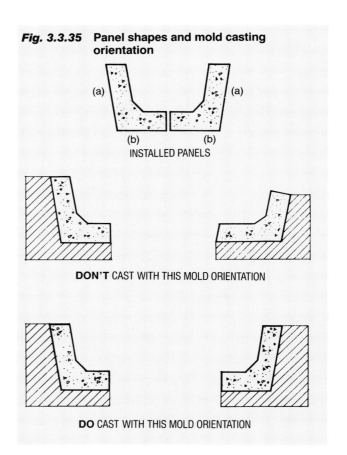

Fig. 3.3.35 Panel shapes and mold casting orientation

(a)   (a)

(b)   (b)

INSTALLED PANELS

DON'T CAST WITH THIS MOLD ORIENTATION

DO CAST WITH THIS MOLD ORIENTATION

Fig. 3.3.36 Separate casting stages of large returns

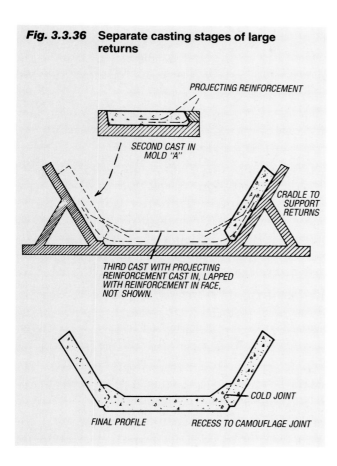

PROJECTING REINFORCEMENT

SECOND CAST IN MOLD "A"

CRADLE TO SUPPORT RETURNS

THIRD CAST WITH PROJECTING REINFORCEMENT CAST IN, LAPPED WITH REINFORCEMENT IN FACE, NOT SHOWN.

COLD JOINT

FINAL PROFILE   RECESS TO CAMOUFLAGE JOINT

Fig. 3.3.37 Alternate casting approaches

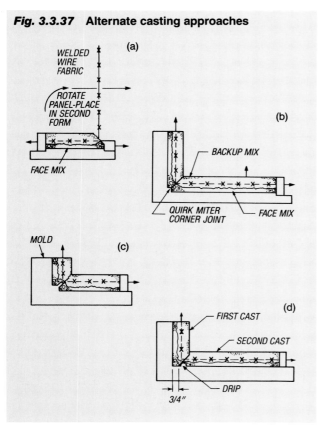

WELDED WIRE FABRIC

(a)

ROTATE PANEL-PLACE IN SECOND FORM

FACE MIX

(b)

BACKUP MIX

QUIRK MITER CORNER JOINT   FACE MIX

MOLD   (c)

(d)

FIRST CAST

SECOND CAST

DRIP

3/4"

mask the joint. Where desired, this joint can be recessed deep enough to allow installation of a small backer rod and placement of a 1/4 in. bead of joint sealant, Fig. 3.3.37c. Sometimes precautions may be necessary to ensure watertightness of the dry joints. The main disadvantage of two-stage precasting is that two or three separate concrete placements are necessary to complete a panel as shown in Fig. 3.3.36. Fig. 3.3.38 shows spandrel panels which were sequentially cast because of large returns.

3.3.9 DIMENSIONS AND OVERALL PANEL SIZE

Panel geometry, referred to in a general sense as shape details which do not affect the architectural concept, can be a major influence on both fabrication economy and engineering requirements. The significant shape details are overall size and configuration. The visual characteristics of a panel will usually be determined by the architect. Preferably, the size of the individual elements and the exact details of the panel geometry should be determined in consultation with the structural engineer and the precaster. Thus, both the architect and the structural designer should be familiar with good production practice as well as production capabilities of the probable fabricators. For overall economy, early coordination of design and erection are essential.

Since many of the manufacturing, handling, and erection costs are independent of the size of a piece, making panels larger can substantially reduce the total cost of the project.

Fig. 3.3.38

The smaller the panel, the greater the number of pieces required for enclosure which means more handling.

Handling of a precast concrete unit constitutes a significant portion of the cost of precast concrete. The cost difference in handling a large rather than a small unit is insignificant compared to the increased square footage of the large unit. In addition to providing savings on erection costs, larger sized panels provide secondary benefits of reduced amounts of caulking, better dimensional controls and fewer connections. Thus, large units are preferable unless they lack adequate repetition, require high tooling costs, or incur cost premiums for transportation and erection. In high seismic areas, very large panels may not be desirable because larger dimensional deformations in the supporting structure must be accommodated. In these areas, normal practice is to use panels that are one story in height, and seldom span more than one structural bay. Where desired, the scale of large panels may be reduced by using false joints (rustications).

The infinite combination of sizes, shapes, materials, and functions that can be incorporated into a precast concrete unit makes it difficult to present a size chart. Overall limiting factors in the physical size of the unit based upon structural requirements are discussed in Section 4.2.9. Other limiting factors may be handling operations during stripping, storage and shipping. To determine the optimum size of architectural panels, a close collaboration between the designer and precast concrete supplier is required. The trend over the past 20 years has been to increase both the size and weight of architectural precast concrete panels.

The most economical element for a project is usually the largest, considering:

1. Production repetition and size of available casting beds.

2. Handling ease and stability and stresses on the element during handling.

3. Transportation size and weight regulations and equipment restrictions.

4. Available crane capacity at both the plant and the project site. Position of the crane must be considered, since capacity is a function of reach.

5. Storage space, truck turning radius, and other site restrictions.

6. Loads imposed on support system. Sometimes multi-story panels can rest directly on foundation walls rather than being hung from the structure. Panels can also be used to span from column to column, thereby reducing floor edge loading.

Limitations of dimensions due to handling and storing vary considerably from plant to plant, but are normally not important considerations for the architect. Tilt tables (Fig. 3.3.39), strong-backs, stiffening trusses or pipe frames for handling, Fig. 3.3.40, and plain ingenuity will often allow the larger pieces to be handled. The precaster may also make use of prestressing to facilitate handling of large units without risk of cracking or damage.

Fig. 3.3.39

Fig. 3.3.40 Temporary strengthening of panels with significant openings

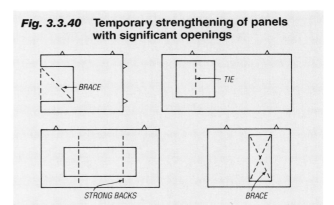

The architect should bear in mind, however, that some finishes, such as bushhammering, honing and polishing, normally require the panels to be turned between casting and finishing. Also, the exposed surface finish may dictate the position of the panel for worker access during removal of surface retarder or sandblasting. This involves special tooling and may add cost for larger panels.

Before deeply sculptured elements are combined into larger units, potential storage problems (weathering or special racks) should be considered, particularly where the period of storage may be long or uncertain.

It is the precast concrete manufacturer's option as to which production and transportation methods will be employed and their responsibility to verify the behavior of the precast concrete units during these operations. However, the designer should be familiar with legal highway load limitations, and with the cost premiums that are associated with transporting members of a particular overheight, overwidth or overlength.

Federal, state (province), and local regulations limit the size and weight of shipping loads. Limitations vary from one locale to another whether the shipment is by truck, barge, or rail. Where large units are to be moved, a thorough check of local statutes is mandatory. Where climatic conditions result in load restrictions during spring thaws, actual timing of the expected transportation becomes significant.

The common payload in many areas is 20 ton with a product size restriction of 8 ft in width, 8 ft in height, and 45 ft in length, Fig. 3.3.41. If a unit will fit within these confines, it can be hauled on a standard flatbed trailer without requiring permits. By use of lowboy (step deck) trailers, the product height can generally be increased to about 10 to 12 ft without requiring special permits, Fig. 3.3.42. However, lowboys (step decks) are not as readily available, and their shorter bed length may restrict the length of precast concrete units. Two triangular frames mounted on an extendable lowboy trailer can carry two architectural precast concrete panels as large as 15 ft 8 in. x 15 ft 2 in. and weighing up to 20,000 lbs. each, Fig. 3.3.43. Finger racks allow architectural precast concrete spandrel beams to be transported in the vertical position. Note the adjustable screw clamps which secure the beams to the racks, Fig. 3.3.44.

In most areas, total heights (roadbed to top of product) of 13 ft 6 in. are allowed without special permit, while in others this limit is 12 ft. On occasion, even such heights require special routing to avoid low overpasses and overhead restrictions. Restrictions generally exist for loads over 8 ft in width; maximum permit widths can vary from 10 ft to 14 ft depending on the area or city. Some areas allow overall lengths over 70 ft with only a simple permit, escorts front and rear, and travel limited to certain times of the day. Beside variations in lengths, heights, and widths, weight restrictions vary widely. Thus, the general load limit without permit (20 to 22 ton) can, in some areas, be increased to 100 ton with special permit, while in others there are very severe restrictions on loads over 25 ton. Permits are issued in most areas only for non-divisible loads. These restrictions will add to the cost of the precast concrete unit, and thus should be evaluated against savings realized by combining smaller units into one large unit.

In determining final dimensions, consideration should be given to utilizing a full truckload. The average payload is 20 tons. A precast concrete unit or several units should approximate this usual payload. For example, an 11 ton unit is not economical, because only one unit can be shipped per load, whereas a 10 ton unit would be economical, because two units per load can be shipped. For quick calculations, standard weight concrete including reinforcement and hardware weighs 150 lbs per cubic foot.

In order to facilitate erection, it is desirable to transport members in the same orientation which they will assume in

Fig. 3.3.41 Common overall trucking volume

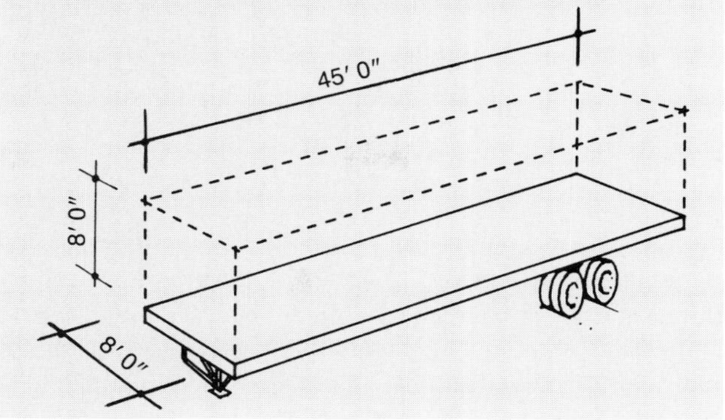

Fig. 3.3.42

Fig. 3.3.43

Fig. 3.3.44

the structure. In many cases this is possible; for example, with single-story wall panels, transportation can be accomplished on an A-frame type trailer with the panels in an upright position from which they can be lifted directly into position. With this type of trailer, good lateral support as well as two points of vertical support are provided to the members, Fig. 3.3.45 a.

Longer units, which are thin compared to their length and width, can be transported in a favorable orientation to reduce tensile stresses. Two or three story panels can be transported on their sides, taking advantage of increased stiffness while supporting the panel on two points, with lateral support along the length of the panel, Fig. 3.3.45 b.

The cost of transportation is not proportional to the distance covered as the cost of loading and unloading (plus protection of the load) becomes less significant on longer hauls. The rate per ton mile is normally reduced for longer hauls. Consequently, long hauls have occurred on competitively bid jobs.

Piggyback transport by rail has been used successfully, but otherwise rail transportation is not common because of the delicate coordination required and the potential damage to units.

Water transportation via barge is inexpensive and safe, but even for plants with navigable water frontage, double handling will occur when the barges reach their destination because the panels will normally have to be transferred to the jobsite by trucks.

3.4 COLORS AND TEXTURES
3.4.1 COLORS

Architectural precast concrete can be cast in almost any color, form, or texture, to meet esthetic and practical requirements of modern architecture.

Design flexibility is possible in both color and texture of precast concrete by varying aggregate and matrix color, size of aggregates, finishing processes, and depth of exposure. Combining color with texture accentuates the natural beauty of aggregates, Fig.3.4.1. Aggregate colors range from white to pastel to red, black and green. Natural gravels provide a wide range of rich warm earth colors, as well as shades of gray.

Color and, consequently, color tone represent relative values. They are not absolute and constant, but are affected by light, shadow, density, time and other surrounding or nearby colors. A concrete surface, for instance, with deep exposed opaque white quartz appears slightly gray. This is due to the fact that the shadows between the particles "mix" with the actual color of the aggregate and produce the graying effect. These shadows in turn affect the color tone of the matrix.

Similarly, a smooth concrete surface will change in tone when striated. Also a white precast concrete window unit with deep mullions will change tone when bronze-colored glass is installed. A change of color tone is constantly going on as the sun travels through the day. A clear sky or one that is overcast will make a difference as will land-

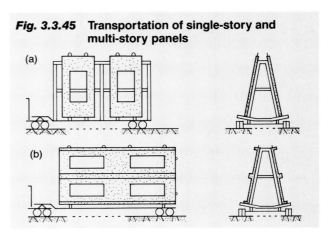

Fig. 3.3.45 Transportation of single-story and multi-story panels

(a)

(b)

scaping and time. And last but by no means least, in large city and industrial environments, air pollution can cause the tone to change.

Color selection should be made under lighting conditions similar to those under which the precast concrete will be used, such as the strong light and shadows of natural daylight. Muted colors usually look best in subdued northern light. In climates with strong sunlight much harder and brighter colors are used with success.

Surface texture also affects color. A matte finish will result in a different color panel than does a smooth finish. Texture helps to determine the visual importance of a wall and hence the color. For example, moderately rough finishes usually are less obtrusive than shiny surfaces. The building's appearance is a function of the designer's success in the use of light, shadow, texture and color.

Cement (plus coloring agent) exerts the primary color influence on a smooth finish because it coats the exposed concrete surface. As the concrete surface is progressively removed and aggregates exposed, the panel color increasingly reflects the fine and then the coarse aggregate colors. Nevertheless, the color of the cement always has an effect on the general tone of the panel.

Cement may be gray, white, buff or a mixture. All cements have inherent color and shading differences depending on their source. For example, some white cements have a buff or cream undertone, while others have a blue or green undertone. In addition, a finely ground gray or white cement is normally lighter in color than a coarse ground cement of the same chemical composition. If color uniformity is essential, white cement from one source only should be specified. Gray and buff cements are generally subject to greater color variation than white cement even when supplied from one source. Normal production variables such as changes in water content, curing cycles, temperatures, humidities, and exposure to climatic conditions at varying strength levels, all tend to cause greater color variations in a gray cement concrete in relation to concrete with white cement. A low water-cement ratio cement paste is almost always darker than a high water-cement ratio paste made with the same cement.

While gray or buff cement can be combined very effectively with many aggregates, the use of white cement, with

Fig. 3.4.1

or without color pigments, greatly extends the range of possible color combinations. Although white cement will give the least amount of color variation, it is important to choose the lightest color aggregates to decrease the shadowing effect of aggregates close to the surface. Gray cement has a greater ability to provide an opaque covering of aggregate, but has color differences that may offset this advantage.

If gray is the desired color of the matrix and the optimum uniformity is essential, a mixture of white and gray is recommended. Uniformity normally increases with increasing percentage of white, but the gray color remains dominant. If such a mixture is desirable, the architect should specify the required percentage of white cement in the contract documents. See also Section 2.3.2 for economy of materials.

Pigments conforming to ASTM C979 often are added to the matrix to obtain colors which cannot be obtained through combinations of cement and fine aggregate alone. Standard colors for integrally colored concrete are red, ivory, cream or buff, yellow, brown, gray, black, green, olive, turquoise, blue and white. Variable amounts of a pigment, expressed as a percentage of the cement content by weight, produce various shades of color. High percentages of pigment reduce concrete strength because of the high percentage of fines introduced to the mix by the pigments. For these reasons, the amount of pigment should be controlled within the limits of strength and absorption requirements. Different shades of color can be obtained by vary-

ing the amount of coloring material or by combining two or more pigments. White portland cement will produce cleaner, brighter colors and should be used in preference to gray cement, especially for the light pastels such as the buffs, creams, and ivories, as well as the bright pale pink and rose tones.

Shades of red, orange, yellow, brown, black and gray are the least costly. Green is quite permanent but expensive, except in light shades. Blue is expensive and some blues are not uniform or permanent. Cobalt blue should be used to avoid problems. Black colors of any kind readily fade to gray because of the thin layer of calcium carbonate that forms on all concrete surfaces when the leached cement hydration products undergo atmospheric carbonation. If the lightening of the color becomes too objectionable, the black can be restored by washing with dilute hydrochloric acid and rinsing thoroughly. Carbon black, due to its extremely fine particle size, has a tendency to wash out of a concrete matrix. As the pigment dissipates the concrete substrate appears increasingly "faded". Synthetic black iron oxide will produce a more stable charcoal color.

Titanium dioxide pigment in quantities of 1 to 3 percent is sometimes used to increase the opacity or to further intensify the whiteness of white concrete. However, it is doubtful that titanium dioxide would overcome the yellowing effect of yellow or brown sands when used in amounts that would not be detrimental to the strength and durability of the concrete. For these same reasons, titanium dioxide cannot be used instead of white cement to produce white concrete. In addition, the practice would not be economically feasible and uniform color would be extremely difficult to obtain.

Architects can best specify the color they desire by referring to a swatch or color card. A cement color card is preferred but one published by a paint manufacturer is acceptable. An excellent color reference is the Federal Color Standard 595 A, published by the U. S. Government Printing Office.

Significant points to consider when color consistency is critical are:

1. Quality and quantity of the coloring agent.

2. Fading characteristics of some coloring agents.

3. Proper batching and mixing techniques and the coloring agent's effect on concrete workability.

4. Quality (freedom from impurities) of the fine and coarse aggregates.

5. Uniform quantities and gradation of fine materials (passing No. 50 sieve and including the cement) in the concrete mix.

6. Careful attention to curing and uniform duplication of curing cycles.

7. Type and color of cement.

8. Constant water-cement ratio in the mix.

9. Consideration of those factors that can contribute to efflorescence. (This is especially important for darker and more intense colors.) Efflorescence deposited on the surface may mask the true color and give the appearance of fading even though the cement paste itself has shown no change. The original color may be restored by washing with a dilute solution of hydrochloric acid and water and rinsing thoroughly. In addition, weathering of the pigmented cement paste exposes more of the aggregate to view. If the color of the aggregate is in contrast to that of the pigment, a change in the overall color of the surface may be noted.

Amounts of pigment in excess of 5% by weight of cement seldom accomplish further color intensity, and in no case should pigment exceed 10% of the weight of cement.

Apparent fading of color can be reduced to a minimal degree by sealing the surface of the panel, when it is cured and cleaned, with a sealer, see Section 3.6.13.

Fine aggregates have a major effect on the color of white and light buff colored concrete, and can add color tones. Where the color depends primarily on the fine aggregates, gradation control is required, particularly where the color tone depends on the finer particles. Where fine aggregates (sand) are manufactured by crushing colored coarse aggregates and bagged by sizes directly from the screening operations, uniformity in gradation can be maintained from one batch to the next. For fine aggregates in bulk, and subject to several rehandling processes, this is not feasible. Consequently, it is recommended that for bulk material, the percentage of the fine aggregates passing the No. 100 sieve should be limited to no more than 5%. This may require double washing of such aggregates, but the premium for this is justified by the increased uniformity of color.

The precast concrete manufacturer should verify that adequate supply from one source (pit or quarry) for each type of aggregate for the entire job will be readily available. If possible, the precaster should obtain the entire aggregate supply prior to starting the project, or have the aggregate supply held by the aggregate supplier. Stockpiling will minimize color variation caused by variability of material and will maximize color uniformity.

For reasons of workability, a percentage of natural sand is preferable in a concrete mix. Manufactured sand, however, often adds valuable color tones, and may be used as part of the fine aggregates. Manufactured sand is generally more expensive than natural sand and may not always be available. Crushed pink granite, for instance, will create a warm colored matrix, but due to the size of sand particles, the pink can hardly be distinguished in the finished unit. With a light to medium exposure, a uniform color appearance may be obtained by using crushed sand of the same material as the coarse aggregate. When maximum whiteness is desired, a natural or manufactured opaque white or light yellow sand is effective. Most naturally occurring sands lack the required whiteness, and the precaster usually must look to the various manufactured aggregates for the white base desired. Generally, these consist of crushed limestone (dolomite, calcite), or quartz and quartzite sands.

The colors in coarse aggregates are multiple, and most precasters will have a supply of aggregate samples at the plant. Selection of aggregates for colors should be governed by the following considerations:

1. Aggregates must measure up to proper durability (soundness and absorption) requirements, be free of impurities (iron oxides) and be available in shapes required for good concrete and appearance (chunks rather than slivers).

2. Aggregate shape affects the appearance of a surface after weathering. Rounded aggregates (pebbles) tend to remain clean, but angular aggregates of rough texture tend to collect dirt, and confine it to the recesses of the matrix. For this reason, as well as architectural appearance, the area of exposed matrix between aggregate particles should be minimized. It may be advisable for the matrix to be darker than the aggregate in structures subject to considerable atmospheric pollution.

3. Final selection of colors should be made from concrete samples that have the proper matrix and are finished similarly to the planned production techniques. Some finishing processes change the appearance of aggregates. Sandblasting will dull the aggregate, while acid etching may increase their brightness. Exposure by retardation normally leaves the aggregates unchanged. The method of exposing aggregate alters the color of the surface by affecting the color of the aggregate and

by the amount of shadow cast by the exposed particles.

4. Aggregates with a dull appearance in a gray matrix may well appear brighter where the matrix is basically light colored.

5. Weathering may influence newly crushed aggregates. When first crushed, many aggregates are bright but may dull slightly with time. Similarly, some of the sparkle caused by acid etching or bushhammering may not survive more than a few weeks. The architect should recognize that samples maintained indoors may not retain their exact appearance after exposure to weather even after a few weeks.

Coarse aggregates are selected on the basis of color, hardness, size, shape, gradation, method of surface exposure, durability, cost and availability. Colors of natural aggregates vary considerably according to their geological classification and even among rocks of one type.

Clear quartz provides a sparkling surface to complement the color created by use of a colored matrix. White quartz ranges from translucent white to a milky white. Rose quartz provides surfaces ranging from a delicate pink to a warm rose color. Green, yellow and gray colors are also available.

Marble offers the widest selection of colors including green, yellow, red, pink, blue, gray, black and white. Blue and yellow marble aggregates are available in pastel hues. Marble is available in many shades running from light to moderately dark. Crushed limestones tend toward white, gray and pink colors.

Granite in shades of pink, red, gray, dark blue-black and white produces a soft mottled appearance when used in concrete. Traprocks, such as basalt, provide gray, black and green colors and are particularly durable.

Some washed and screened gravels supply brown or reddish brown finishes. Yellow ochers, umbers, and sandy (buff) shades abound in river bed gravels. Also, an almost pure white gravel occurs in several sedimentary formations.

Ceramic aggregates and vitreous materials such as glass offer the most brilliant and varied colors. Almost any color can be produced. Glass aggregates must have low reactivity with portland cement. While the colors of ceramic aggregates are bright and clear, the aggregates are characteristically soft.

The local architectural precasters are familiar with available aggregates and usually have concrete samples made with the different materials on display, Fig. 3.2.2.

Coarse aggregates should be reasonably uniform in color. However, surfaces consisting of a single color lack clarity and, strangely enough, purity. Light and dark coarse aggregates require care in blending to provide color uniformity within a single unit. With a small color difference between the light and the dark aggregates and a small variance in total amounts of each aggregate, the chances of uniformity are enhanced. It is advisable to match the color

or tone of the matrix to that of the coarse aggregate so minor segregation of the aggregate will not be noticeable. Panels containing aggregates and matrices of contrasting colors will appear less uniform than those containing materials of similar colors (as the size of the coarse aggregate decreases, less matrix is seen and the more uniform the color of the panel will appear).

The choice of aggregates becomes more critical in smooth white concrete. Due to the greater difference in color between the white cement and aggregates, the white cement has less ability than gray cement to form an opaque film over the aggregates and prevent the aggregate color from showing through. Thus, special consideration must be given to the selection of suitable aggregates to ensure against variations in color and in its intensity on the finished surface. A light-colored aggregate is preferable to a dark aggregate to avoid the possibility of getting shaded or toned areas.

Two concrete mixes with differently colored matrices exposed at the face of the same panel should not be specified. No demarcation feature can prevent a white cement paste from leaking into a gray or vice versa. Such mixtures are acceptable (although at added cost) where one part of the panel (for instance a spandrel part) is cast first of the one mix and, after curing, is cast into the total panel. See Fig. 3.3.36. This method has been successfully employed by proper sealing of the cold joint and by providing special protection to the initially cast concrete element, See Fig. 3.5.87.

The ease of obtaining uniformity in color is directly related to the ingredients supplying the color. Whenever possible, the basic color should be established using colored fine or coarse aggregates (depending on depth of exposure) rather than pigments or colored cements, and extreme color differences between aggregates and matrix should be avoided. In all cases, color should be judged from a full-sized sample that has the proper matrix and has been finished in accordance with planned production techniques.

The sample should be assessed for appearance during both wet and dry weather. The difference in tone between wet and dry panels is normally less with white concrete. In climates with intermittent dry and wet conditions, drying-out periods often produce blotchy appearances in all-gray cement facades. This is particularly true of fine-textured surfaces. On the other hand, dirt (weathering) will normally be less objectionable in gray panels. These comparisons are based on similar water absorption or density of white and gray concrete.

3.4.2 TEXTURES

Textures allow the naturalness of the concrete ingredients to be expressed, provide some scale to the mass, express the plasticity of the concrete, and normally improve its weathering characteristics. A wide variety of textures is possible, ranging from a honed or polished surface to a deeply exposed, 3 in. diameter uncrushed aggregate.

The surface finish enhances the basic design of the building by contributing a tone or feel to the precast concrete.

However, a small concrete sample in its isolation can actually mislead the architect in the value of a finish, which may in reality contribute very little to the total building design when seen in its entirety or from a distance.

As a general rule, a textured surface is esthetically more satisfactory (greater uniformity) than a smooth surface. The surface highlights and natural variations in aggregate color will, to a large extent, camouflage subtle differences in texture and color of the concrete.

A texture may be defined, in comparison with a smooth surface, as an overall surface pattern. The range of textured finishes for concrete direct from the mold include the characteristic imprint or patterns from a form liner or mold. Textured finishes of a different type may be produced by removing the surface mortar to expose the coarse aggregate in the mix, either before or after the concrete has hardened, and also by tooling the hardened concrete.

A profile may be defined in comparison with a flat surface as a shape, rather than a texture, produced from specially made mold or form liner. One well known example is the striated or ribbed finish.

Profiled surfaces can be either smooth or textured; in a similar way flat surfaces can be either smooth or textured. This gives four possible combinations.

It is also possible for part of a surface to be given one finish and part another, so there is a wide range of options from which to choose.

A detailed description of the more common textures is given in Section 3.5 based on the techniques for obtaining such textures.

Four major factors should be considered in choosing a texture:

1. **The area of the surface.** This affects the scale of the texture. Coarse textures usually cannot be used effectively for small areas. Dividing large flat areas into smaller ones by means of rustications tends to deemphasize any variations in texture.

2. **The viewing distance.** The designer may seek a visually pronounced texture or may use texture as a means to achieve a particular tone value. The visual effect desired at the normal viewing distance influences the texture and size of aggregate chosen for the panel face. Fig. 3.4.2 shows different size aggregates viewed at 30 ft and 75 ft.

Suggested Visibility Scale	
Aggregate size, in.	Distance at which texture is visible, ft
1/4 - 1/2	20 - 30
1/2 - 1	30 - 75
1 - 2	75 - 125
2 - 3	125 - 175

These viewing distances are based on the use of aggregates of one color. They will require modifications when the aggregate contains both light and dark particles. Further modifications may be required to include the effects of panel orientation. For example, the contrast caused by shadows from aggregate particles will vary with conditions of lighting.

3. **The orientation of the building wall elevation.** This determines the amount and the direction of light on the surface and how the panel will weather.

4. **Aggregate particle shape and surface characteristics.** For exposed aggregate textures, the aggregate particles may be rounded, irregular, angular or flat. Their surface may be glossy, smooth, granular, crystalline, pitted or porous. Both the shape and surface characteristics determine how the surface will weather and reflect light. Flat, pitted or porous aggregates should only be considered in special applications not exposed to weather.

In addition to the visual effect of texture within reasonable distances, textures may be used to achieve colors based on the natural colors of the exposed aggregates and matrix.

The size of the aggregate should be related to the configuration of the panels. The larger the aggregate, the more difficult it will be to accommodate edges and returns. The smaller the aggregates, the more cement and water is required in the mix increasing the shrinkage of the con-

Fig. 3.4.2

crete and thus the possibilities of hairline cracking.

Exposed aggregate finishes are popular because they are reasonable in cost and provide a good variety in appearance. This variety is achieved by varying the type, color, and size of aggregate, color of matrix, method of exposure, and depth of exposure.

The different degrees of exposure are:

a. **Light Exposure** — where only the surface skin of cement and sand is removed, just sufficiently to expose the edges of the closest coarse aggregate.

b. **Medium Exposure** — where a further removal of cement and sand has caused the coarse aggregate to visually appear approximately equal in area to the matrix.

c. **Deep Exposure** — where cement and sand have been removed from the surface so that the coarse aggregate becomes the major surface feature.

The extent to which aggregates are exposed or "revealed" is largely determined by their size. Reveal should not be greater than one-third the average diameter of the coarse aggregate particles or one-half the diameter of the smallest sized coarse aggregate.

3.5 FINISHES

3.5.1 GENERAL

Finishes, in terms of colors and textures, are discussed in Section 3.4. This section describes the various methods of obtaining these finishes. Since surface finishes depend on properly fabricated molds, the designer should clearly understand the capabilities and limitations of mold production, refer to Section 2.2.

The appropriate finish, which may vary from face to face, should be carefully chosen and clearly specified. The designer should base the final choice of finish on a balance between appearance, quality, and cost. The appearance can be judged using a combination of samples and reduced scale or full scale mockups, see Section 3.2. These samples or mockup panels can then be made available at the precast concrete plant so that all concerned can be assured standards of finish and exposure are being maintained.

The quality of the finish includes the durability of both the individual ingredients and the end product. The quality assessment should also include the likelihood of maintaining a reasonable level of uniformity from start to finish of production. For instance, it is not too difficult to get a uniform distribution of two differently colored aggregates in a small sample produced under laboratory conditions, but it is quite a difficult task to produce the same uniform appearance on a daily basis. Generally speaking, if two differently colored aggregates are contemplated, the difference in appearance (colors) should not be too prominent, and similarly the color difference between aggregates and matrix should also be weighed against the practicality of obtaining a uniform appearance.

A compromise may be required between the finish and the shape of a precast concrete panel. The surface finish should be chosen before the shape of the unit is finalized. Sculptured panels may have visible air voids on the returns which become accentuated when the surface is lightly sandblasted. Normally, smooth finishes also will have air voids on return surfaces. If air holes are of a reasonable size (1/8 to 1/4 in.) it is recommended that they be accepted as part of the texture. Filling and sack-rubbing will eliminate the voids, but this method is expensive and may cause color differences. Exposed aggregate finishes often have variations between faces and returns. To minimize differences, concrete mixes should contain reasonably spherical or cubical aggregates. For large returns, or other situations where variations in appearance must be minimized, concrete mixes should have fully graded aggregates. Also, exposure of aggregate should be medium to deep with minimal color differences between mix ingredients.

Color and texture may be achieved at different stages of manufacture. These stages are described in the sequence of precast concrete operations.

Before Cast-finish is established by mold surface (before concrete is cast).

Treated After Cast-finish is accomplished after the concrete is cast but during precasting operations.

Finish After Hardening-finish is accomplished any time after concrete has hardened.

The unique versatility of precast concrete permits color, pattern and texture to be accomplished during any one of these three stages. The final decision should be based on the results desired and by the economics of the individual precasting operation. The precaster is usually the best judge, and should be allowed to decide the most efficient methods for achieving the desired effect.

Finishing techniques used in individual plants may vary considerably from one part of the continent to another, and between individual plants. Many plants have developed specific techniques supported by skilled operators or special facilities. It is difficult to provide general cost figures for different finishes since individual plants may price them differently. Many plants, for instance, consider acid etching an expensive finish. Some precasters discourage its use, while others, having used this finish for years may, in fact, prefer it over more common textures.

Finishing costs cannot be analyzed without considering the cost of concrete materials and labor because these normally change with different finishes. Repetition, shape and depth of exposure also play a major role in finishing costs. With these reservations in mind, an attempt has nevertheless been made to list finishes in groups, indicating increasing cost as finishes are chosen from groups 1 to 4.

1. Smooth (subject to qualifications stated in Section 3.5.2 or cost increases to Group 3)

2. Aggregates exposed by retarders or water washing
   Form liners
   Sandblasting or abrasive blasting

3. Aggregates exposed by acid etching

Tooling or bushhammering
Hammered ribs (fractured fins)
Sand embedment
Clay product (brick, tile, or terra cotta) facing

4. Honing or polishing
   Stone veneer-faced precast concrete

These groups include only the more common finishes; deviations from and combinations of these finishes are numerous. For actual projects be sure to confer with the local precaster about relative costs. The following sections discuss each of the finishing techniques. Description of the production techniques should reveal certain limitations. Most of the finishes require uniform plant-controlled conditions to assure satisfactory results.

3.5.2 SMOOTH OFF-THE-MOLD

A smooth off-the-mold or as-cast finish shows the natural look of the concrete without trying to simulate any other building product. Fine surface details and sharp arrises can be achieved with a smooth finish. This finish can be one of the most economical. Unfortunately, smooth off-the-mold finishes are seldom esthetically acceptable due to misunderstandings concerning surface imperfections. There is also the question of how the surface will change when exposed to the weather, see Section 3.6.4.

Smooth concrete makes the maximum demands on the quality and maintenance of the mold and on the concrete itself. Color variations tend to be most pronounced when the mold face is glassy and impermeable. Color uniformity is difficult to achieve on gray, buff, and pigmented concrete surfaces. The use of white cement will give better color uniformity than gray cement. Allowable color variation in the gray cement is enough to cause noticeable color differences in precast concrete panels. The slightest change of color is readily apparent on the uninterrupted surfaces of smooth off-the-mold concrete, and any variation is likely to be regarded as a surface blemish. Fig. 3.5.1 shows smooth gray sculptured panels and gray cast-in-place concrete shear walls. Fig. 3.5.2 shows smooth white panels with one-foot high horizontal strips of glass enclosing a 200,000 sq ft office building.

Smooth off-the-mold precast concrete panels have a smooth film of hardened cement paste, and the finished color is therefore determined primarily by the color of the cement. In some instances the sand may also have some effect. Initially, this is unlikely to be significant unless the sand contains a high percentage of fines or is itself highly colored. As the surface weathers, the sand will become more exposed, causing its effect to become more marked.

The color of the coarse aggregate should not be significant unless the particular panel requires extremely heavy consolidation. Under this circumstance, some aggregate transparency may occur, causing a blotchy, non-uniform appearance. Aggregate transparency or "shadowing" occurs when a dark or deeply colored aggregate shows like a shadow through a light colored cement skin. The effect can be reduced by using light colored coarse aggregates and white cement.

Fig. 3.5.1

Smooth concrete may be susceptible to surface crazing when exposed to wetting and drying cycles. This is merely a surface phenomenon (penetrates only as deep as the thin layer of cement paste at the surface of a panel) and does not affect structural properties or durability. In dirty atmospheres, crazing will be accentuated by dirt collecting in the tiny lines. This will appear more in white than in gray finishes and more on horizontal than on vertical surfaces. Crazing can be minimized by proper mix design, using a minimum water and cement content.

Precast concrete panels with a smooth finish will normally have air voids, particularly on any return surfaces. If these air holes are of reasonable size, 1/8 to 1/4 in., it is recommended that they be accepted as part of the surface texture rather than sack-rubbed. Filling and sack rubbing will eliminate the voids, but may cause increased color differential. Samples or the mockup panels should be used to establish the acceptability of color variation and air voids with respect to frequency, size, and uniformity of distribution.

For true economy, units with smooth surfaces should be produced without additional surface treatment after stripping from the molds, except possible washing and cleaning. This, in turn, demands the following precautions:

1. Provide architectural relief to flat exposed surfaces. Some sculpturing of the panel is highly desirable. Careful attention to detailing is essential. Make provisions for ample draft, proper edges and corners, see

Fig. 3.5.2

Sections 3.3.2 and 3.3.6, suitable water drips, and other weathering details, see Section 3.6. Design shapes to minimize stripping damage.

2. Non-porous molds should be well constructed so that no water or grout leakage will occur. Avoid joints in the mold face, and finish the mold surface so that minor imperfections in the mold will not be mirrored in the surface of the panel. Molds must also be able to withstand high energy consolidation. Use form vibration whenever possible. If immersion vibration is necessary, ensure that the vibrator head does not touch the sides of the mold.

3. Concrete mix designs that combine a minimum cement content and a constant, low water-cement ratio with high density are desirable to minimize crazing, entrapped air voids and color variations.

4. Uniformity of manufacturing procedures is most important. Procedures that ensure consistent techniques for cleaning the mold, application of mold release, uniformity of concrete quality, consistent curing processes and careful storage procedures will all contribute to produce a uniform quality product. Proper consolidation and curing will minimize uniformity shortcomings and air voids which show easily on smooth surfaces, particularly with gray concrete mixes. Uniform slow curing with minimum loss of moisture from the smooth surface will help to minimize crazing tendencies.

5. Efforts to minimize chipping or other damage are very worthwhile because smooth off-the-mold finishes are the most difficult of all precast concrete finishes to repair and match in appearance because of possible variations in color and texture. Repairs to this finish tend to be even more noticeable after weathering.

Smooth finishes can be very economical when there is a reasonable repetition to absorb the high mold costs and when both designer and precaster accept its limitations and comply with the listed requirements. Smooth off-the-mold precast concrete panels usually have some surface imperfections. Minor variations in surface texture of mold surface reflected on the smooth concrete surface, color variations, air voids (bug/blow holes), and minor surface crazing are to be expected. Both designer and precaster must be aware of the realistic surface finish that will be obtained. Of all precast concrete finishes, this finish is the most misunderstood when it comes to acceptability. An acceptable smooth finish can be very difficult and expensive to achieve if a high degree of uniformity is anticipated by the architect or owner. If the surface is to be painted, this finish will provide an excellent surface, while keeping costs to a minimum.

Many of the esthetic limitations of smooth concrete may be masked by the shading and depth effects induced by creating profiled surfaces (fluted, sculptured, board finishes, etc.), by subdividing the panels into smaller surface areas by means of vertical and horizontal rustications, or by using white cement, Fig. 3.5.3. Any introduction of shapes to provide shadow effects will enhance the final finish.

Fig. 3.5.3

3.5.3 EXPOSED AGGREGATE BY CHEMICAL RETARDERS OR WATER WASHING

Chemical surface retarders provide a non-abrasive process which is very effective in bringing out the full color, texture, and natural beauty of the coarse aggregate. The aggregate is not damaged or changed by this exposure method. The application of a chemical retarder to the concrete surface (normally the mold surface bottom and/or sides) delays the surface cement paste from hardening within a time period and to a depth dependent on the type of retarder. After hardening of the concrete mass (normally overnight), the retarded outer layer of cement paste is removed by brushing, by high pressure water washing or a combination of both, exposing the aggregate to the desired depth. This process should take place immediately after the panel is stripped from the mold, and if possible at a predetermined time after casting.

The aggregate exposure obtained is controlled by the retarder; therefore, any variations in exposure are not as correctable as in sandblasting. Furthermore, deep retarded surfaces clearly do not allow for sharp profiles.

The action of a chemical retarder specifically affects the cement mortar fraction of the concrete. The shape of the coarse aggregate, its position after consolidation, and depth of etch will determine the surface appearance. Appearance will therefore vary to some degree with surface orientation and aggregate shape. This is especially critical in units with returns, where vertical sides are

expected to reasonably match the bottom face. Refer also to Section 3.3.7.

Special detailing of the exposed aggregate panels in Fig. 3.5.4 incorporated exaggerated beveled edges and form liners to creat a decorative, monolithic appearance emphasizing vertical columns and minimizing apparent width of spandrels. In Fig. 3.5.5, local aggregates and gray cement were used to produce a light buff exposed aggregate panel.

Depending on the particular retarder and mold configuration (vertical, radius, or complicated shapes) the placing of concrete may scour the retarder applied to sloping surfaces and affect the finish of the concrete. These factors may compound the problem of matching bottom and returns. See Section 3.3.8.

If the bright, natural colors of the aggregate are the prime concern, exposed aggregate from retarded surfaces are the best way to achieve this result. The mix design, aggregate gradation and physical characteristics of the aggregate, and matrix/aggregate color compatibility are important. Aggregate should be round or aggregate fracture should be conical and not slivered. Slivered coarse aggregate will tend to be dislodged during high pressure washing and may not give a consistent appearance. It is advisable to vary the color or tone of the matrix wherever possible to match or blend in with the color of the aggregate. This match can be achieved by careful selection of cement and sand colors, and the use of coloring agents.

Fig. 3.5.4

Fig. 3.5.5

Good matrix to aggregate match will prevent "patchy" effects (minor segregation of aggregate) from being noticeable.

Chemical retarders are also available for the face-up method of casting concrete. Retarder is sprayed on the concrete surface following consolidation and finishing operations. Since the consolidation of concrete brings an excess of cement mortar and water to the surface, an exposure of this surface will normally fail to show as dense a coarse aggregate distribution as a down-face mold surface. This may be overcome by seeding the surface with coarse aggregates following the initial finishing of the surface, then tamping or rolling the aggregates into the surface, refinishing the surface, and applying the retarder. The spandrels and sunscreen panels shown in Figs. 3.5.5 and 3.5.6 are exposed on all sides.

Water washing and brushing without the use of chemical retarders are normally performed on unformed surfaces or on surfaces from which the molds have been removed prior to hardening of the paste. These molds should be designed with care to allow the mold sections to separate easily with a minimum of aggregate dislodgement prior to washing. Angular aggregates that will rotate during washing and brushing should be avoided when designing the mix. Water washing will remove the cement mortar to the required depth. This method of exposing the aggregate usually requires some seeding of coarse aggregate to obtain a uniform dense surface. The seeded aggregate is embedded by troweling, tamping or rolling into the plastic concrete. A magnesium float or darby should be used for final embedment of the seeded aggregate so that mortar completely surrounds the aggregate. Seeding and water washing requires skilled workmen. Thus, the process is not as common as the use of chemical retarders.

The appearance of aggregates in precast concrete units subjected to retardation or water washing will not change from the natural appearance of these aggregates prior to incorporation in the concrete mix. These methods may be used for all three degrees of exposure, but they are most commonly used on medium or deep exposure. The three exposures shown in Figs. 3.5.7, 3.5.8 and 3.5.9, illustrate the use of light, medium, and deep exposure of a quartz aggregate in a white cement matrix, while Fig. 3.5.10 shows deep exposure of a river gravel in a gray cement matrix. Retarded and water washed finishes are relatively easy to repair, a major advantage. Also, the mold surface is not as critical if the aggregate is to be exposed. Figs. 3.5.11 and 3.5.12 show the different appearances resulting from a medium retarded finish (3.5.11) and a medium sandblast finish (3.5.12) with the same concrete mix design.

After the aggregate is properly exposed, and the panel is well cured, the exposed faces of the precast concrete units and all edges where sealant material is to be applied may be given one or more washings with a 5 to 10 percent solution of muriatic acid to thoroughly clean the exposed aggregate and to remove any retarded cement paste and foreign materials. See Sections 3.5.5 and 3.6.14 for further discussions on acid cleaning.

Fig. 3.5.6

Fig. 3.5.7

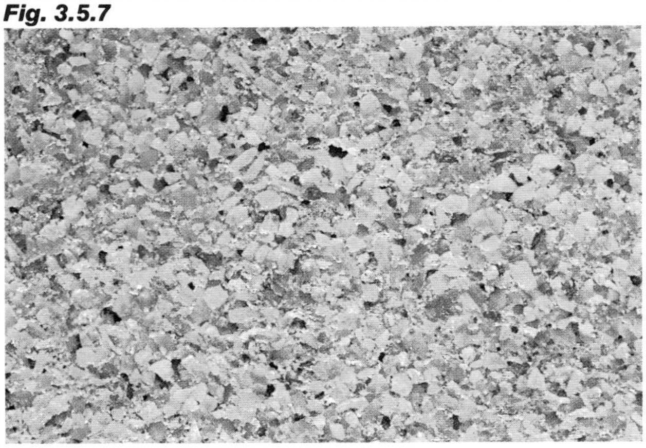

Fig. 3.5.8

Fig. 3.5.9

Fig. 3.5.10

Fig. 3.5.11

Fig. 3.5.12

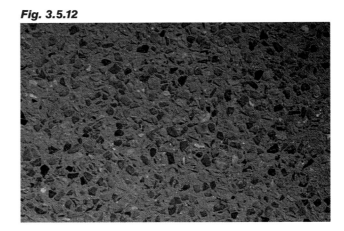

3.5.4 FORM LINERS

An almost unlimited variety of attractive patterns, shapes, and surface textures can be achieved by casting against wood, steel, plaster, elastomeric, plastic, or foam plastic form liners, Fig. 3.5.13. These form liners may be incorporated in or attached to the surface of a form. The faithful reproduction of the form materials is due to the plasticity of the freshly placed concrete. However, the concrete finishes are only as good as the quality of the form liner itself. A large pattern offers ever-changing details due to light shading; a fine texture offers a muted appearance that is subtle but not drab; and smooth surfaces bring out the elegance and richness of simplicity. A form liner's texture can make smooth off-the-mold surfaces appear more uniform.

Fig. 3.5.13

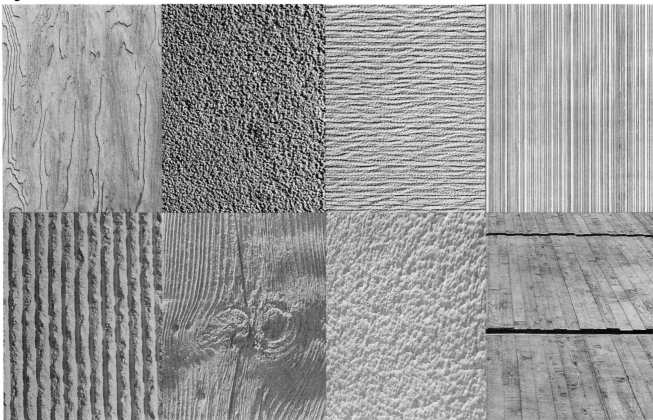

Liners can highlight large concrete surfaces with low relief patterns at a reasonable cost. Also, the light and shade created by modeling or sculpturing with liners, and the weathering patterns they produce, tend to mask color differences. Use of this type of finish requires close coordination between the design team and the precaster to be sure the end result is achievable.

Figs. 3.5.14, and 3.5.15 show simulated bushhammered rib, and brick patterns, respectively. All panel surfaces were given a light acid etch to brighten the surface.

Fluted panels demand considerable attention to detailing as panel sizes and distances between openings must be a multiple of the rib spacing. Panel joints should normally also be in the bottom of a groove unless the ribs are wide enough to accommodate a joint in the raised rib and still have adequate rib thickness to minimize breakage.

An important consideration is selecting the texture and/or type of form liner best suited to the project. Material selection should depend on the amount of usage and whether or not the pattern has undercut (negative) drafts. Concrete can be produced with vertical ribs or striations in a range of sizes to suit a particular structure and the distance from which it will most often be seen. Overall, the cost of liners depends upon the ease of use and the number of uses actually obtained. Regardless of the type of form liner used, draft must be considered to prevent chipping or spalling during stripping of the unit from the mold.

If the concrete is to be left as-cast, that is without further

Fig. 3.5.14

Fig. 3.5.15

Fig. 3.5.16

Because of the difficulty in matching joints between liners, this technique should either be limited to widths less than the available width of the liner, or liner joints should be at form edges or be detailed as an architectural feature in the form of a groove, recess or rib. In Fig. 3.5.16, accent grooves are placed between random width barnwood form liners. Joints are placed in the groove and caulked with material matching the color of the concrete. In Fig. 3.5.3, the precast concrete panels were cast against wood clapboard siding for both the exterior facade and interior lobby. All of the vertical joints in the form liner were concealed in the corner joint. All of the horizontal joints occurred in the regular 6 in. o.c. V-shaped grooves.

Sandblasted wood, textured plywood and rough-sawn lumber are useful in creating rugged textures. (Resultant surface texture may also be obtained by use of other liners reproducing this finish). Rough-sawn lumber is often used for board surface textured finishes where concrete color variations and rough edges are acceptable. The weathering of lumber can also affect the outcome of the concrete finish. If rough sawn lumber is being used, it is important to cast samples to determine the effect the lumber will produce. The lumber selected should all be purchased at one time from one source to minimize the possibility of variations. In Fig. 3.5.17, the soft green colored panels were cast from sandblasted pine forms designed to emulate wooden tobacco barns previously on the site. The panels were 6 in. thick and included column covers, spandrels with soffit, corner spandrels and curved sections.

treatment, its appearance will be determined by the surface characteristics of the liner material as well as by the chosen pattern or texture. Variations in the absorbency of the form surface will tend to produce corresponding variations in the color of the concrete, a dark color being associated with water loss.

The method of attaching form liners should be studied for resulting visual effect. Form liners should be secured in molds by gluing or stapling, rather than by methods which permit impressions of nail heads, screw-heads, rivets, etc. to be imparted to the concrete surface finish, unless desired. The surfaces of wood form liners should be sealed to minimize the discoloration of the concrete caused by differential absorption of mix water by the form liner.

Fig. 3.5.17

Fig. 3.5.18

Virtually any design can be achieved with plastic form liners when the following rules are observed: limit depth of design to 1/2 in. to 1 in. In most cases, maintain a 10 deg. draft on all indentation sides to prevent chipping and spalling during stripping of the panel from the mold; and keep all edges and corners rounded or chamfered. Relief may be more than 1 in. if the depressed area is sufficiently wide.

The 8 ft by 28 ft insulated sandwich panels in Fig. 3.5.18 simulate the hand-hammered look of fractured fins. The panels were produced by first building a smooth-ribbed mold of wood and casting from it one master panel with alternating directions of diagonal fractured ribs. That panel was then hammered and sandblasted, and an elastomeric mold was made of its hand-finished surface. This second mold was used to cast the final panels. After demolding, the rib surface was sandblasted to expose the aggregate to the desired texture.

Whenever reuse of liner is not required, sculptural designs have been produced using smooth or irregular sections of foamed polystyrene or polyurethane as form liners or inserts. Abstract patterns and deeply revealed designs with undercut edges can be shaped easily in these materials. An example of a large sculptured relief developed with the use of inserts is shown in Fig. 3.5.19. Figs. 3.5.20 and 3.5.21 show how this was accomplished.

Elastomeric liners are useful for detailed textured or profiled surfaces, because they greatly facilitate stripping.

Fig. 3.5.19

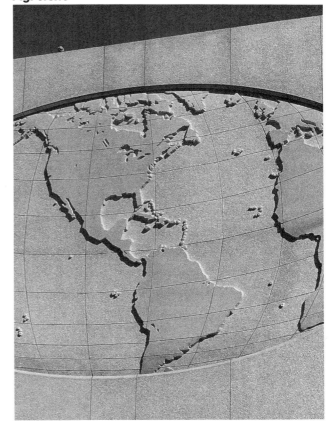

Fig. 3.5.20

Fig. 3.5.21

Fig. 3.5.22

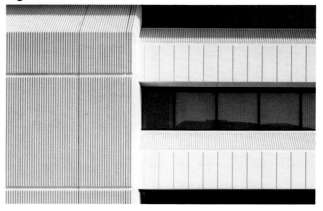

If other materials were used for such detail, the forms would be virtually impossible to strip. Close patterns should be avoided, since the removal of air trapped between the concrete face and the elastomeric liner tends to be difficult, and the resulting air voids mar the finished appearance of the concrete. The surface of the panel is usually difficult to repair, particularly if the surface is left as-cast.

Combination finishes, involving the use of one or more basic finishing methods, together with form liners are almost infinite. One common combination is the ribbed form liner and a light sandblast finish, see Fig. 3.5.22 and Section 3.5.13.

3.5.5 SAND OR ABRASIVE BLASTING

Sand or abrasive blasting of surfaces can provide all three degrees of exposure. Fig. 3.5.23 is typical of a light sandblasting, while Fig. 3.4.24 shows the medium sandblasting finish and Fig. 3.5.25 illustrates a deep (or heavy) sandblasting of a river gravel aggregate mix. Fig. 3.5.26 shows the same treatment applied to a soft marble aggregate. This process is suitable for exposure of either large or small aggregates.

Generally, the technique is used when a light exposure is desired, because sandblasting costs increase with the depth of the exposure. A light sandblast can be specified for the purpose of obtaining a sand finish resembling limestone. By blending yellow and black sands and lightly sandblasting the panels, a veining effect resulted, which closely simulated natural limestone, Fig. 3.5.27. Such resemblance may be obtainable on a regular production basis, but the importance of developing and approving realistic samples with respect to shape, thickness of the concrete, and the actual production methods cannot be overemphasized for this finish.

Sometimes a panel surface is given a brush-blast, which is little more than a uniform scour cleaning that lightly textures the surface skin, to remove minor surface variations such as surface laitance. A brush-blast surface seldom appears uniform at close inspection and should be viewed at a distance for uniformity.

Uniformity of depth of exposure between panels and within panels is essential in abrasive blasting, as in all other exposed aggregate processes, and is a function of the skill and experience of the operator. Different shadings and to

Fig. 3.5.23

Fig. 3.5.24

Fig. 3.5.25

Fig. 3.5.26

Fig. 3.5.27

some extent, depth of color will vary with depth of exposure.

The degree of uniformity obtainable in a sandblasted finish is generally in direct proportion to the depth of sandblasting. A light sandblasting may look acceptable on a small sample, but uniformity is rather difficult to achieve in reality. The deeper the exposure, the more costly the finish.

The lighter the sandblasting, the more critical the skill of the operator, particularly if the units are sculptured. Small variances in concrete strength at the time of blasting may further complicate results. Sculptured units will have air holes on the returns which might show strongly in a light sandblasted texture. If such air holes are of reasonable size, 1/8 to 1/4 in., it is strongly recommended that they be accepted as part of the texture, because sack-rubbing is expensive and will nearly always cause color differences. A few inches of concrete can be consolidated with only small air holes likely to show; thicker layers will increase the possibility of larger air holes. For this reason, the use of a face mix or a specially placed mix of only 1 to 2 in. thick may be advisable.

To improve uniformity, but at increased cost, the light sandblast finish may be obtained by blasting with materials softer than silica sand, such as walnut shells. Walnut shells will cut the cement and sand but only slightly affect most coarse aggregates. As an additional step towards uniformity, the cement and sand color should be chosen to blend with the slightly "bruised" color of the sandblasted coarse aggregate as the cement-sand matrix color will predominate when a light sandblast finish is desired. With a light sandblasting only some of the coarse aggregates near the surface will be exposed, so a reasonable uniform distribution of such aggregates is not controllable.

Blasting will cause some etching of the face of the coarse aggregate, and softer aggregates will show this to a greater extent beyond a medium exposure. Etching of the aggregate surface is more noticeable on dark-colored aggregates which have a glossy surface texture. This will produce a muted or frosted effect which tends to lighten the color and subdue the luster of the aggregate. For example, white concrete tends to become whiter when blasted. Depth of sandblasting should also be adjusted to suit the aggregate and abrasive hardness. For example, soft aggregates might be eroded at the same rate as the mortar; exposed aggregates lose their sharp edges when blasted.

Type and grading of abrasives determine the surface texture and should remain the same throughout the entire project. Since some aggregates change color after sandblasting, trials of different abrasive materials with sample panels are desirable to check the texture and color tone.

Although sandblasting is generally specified as an overall treatment, it may be used to develop textured patterns by means of special templates. Portions of a panel can be left unblasted by making a shield of wood, rubber or sheet metal to fit over the panel and cover those areas. Masking may be adopted for geometric patterning, or the technique can be employed by artists in producing murals in concrete.

The time when sandblasting should take place is determined by scheduling, economics, visual appearance desired, and hardness of the aggregate. The concrete matrix will be easier to cut in the first 24 to 72 hrs after casting. As the concrete cures and gains strength, it becomes more difficult to blast to any appreciable depth, thus increasing the cost of the operation. Softer aggregates tend to abrade more when concrete strengths are high and the concrete surface will have a duller appearance, Fig. 3.5.26. In some cases, the higher cost of deferred blasting may be justified by avoiding scheduling problems. However, all surfaces should be blasted at approximately the same age or compressive strength for uniformity of appearance. The concrete mix used, and the matrix strength at time of blasting, will affect the final exposure, as will the gradation and hardness of the abrasive.

Fig. 3.5.22 shows a lightly-blasted fluted surface on a white concrete panel. The deeply articulated beams and columns, in Fig. 3.5.28, were lightly sandblasted to expose various colored granite aggregates to match the appearance of granite. The medium sandblasted panels in Figs. 3.5.29 and 3.5.30 have two different coarse aggregates to achieve a mix that is compatible with both the granite insets and champagne colored glass. Limestone aggregate panels produced with white cement and buff pigment were given a deep sandblast finish, Fig. 3.5.31. The half round string courses and the feature strips on the panels in Fig. 3.5.32 received light sandblasting opposed to a central portion of a heavy sandblasted texture. A sandblasted

Fig. 3.5.28

Fig. 3.5.29   *Fig. 3.5.30*

Fig. 3.5.31

Fig. 3.5.32

finish is not widely used to achieve a deep, heavy texture because of the time and labor associated with deep exposure. Unless it is the intent of the architect to achieve a severely weathered look, deep exposed aggregate finishes are more readily achieved with other methods.

For a medium or deep exposure with a sandblasted appearance, retarders may be used initially followed by sandblasting to dull the finish. This approach reduces blasting time and lessens the abrasion of softer aggregates. Using sandblasting to achieve the final texture allows for correction of any variations in exposure, so this method can result in a very uniform surface. Extra care, however, should be taken to avoid non-uniform exposure that may be caused by the presence of soft and hard spots on the retarded surface. This is especially true and more noticeable on large flat surfaces. Small flat areas or surfaces that are divided by means of rustications will tend to call less attention to these texture variations.

When sandblasting after applying a retarder, the selected chemical retarder strength should only give 50 to 75 percent of the expected reveal. Abrasive blasting then removes the retarder and part of the matrix to obtain the desired surface texture. The end result is a matte finish, as opposed to a brighter finish achieved with water blasting.

Exposed aggregates can be brightened by washing with a 5 to 10 percent solution of hydrochloric acid, which removes the dull cement film remaining from some exposure techniques, such as sandblasting, retardation, washing and brushing. The acid solution should not materially affect the cement or aggregate. The acid normally is applied to a prewetted surface by brush or it can be sprayed. The surface is wetted to reduce acid penetration. Immediately after each washing with the acid solution, the precast concrete units should be thoroughly rinsed with fresh, clean water, to completely remove all traces of the acid.

Acid cleaning is best used after a 1 to 2 week delay to ensure more complete hydration but often is done immediately after initial curing to reduce handling operations. The surface should have at least 3 or 4 days of curing. Acid washing too soon contributes to formation of white deposits on units with a gray cement matrix or with dark colored aggregates.

3.5.6 ACID ETCHING

Acid etching dissolves the surface cement paste to reveal the sand with only a small percentage of coarse aggregate being visible. It is most commonly used for light or light to medium exposure. The acid etch finish is typically used to produce a fine sand texture closely resembling natural stones such as limestone or sandstone. Only acid resistant siliceous aggregates, such as quartz and granite should be used. Carbonate aggregates, e.g. limestone, dolomite and marble, will discolor or be dissolved due to their high calcium content. The aggregates on an acid etched surface present a clean or bright look. However, after normal weathering, the aggregates lose this brightness and will closely resemble their original condition.

Concrete should be thoroughly wetted with clean water prior to acid treatment because concrete is porous and acids will penetrate faster and deeper into dry concrete. The water fills the pores and capillaries and prevents the acid from etching too deeply.

The best results with acid etching of surfaces are obtained using low concentrations of solution after a 2-week delay, but often this finishing operation is done immediately after curing to reduce handling operations. The units to be treated should have uniform temperatures and strength levels (preferably about 4500 psi) when acid is applied.

Acid etching may be accomplished by brushing the surface with a stiff bristled fiber brush immersed previously in the acid solution; by spraying acid and hot water onto the panel surface using specially designed pumps, tanks and nozzles; or by immersing the unit for a maximum of 15 minutes in a tank containing from 5 to 35 percent hydrochloric acid. Only the first two methods should be used with sandwich wall panels where the insulation goes to the edges of the panels and may be damaged by the acid.

Within 15 minutes of completion of acid etching, the unit must be thoroughly flushed with large amounts of clean water. Acid solutions lose their strength quickly once they are in contact with cement paste or mortar; however, even weak, residual solutions can be harmful to concrete due to possible penetration of chlorides. Failure to completely rinse the acid solution off the surface may result in efflorescence or other damaging effects.

Fig. 3.5.33

Fig. 3.5.34

Fig. 3.5.35

Prior to acid etching, all exposed metal surfaces, particularly galvanized metal, should be protected with acid-resistant coatings. These include vinyl chlorides, chlorinated rubber, styrene butadiene, bituminous paints and enamels and polyester coatings. Touch-up of all exposed galvanized metals cast in the precast concrete units and affected by acid washing or etching should be done utilizing a 3-mil thickness of a single component zinc-rich compound with 95 percent pure zinc in the dried film. Acid etching may also damage the reinforcement if less than the recommended cover is used.

Acid etching of concrete surfaces will result in a pleasing fine, sandy texture with retention of detail if the concrete mix and its consolidation have produced a uniform distribution of aggregate particles and cement paste at the exposed surfaces. Concentrations of cement paste and under and over etching of different parts of a panel, or variation in sand color or content may cause some uniformity problems, particularly when the acid etching is light or used for large, plain surfaces.

Figs. 3.5.33, 3.5.34 and 3.5.35 show a light, medium and deep acid etch finish on panels produced with gray granite aggregate and gray cement.

Color compatibility of the cement and the aggregates becomes more important, the lighter the texture, to avoid a blotchy effect. Brown, buff, yellow or pink colors are forgiving to the eye and increase the likelihood of better color match from panel to panel. Using a light acid etch on the cladding panels in Fig. 3.5.36, produced with a rose tone feldspar aggregate, produced an attractive finish. A close-up of the texture is shown in Fig. 3.5.37. Gray is the worst color to select for uniformity unless both the fine and coarse aggregates are also gray. In any case an acid etch finish is not going to be as uniform as an exposed aggregate finish, particularly on returns. An acid etch finish is also more difficult to patch than many of the deeper texture finishes, however, minor air voids are fairly easy to grout and refinish.

Vertical rustications or reveals extend to the bottom of the panel to avoid the potential of a deeper etch on the bottom flat band of the panel. If not extended, acid collects in the reveal and can run down and streak the bottom band.

The exterior of the hotel in Fig. 3.5.38 was designed to have the appearance of an English manor house. The buff colored precast concrete panels were acid etched to closely resemble Indiana limestone. Note the extremely fine detail on the Tuscan arch reproduced from the formwork, Fig. 3.5.39. Precast concrete was also used as a backup suport for the brick veneer, which was laid up in the field and tied back into the precast concrete wall with steel connectors. This approach allowed for efficient scheduling of construction trades, and cleaner brick detailing.

3.5.7 TOOLING OR BUSHHAMMERING

Any tooling technique is usually called "bushhammering." Concrete can be mechanically spalled or chipped with a variety of hand and power tools to produce an exposed aggregate texture. Each type of tool produces a distinctive surface effect and a unique shade of concrete color. All tooling removes a layer of hardened concrete while fracturing the larger aggregates at the surface. It produces an appearance somewhat different from other forms of exposed aggregate. The color of the aggregate, but not necessarily aggregate shape, is revealed. The technique is most suitable for flat or convex surfaces.

Pneumatic or electric tools may be fitted with a bush-hammer, comb chisel, crandall, or multiple pointed attachments, Fig. 3.5.40. To decrease the cost of a tooled finish, tools with multiple impact heads are normally used. The finish obtained can vary from light scaling to the deep bold texture achieved by jack-hammering with a single pointed chisel. The bushhammered surface in Fig. 3.5.41 was achieved with air hammers equipped with 1-1/2 to 2-in.

Fig. 3.5.36

Fig. 3.5.37

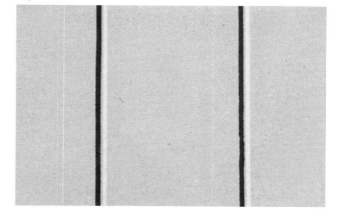

wide, carbide-tip, single blade chisel bits. Bushhammering was done with a horizontal motion only. A heavily chiseled recess surrounds a striated center panel in Fig. 3.5.42. Bushhammered finishes affect the appearance, color and brightness of the aggregate. Color tends to be lightened by the fracturing, which on dark materials has a dulling effect, but it often improves the lighter silver grayish and, in particular, the white tones. By increasing or decreasing the shadow content of the texture, tooling alters the panel reflectancy and changes the tone value.

Scaling is the lightest texture in tooled finishes. It is achieved by passing a triple-pronged scaler (originally developed to remove scale from steel prior to painting) singly or in gangs over the surface to remove only a thin

Fig. 3.5.38

Fig. 3.5.39

Fig. 3.5.40

Fig. 3.5.41

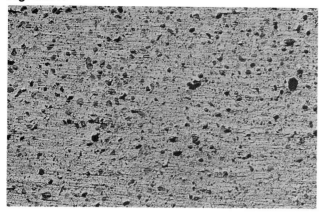

Fig. 3.5.42

skin. Some aggregate is exposed and fractured in the process. Under certain conditions, almost the same result can be achieved less expensively by light abrasive blasting.

Although a dense, fully-graded concrete mix is desirable, bushhammering has been successfully applied to gap-graded concrete. Natural gravels are inclined to shatter, leading to bond failure and loss of aggregate particles. Aggregates such as quartz and granite are difficult to bushhammer uniformly because of their hardness, and they may fracture into rather than across the concrete surface. Aggregates such as dolomite, marble, calcite and limestone are softer and more suitable for bushhammered surfaces. The comb chisel should be used only with the softer aggregates. Concrete containing soft aggregates cannot be satisfactorily point-tooled.

Bushhammering at outside corners may cause jagged edges. If sharp corners are desired, bushhammering should be held back from the corner a distance of 1 to 2 in. It is quite feasible to execute tooling along specific lines. If areas near corners are to be tooled, this must normally be done by hand since tools will not reach into inside corners, making this operation more expensive. Chamfered corners are preferred with tooled surfaces and with care a 1 in. chamfer may be tooled.

In Fig. 3.5.43, the 3-1/2 in. deep, 6 in. on center ribs are hammered on the extended face to soften the sharpness of the vertical ribs. Horizontal ribs are subdued by tapering off the ribs at and above the joints. Fig. 3.5.44 shows a closeup of a bushhammered rib panel made with a yellow marble aggregate and white cement.

Exposing the aggregate by tooling requires trained operators in order to produce a uniformly textured surface (if uniformity is what is desired). Orientation of equipment and direction of movement for tooling should be kept uniform throughout the tooling process as tooling produces a definite pattern on the surface. Variations due to more than one person working on the panels may occur with this finish.

Tooling removes a certain thickness of material, 3/16 in. on an average, from the surface of the concrete and may fracture particles of aggregate causing moisture to penetrate the depth of the aggregate particle. For this reason the minimum cover to the reinforcement should be somewhat greater than normally required. It is sometimes recommended that 2 in. of cover be provided (prior to tooling).

Fig. 3.5.43

Fig. 3.5.44

3.5.8 HAMMERED RIB OR FRACTURED FIN

A hammered rib or fractured fin finish may be produced by casting ribs on the surface of the panels and then using a hammer or bushhammer tool to randomly break the ribs, and expose the aggregate. Fig. 3.5.45 shows a closeup of a fractured fin panel made with a yellow marble aggregate and white cement. The effect is a bold, deeply textured surface as shown in Fig. 3.2.8. Rib size measured at the outer face should be a maximum of 1 in. as larger sections are difficult to fracture. The ribs should not be narrower than 5/8 in. or they may break off at their base without leaving any of the rib projecting. The ribs may be hammered from alternate sides, in bands, to obtain uniformity of cleavage, or randomly, depending on the effect required. There should be a definite plan, even with a so-called random pattern, because unless care is exercised, an uneven shading effect on the concrete surface may be produced. This finish is expensive but is justified if the panels will be viewed at close hand on ground level walls or interior walls.

The diagonal striated pattern of the panels in Fig. 3.5.46 were designed for maximum color and texture. Light gray limestone and white river gravel were used for color control. A warm buff sand added color to the gray matrix. Before finishing, the ribs were 3/8 in. tall and 3/8 in. wide at the base and repeated at 1/2 in. intervals. The striated ribs were bushhammered to expose the aggregates. The size of the ribs and the size of the aggregate were carefully coordinated to achieve maximum coarse aggregate exposure. Fig. 3.5.47 shows a closeup of the fractured fin.

Fig. 3.5.45

Fig. 3.5.46

Fig. 3.5.47

An effect similar to a fractured rib finish may be achieved less expensively using forms or form liners that simulate the fractured ribs. The exposed, weathered look can then be achieved by chemical retardation or by sandblasting the rib surface. See Section 3.5.4.

3.5.9 SAND EMBEDMENT

When bold and massive architectural qualities are desired, stones ranging from 1 to 8 in. in diameter, Fig. 3.5.48, or flagstones, Fig. 3.5.49, may be exposed by the sand embedment technique. These large stones must be hand placed in a sand bed, or other special bedding material, at the bottom of the mold to a depth which keeps the backup concrete 25 to 35 percent of the stone's diameter from the face. This technique reveals the facing material and produces the appearance of a mortar joint on the finished panel.

Special care should be taken to ensure that adequate aggregate density is obtained around corners, edges and openings as well as on the flat surfaces. All aggregate to be exposed should be of one size gradation for a uniform exposure. Where facing materials are of mixed colors, their placement in the mold must be carefully checked for the formation of unintended patterns or local high incidence of a particular color. If it is the intention, when using some aggregates, to expose a particular facet of the stone, placing should be checked with this in mind before the backup concrete is placed.

If a white or colored mortar joint is desired, the mortar consisting of one part white air-entrained cement (plus pigment) to 2-1/2 parts well-graded white or light-colored sand with sufficient water to make a creamy mixture can be placed over the aggregate. If a mortar facing mix is used, usually the backup mix should be of a low slump (a maximum of 1 in.), to absorb excess water from the facing mix. Otherwise, the backup mix should be standard structural concrete with a slump of 2 to 4 in. Care must be exercised during vibration not to disturb the sand or aggregates causing uneven aggregate distribution.

When the panels have cured, they should be raised and any clinging sand removed by brushing, air blasting, or washing with a stream of water. Some sand bonds to the

Fig. 3.5.48

Fig. 3.5.49

concrete; therefore, the color of the bedding sand should be carefully chosen to harmonize with the aggregate to be exposed.

3.5.10 CLAY PRODUCT-FACED PRECAST CONCRETE PANELS

Clay product-faced precast concrete allows the architect to combine the pleasing visual appearance of traditional clay products with the strength, versatility, and economy of precast concrete. Clay products which have been bonded directly to precast concrete include brick, ceramic tile, and architectural terra cotta. The clay product facing may cover the entire exposed panel surface or only part of the face, serving as an accent band. Also, marble, glass and ceramic mosaics have been bonded directly (preferred) or applied to the hardened concrete.

The exterior of the multi-purpose sports and entertainment building, in Fig. 3.5.50, is wrapped with 9 in. thick precast concrete panels inlaid with a contrasting basketweave pattern of 8 in. square brick units and a rectilinear grid of blue-green glazed brick. Two in. of rigid insulation is embedded in the concrete backup. Each panel is 20 ft wide by 10 ft high with a false joint in the middle of each panel. The panel edges are exposed concrete, lightly sandblasted for uniformity and deeply chamfered to provide a vertical and horizontal shadow line, Fig. 3.5.51.

The ability of architectural precast concrete to create the new and restore the old is demonstrated by the new 22-

Fig. 3.5.51

Fig. 3.5.50

story tower, Fig. 3.5.52, and, the barely visible, old 6-story adjacent building. The 22-story tower has three different surface finishes: (1) lightly sandblasted white concrete, (2) 6 in. sq. glazed white ceramic tiles and some additional 10 in. sq. accenting blue tiles, and (3) a combination of the first two.

Built in 1906, the 6-story building is considered one of San Francisco's architectural landmarks. For that reason, it was decided the building's terra cotta facade would be preserved on an otherwise all-new structure of slightly taller height. The terra cotta was taken off the building, piece by piece and identified for subsequent reassembly on new precast concrete panels, Fig. 3.5.53. Stainless steel wires were looped through the back ribs of the terra cotta pieces and projected into the backup concrete, in order to anchor the pieces to the concrete.

The combination of precast concrete and clay products has several advantages over site laid-up masonry. By using precast concrete panelized construction, the need for on-site scaffolding is eliminated, which can be a significant cost savings over masonry construction.

The panels can be produced during foundation work or superstructure construction, to permit panel erection when needed. The panels can be quickly erected as they are delivered, eliminating any need for panel storage at the site. One of the distinct advantages of plant production is that it permits year-round work under controlled temperature conditions. Cold and inclement weather do not cause construction delays. Fabrication can take place during cold wintery months when outside temperatures are too low for installing clay products and in hot summer months or on windy days when special precautions would be necessary to prevent rapid moisture loss and excessive shrinkage in the mortar. The use of clay product-faced precast concrete panels may eliminate the need for, or actually provide, the means of winterizing the structure, permitting floor topping and finishing trades to continue without any weather delays.

Precasting allows for the use of stringent quality control. Concrete and mortar batching systems and curing conditions can be tightly controlled. Use of clay product-faced precast concrete has the advantage of reducing efflorescence, since precasting techniques do not require the use of chloride accelerators in the mortar and better curing is obtained through plant production.

Panel configuration may range from flat panel sections through C-shaped spandrels, soffits, arches, and U-shaped column covers, Fig. 3.5.54. Repetitive usage of these shapes can lower costs appreciably. Returns on spandrels or column covers may be produced by the sequential (two-stage) casting method or as a single cast depending on the height of the return. Panels may serve only as cladding or may be loadbearing panels supporting floor and roof loads.

General Considerations. Structural design, fabrication, handling and erection considerations for clay product-

Fig. 3.5.52

Fig. 3.5.53

Fig. 3.5.54

faced precast concrete units are similar to those for other precast concrete wall panels, except that special consideration must be given to the clay product material and its bond to the concrete. The physical properties of the clay products must be compared with the properties of the concrete backup. These properties include the coefficient of thermal expansion and volume change.

Reinforcement of the precast concrete backup should follow recommendations for precast concrete wall panels relative to design, cover and placement. Cover depth of reinforcement must be a minimum of 3/4 in. to the back of the clay product. This cover is maintained by non-corrosive spacers, such as plastic. Galvanized or epoxy-coated reinforcement is recommended at cover depths of less than 3/4 in., although brick thickness, joint depth and weather exposure affect the cover requirements.

Because of the difference in material properties between the facing and concrete, clay product-faced concrete panels are more susceptible to bowing than all-concrete units. However, panel manufacturers have developed design and production procedures to minimize bowing. Many manufacturers compensate by using cambered forms, e.g., 1 in. for 40 ft, to produce panels initially bowed inward. Bowing is also a consideration in reinforcement design. If thickness is sufficient, two layers of reinforcement should be used, as this helps to reduce bowing caused by differential shrinkage or temperature changes. In some cases, reinforcing trusses are used to add stiffness; in others, concrete ribs are formed on the back of the

panel, but this may require backforming and is more costly. Minimum thickness of backup concrete of flat panels to control bowing or warping is usually 5 to 6 in. but 4 in. has been used where the panel is small or it has adequate rigidity obtained through panel shape or thickness of clay product.

Prestressing of panels has been employed on several projects and has been effective in controlling bowing of long, flat, relatively thin panels. Such panels are generally more susceptible to bowing. As with any multi-layer panel, trial runs may be necessary to verify an analysis as to the best prestressing strand location in order to avoid bowing. Prism tests should be conducted on the proposed brick to establish the modulus of elasticity. The precaster can then design the prestressed cross section for a transformed cross section based upon the ratio of the established moduli. This usually results in the prestressing strand being moved laterally off center to compensate for the transformed section. Prestressed panels, 8 in. thick, have been produced up to 60 ft in length, with minimum sweep (bowing). It is recommended that a control joint be introduced through the clay product face thickness any time the panel length exceeds 30 ft.

Concrete after initial set begins to shrink as it loses excess water to the surrounding environment. Since the clay products are bonded, the shrinkage of the concrete is restrained by the facing. This results in compressive stresses in the facing and tensile stresses in the concrete at the interface. The deformation resulting from these stresses may cause an outward bowing of the clay product surface. Mid-point tie back connections can help minimize convex bowing.

Control of concrete shrinkage necessitates close attention to concrete mix design, continuous control of water and cement content in the mix, and prolonged curing under proper humidity conditions or the use of curing compounds on all exposed concrete surfaces, e.g., back surface and panel edges.

In the case of differential thermal expansion, the direction of deflection depends on the coefficients of expansion of both the facing and concrete. If the facing has a larger coefficient of expansion than the concrete, stresses and deformations in the panel under decreasing temperature will be the reverse of those due to shrinkage. For rising temperatures, the stresses and deformations will be added to that of shrinkage. The reverse situation exists for conditions where the coefficient of expansion of the facing is less than that of the concrete.

It is desirable, therefore, to have a backup concrete with low shrinkage, and a thermal expansion coefficient that closely approximates that of the clay product facing. The coefficient of thermal expansion of concrete can be varied by changing aggregate type.

Panel design must also include "in-service" requirements—in other words, the conditions that panels will encounter when in final location in the structure and subjected to the wide range of seasonal and daily temperatures. Generally the interior surfaces of panels are subjected to a very small temperature range while the exterior surfaces may be exposed to a large daily or seasonal range. The precaster and designer should consider the following in design and production in order to minimize or eliminate panel bowing:

1. The temperature differential (exterior to interior).

2. Coefficients of expansion of the materials.

3. Ratio of cross sectional areas of the materials and their moduli of elasticity.

4. Amount, location and type of reinforcement in the concrete panel.

5. The use of prestressing.

6. Type and location of connections to the structure.

7. Shrinkage of the concrete and expansion of the facing..

Clay Product Properties. Physical properties of brick vary considerably depending on the source and grade of brick. Table 3.5.1 shows physical properties of brick from four different regions of the U.S.

As the temperature or length of burning period is increased, clays burn to darker colors, and compressive strength and modulus of elasticity are increased. The modulus of elasticity of brick ranges from 1.4 to 5.0 x 10^6 psi and Poisson's ratio from 0.04 to 0.11. In general, the modulus of elasticity of brick increases with compressive strength to a compressive value of approximately 5000 psi, after which, there is little change. The average modulus of elasticity is 3.2 x 10^6 psi.

The average coefficient of thermal expansion of brick is 3.6 x 10^{-6} in./in./deg. F. The thermal expansions of clay units is not the same as the thermal expansion of clay product-faced precast concrete panels due to joints.

The compressive strength of clay tile varies from 2000 to 10,000 psi, the modulus of elasticity from 1.8 to 8.0 x 10^6 psi and Poisson's ratio from 0.05 to 0.10. Tile has an average coefficient of thermal expansion of 3.3 x 10^{-6} in./in./deg F.

The compressive strength of terra cotta units usually ranges from 8000 to 11,000 psi and the average coefficient

of thermal expansion is 4.0 x 10^{-6} in./in./deg. F. Since clay products are subject to local variation, the designer should seek property values from suppliers that are being considered.

Clay Product Selection. Clay product manufacturers or distributors should be consulted early in the design stage to determine available colors, textures, shapes, sizes and size deviations as well as manufacturing capability for special shapes, sizes and tolerances. In addition to standard facing brick shapes and sizes (conforming to ASTM C216), thin brick veneer units 3/8 to 3/4 in. thick are available in various sizes, colors and textures. Thin brick units 1 in. thick may also be available. The thin brick units should conform to Type TBX of ASTM C1088, *Specification for Thin Veneer Brick Units Made from Clay or Shale.* Stretcher, corner or 3-sided corner units are typically available in a variety of color ranges, Fig. 3.5.55. The face sizes are normally the same as conventional brick and therefore, when in place, give the appearance of a conventional brick masonry wall.

The most common face size is the standard modular with nominal dimensions of 2-2/3 in. by 8 in. The actual face dimensions vary slightly among manufacturers, but are typically 3/8 in. to 1/2 in. less than the nominal dimensions. The economy size is 50 percent longer and higher, but this difference goes virtually unnoticed since the aspect ratio (length to height) is the same for both the standard and the

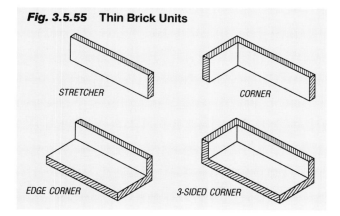

Fig. 3.5.55 Thin Brick Units

STRETCHER

CORNER

EDGE CORNER

3-SIDED CORNER

Table 3.5.1 Average physical properties of brick

Source of Brick	Compressive Strength*		Modulus of Rupture		Tensile Strength, psi	Shear Strength, psi
	Flatwise, psi	Edgewise, psi	Flatwise, psi	Edgewise, psi		
Chicago	3280	3350	1225	1340	417	1100
Detroit	3540	3270	670	680	222	1165
Mississippi	3410	3625	820	760	317	1590
New England	8600	11470	1550	1640	601	3550

*The ratio of compressive strength of clay and shale brick tested on edge to the compressive strength tested flatwise ranges from 0.74 to 2.3.
Source: Plummer, Harry C., "Brick and Tile Engineering," Brick Institute of America, McLean, Virginia, 1962

Table 3.5.2 Nominal Modular Face Sizes of Brick

Unit Designation	Face Dimensions		Number of Courses in 16 in.
	Height in.	Length in.	
Standard	2 2/3	8	6
Engineer	3 1/5	8	5
Economy 8 or Jumbo Closure	4	8	4
Double	5 1/3	8	3
Roman	2	12	8
Norman	2 2/3	12	6
Norwegian	3 1/5	12	5
Economy 12 or Jumbo Utility	4	12	4
Triple	5 1/3	12	3

economy modular units. The economy modular face size, 4 in. by 12 in., is popular for use in large buildings because productivity is increased, and the unit's size decreases the number of visible mortar joints, thus giving large walls a more pleasing appearance by reducing the visual scale of the wall. Other sizes, such as Norwegian, 3-in., non-modular, oversize, etc., may be available. Table 3.5.2 contains face sizes of several modular brick units; however, thin brick may not be available in each size.

Many bricks are too dimensionally inaccurate for precast concrete panel applications. They conform to an ASTM specification suitable for site laid-up applications yet are not manufactured accurately enough to permit their use in a preformed grid used to position bricks for a precast concrete panel. Tolerances in an individual brick of ± 3/32 in. cause problems for the precast concrete producer. Brick may be available from some suppliers to the close tolerances (+ 0, – 1/8 in.) necessary for precasting, or tolerances must be changed by saw cutting each brick thereby increasing costs.

Whole brick are typically not used in precasting. For example, extruded, saw cut brick have kerf lines connecting the extruded holes. The purpose of these lines is to enable the opposing brick faces to be split apart by simply tapping the end of the brick with a mason's hammer. Both sides of the brick are used as facing veneer. Special bricks with a sloping face are used at soldier courses or at the junction with a sloping face. The side cuts on these units are made with a masonry saw and the brick is tapped on its end to remove the waste section.

Brick should preferably be a minimum of 3/4 in. thick (many precasters have used 1/2 in. thick brick) to ensure proper location and secure fit in the template during casting operations, and to minimize the misalignment or tilting of individual units. In site laid-up masonry, each brick is tilted a little bit in one direction or another so it's not quite as noticeable, particularly with bricks having surface imperfections (rough finish). On smooth or glazed face brick, it becomes more noticeable. When placing the units down on a flat form liner, all the brick are perfectly flat, so any tilted brick is more obvious than it would be in site laid-up masonry.

Glazed and unglazed ceramic tile units should conform to American National Standards Institute (ANSI) A137.1, *Standard Specifications for Ceramic Tile*, which includes American Society for Testing and Materials (ASTM) test procedures and provides a standardized system for evaluating a tile's key characteristics. Tiles are typically 1/2 in. thick with a 1 percent tolerance on the length and width measurements. Within one shipment, the maximum tolerance is ± 1/16 in. When several sizes or sources of tile are used to produce a pattern on a panel, the tiles must be manufactured on a modular sizing system in order to have grout joints of the same width.

Glass mosaic tiles require special attention to avoid alkali reaction and subsequent bond failure. They should be painted or sprayed with an epoxy resin and fine sand broadcast over the surface while the resin is wet. In addition, to determine tile products suitable for freeze-thaw conditions tile manufactures should be consulted.

Architectural terra cotta is a custom product and, within limitations, is produced in sizes for specific jobs. Two thicknesses and sizes of units are usually manufactured: 1-1/4 in. thick units, including dovetails spaced 5 in. on centers, size may be 20 x 30 in.; 2-1/4 in. thick units including dovetails spaced 7 in. on centers, size may be 32 x 48 in. Other sizes used are 4 or 6 ft x 2 ft. Tolerances on length and width are a maximum of ± 1/16 in. with a warpage tolerance on the exposed face (variation from a plane surface) of not more than 0.005 in. per in. of length.

Variations in brick or tile color may occur. The clay product supplier must preblend any color variations and provide units which fall within the color range selected by the architect. Also, clay products, like most materials used for architectural facings, suffer from various surface defects. Defects such as chips, spalls, face score lines, cracks and edge "finger marks" are common and the defective units have to be culled from the bulk of acceptable units according to the architect's and applicable ASTM specifications. If these requirements are not met, the visual appearance of the completed panel or a building elevation will be unattractive and may be cause for rejection.

Usually the precast concrete producer buys the clay products but the general contractor or owner may also purchase the units.

Bond. The nature of the surfaces of brick are important for bond to the backup concrete. Smooth, dense, heavily sanded or glazed surfaces are not usually satisfactory where high bond is required. Textures which give good bond include: scored finish, in which the surface is grooved as it comes from the die; combed finish where the surface is altered by parallel scratches; and roughened finish produced by wire cutting or wire brushing to remove the smooth surface or die skin from the extrusion process.

The backside of clay product units should preferably have a keyback or dovetail configuration in order to develop adequate bond to concrete. Grooved or rib back units will also develop adequate bond. Shear test data for terra cotta units are shown in Fig. 3.5.56.

With most clay products used to face precast concrete panels, metal ties are not required to attach them to the concrete. If die skin or heavily sanded brick must be used, a mechanical anchor with corrosion-resistant metal ties is required, Fig. 3.5.57.

Where ties are required, there should be one for each 4-1/2 sq ft of wall area. Ties in alternate courses should be staggered. The maximum vertical distance between ties should not exceed 24 in. and the maximum horizontal distance should not exceed 36 in., See Fig. 3.5.58.

Additional ties should be provided at all openings, spaced not more than 3 ft apart around the perimeter and within 12 in. of the opening. Ties should be of corrosion-resistant metal or should be coated with a corrosion-resistant protective coating.

Corrosion resistance is usually provided by coating the metal with copper or zinc. To ensure adequate resistance to corrosion, coatings should conform to the following ASTM specifications:

1. Zinc coated ties–A153, class B1, B2 or B3.
2. Zinc coated wire–A116, class 2 or class 3.
3. Copper coated wire–B277, grade 30HS.

When ties are used, the brick joints are grouted and the ties placed into the horizontal joint as the wet grout is placed. The required concrete reinforcing steel is anchored in place after the brickwork has received initial set. The concrete is then placed and cured.

Fig. 3.5.57 Typical wall ties

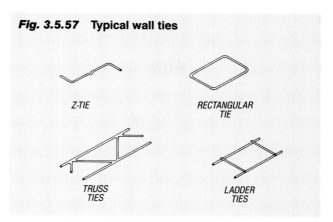

Z-TIE

RECTANGULAR TIE

TRUSS TIES

LADDER TIES

Fig. 3.5.58 Spacing and staggering of metal ties (where required)

24 " MAXIMUM

36" MAXIMUM   MAX. AREA 4-1/2 SQ FT

Fig. 3.5.56 Load tests of terra cotta units with structural lightweight concrete

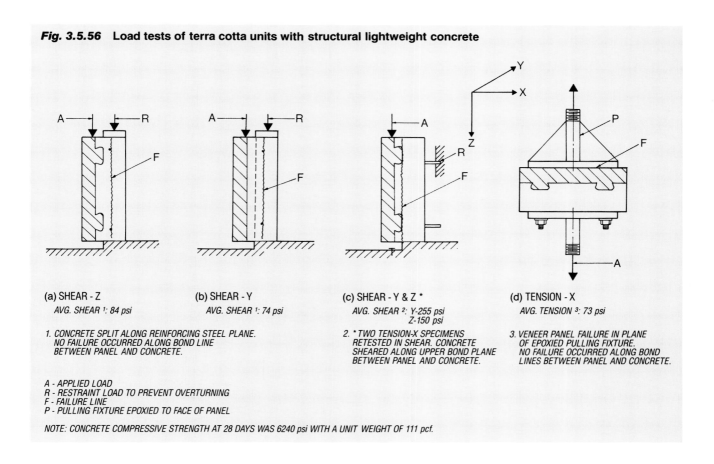

(a) SHEAR - Z
AVG. SHEAR [1]: 84 psi

(b) SHEAR - Y
AVG. SHEAR [1]: 74 psi

(c) SHEAR - Y & Z *
AVG. SHEAR [2]: Y-255 psi
Z-150 psi

(d) TENSION - X
AVG. TENSION [3]: 73 psi

1. CONCRETE SPLIT ALONG REINFORCING STEEL PLANE. NO FAILURE OCCURRED ALONG BOND LINE BETWEEN PANEL AND CONCRETE.

2. * TWO TENSION-X SPECIMENS RETESTED IN SHEAR. CONCRETE SHEARED ALONG UPPER BOND PLANE BETWEEN PANEL AND CONCRETE.

3. VENEER PANEL FAILURE IN PLANE OF EPOXIED PULLING FIXTURE. NO FAILURE OCCURRED ALONG BOND LINES BETWEEN PANEL AND CONCRETE.

A - APPLIED LOAD
R - RESTRAINT LOAD TO PREVENT OVERTURNING
F - FAILURE LINE
P - PULLING FIXTURE EPOXIED TO FACE OF PANEL

NOTE: CONCRETE COMPRESSIVE STRENGTH AT 28 DAYS WAS 6240 psi WITH A UNIT WEIGHT OF 111 pcf.

The bond between the facing and the concrete varies depending on the absorption of the clay product. Low absorption will result in poor bond, as will high absorption due to the rapid loss of the mixing water preventing proper hydration of the cement and the development of good bond strength. Bricks with a water absorption by boiling (ASTM C216) of about 6% to 9% provide good bonding potential. Bricks with an initial rate of absorption (suction) less than 20 g per min. per 30 sq. in., when tested in accordance with ASTM C67, are not required to be wetted. However, brick with high suction or with an initial rate of absorption in excess of 20 g per min. per 30 sq. in. should be wetted, prior to placement of the concrete, to reduce the amount of mix water absorbed, and thereby improve bond. Terra cotta units should be soaked in water for at least one hour prior to placement in order to reduce suction. They should be damp at the time of concrete placement.

Latex additives in the concrete or latex bonding materials provide high bond and high strength, but they have limitations. They are water sensitive, losing as much as 50% of their strength when wet (although they regain that strength when dry). The lowered strength is usually sufficient to sustain low shear stress like the dead weight of the clay product, but when differential movements cause additional stress, problems can occur.

Generally, clay products cast integrally with the concrete have bond strengths exceeding that obtained when laying units in the conventional manner in the field (clay product to mortar). It is necessary, in either case, to use care to avoid entrapped air or excess water-caused voids which could reduce the area of contact between the units and the concrete, thereby reducing bond.

Clay bricks, when removed from the kiln after firing, will begin to permanently increase in size as a result of absorption of atmospheric moisture. The design coefficient for moisture expansion of clay bricks as recommended by the Brick Institute of America is 0.0003 to 0.0004 in. per in. while 0.0006 is recommended for tile by the Tile Council of America. The environmental factors affecting moisture expansion are:

1. Time of exposure. Expansion increases linearly with the logarithm of time. It is estimated that approximately 40% of the total potential moisture expansion of bricks will occur approximately three months after the bricks have been fired and that approximately 50% of the total potential moisture expansion will occur approximately one year after the bricks have been fired.

2. Time of placement. How much the brick will expand subsequent to placement in the panel depends upon how much expansion has already occurred and what proportion this represents of the total potential for expansion.

3. Temperature. The rate of expansion increases with increased temperature when moisture is present.

4. Humidity. The rate of expansion increases with an increase in relative humidity. Bricks exposed to a relative humidity of 70% has a moisture expansion two to four times as large as those exposed to RH of 50% over a 4-month interval. The 70% RH bricks also exhibit almost all of their expansion within the first twelve months of exposure, while the 50 % RH bricks generally exhibit a gradual continuous moisture expansion.

In addition to continuous permanent growth due to moisture absorption, seasonal reversible expansion and contraction of clay bricks will occur due to changes in the ambient air temperature. It is not uncommon for the exterior surface to reach temperatures of 165 deg. with dark colored brick, 145 deg. F for medium color, or 120 deg. F for light colored brick on a hot summer day when directly exposed to solar radiation. Likewise surface temperatures as low as – 30 deg. F can be reached on a cold winter night.

Field experience and analysis has shown that when the moisture expansion potential of clay products is combined with thermal expansion and both are acting opposite to the shrinkage of the concrete backup, the resulting stresses do not cause bond failure, bulging or cracking.

The expansion of the clay products can be absorbed by four simultaneously occurring negative dimensional changes of the clay product and grout (mortar) or concrete:

1. Drying shrinkage of the grout.

2. Elastic deformation of the grout under stress.

3. Creep of the grout under stress.

4. Elastic deformation of the clay product under stress.

In general, strains imposed slowly and evenly will not cause problems. Consider the first 6 months to a year after panel production, see Fig. 3.5.59. Tile expansion is small (rate of strain application is slow) but mortar shrinkage is nearly complete. The mortar or concrete creeps under load to relieve the tensile stress generated in the tile by the mortar or concrete shrinkage since the tile are relatively rigid (E of tile/E of mortar or concrete). After this time period, the tile have years to accommodate the additional moisture expansion.

Failures occur when strain rates exceed creep relief rates. This can occur when:

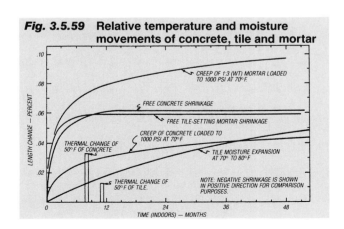

Fig. 3.5.59 Relative temperature and moisture movements of concrete, tile and mortar

1. Total shrinkage is higher than normal because overly rich concrete or mortar was used.

2. Sudden rise in temperature or drop in humidity causes shrinkage to proceed faster than creep relieves the stresses that are generated.

3. Bond between clay product and concrete was never adequately achieved.

4. Sudden temperature drop imposes a sudden differential strain because the clay product and mortar (or concrete) have different thermal coefficients of expansion.

The difference in creep characteristics between concrete and brick, along with the differences in their respective moduli of elasticity, do not pose a problem to the production of small (less than 30 ft) panels when good quality brick is used. This observation is based on static load tests simulating differential creep.

Design and Production Considerations. Overall size and weight of panels are generally limited to what can be conveniently and economically handled by available transportation and erection equipment. Generally, panels span between columns, usually spaced 20 to 30 ft on centers, although spandrel panels have been as large as 6 x 60 ft and 8 x 55 ft.

The standard nominal widths and heights of the panels should be in multiples of nominal individual masonry unit heights and lengths. The actual specified dimensions may be less than the required nominal dimensions by the thickness of one mortar joint, but not by more than 1/2 in. The precast concrete plant should be able to use uniform, even coursing without cutting any units vertically or horizontally.

The appearance of clay product-faced precast concrete panels is principally achieved by the selected clay product with type, size and texture contributing to overall color. Also, the clarity with which the clay product units are featured will depend upon the profile of the joint between units, thus requiring a choice between recessed, flush or concave-tooled joints. Most brick faced panels have recessed joints, Fig. 3.5.60. The joints between panels are usually butt joints. A quirk miter joint (with the thin brick thickness being the quirk dimension) may be used if the joint sealant covers the brick/concrete interface. This precaution is necessary to prevent moisture penetration into the interface and possible brick separation under freeze-thaw conditions. The final element in the appearance of the panel is the grout, mortar or concrete used in the joints.

Clay product-faced units have joint widths controlled by locating the units in a suitable template or grid system set out accurately on the mold face, Fig. 3.5.61. The most popular grid system consists of either an elastomeric (or rubber) form liner or a wooden base tray with shaped wood joint sections. Liner ridges may be shaped so that joints between units may simulate raked (recessed) or tooled joints. The elastomeric form liner can be produced to a tolerance of ± 1/64 in. In some cases, this grid system has successfully incorporated loose rubber joint strips of Shore A, 60 durometer hardness to form a "build-as-you-apply-brick" grid or has used urethane, plaster or sand to ensure

Fig. 3.5.60

proper location and secure fit during casting operation, and prevent grout leakage to the exposed face of the panel.

Both stack and running bond patterns have been widely used in precast concrete panels. With running bond it is essential that courses start and finish with half or full bricks to avoid cutting and to match adjacent spandrels or column covers, see Fig. 3.5.60. Attempts to make the finished exterior look like running bond brick have not always been successful where the pattern is carried unbroken from panel to panel to form large walls. Where the perimeter of the panels is recognized and a narrow strip of exposed concrete is left at the ends, the running bond pattern becomes successful within individual panels from a design point of view, Fig. 3.5.62. Vertical alignment of joints, especially with stack bond requires close tolerances or cutting of brick to the same length.

Joints between the clay products contribute to the appearance and should not be less than 1/4 in., with 3/8 in. preferred. A 1/4 in. joint may be satisfactorily used when the joint faces are smooth and well defined (such as a wall of flush-pointed, smooth-faced clay products). If the unit profiles or the mortar beds in the sides of the joint are rough, then joints must be wider than 1/4 in. If conditions exist which increase the rate of clay product moisture expansion, it may be desirable to use 1/2 in. grout joints to provide maximum accommodation of moisture expansion strains.

Fig. 3.5.61

Fig. 3.5.62

Fig. 3.5.63

Using a grid system, the joints are recessed, usually 1/4 in., and are filled either with a 1:4 cement-sand mortar or the structural concrete used in the structural component of the panel. Colored concrete in the joint is not necessary, although such an architectural choice can be readily accommodated. The aggregate in the concrete for the joints should have a maximum size less than the joint width. Joints narrower than 3/8 in. may have to be filled from the front after the panel has been stripped.

Steel reinforcement is positioned in the panel, and connection hardware and handling inserts are located and secured. Then the backup concrete is placed in a normal manner. Care must be taken during concrete placing and consolidation to prevent movement of the individual facing materials and thus upset the appearance of the finished surface.

Tiles, measuring 2 x 2 in. or 4 x 2 in., may be supplied face-mounted on polyethylene or paper sheets and secured to the mold by means of double-faced tape or a special paste. The space between the tiles is filled with a thin grout film and then the backup concrete is placed prior to initial set of the grout. Fig. 3.5.63 shows a project that used 2 x 2 in. tile with the placing method described.

After the concrete cures and the panel is removed from the mold, joints may be filled, if necessary, with pointing mortar or grout carefully formulated as to color and texture. Before pointing, joints should be saturated with clean water. After joints are properly pointed and have become thumb print hard they may be (1) tooled to a smooth concave surface, or (2) struck and troweled flush with the face of the clay units. Initial grout cleanup should be done within 15 minutes of pointing to avoid hard setting of the grout on the units. Final cleanup should be completed within 60 to 90 minutes.

When tile or terra cotta joints are grouted after the panel is stripped, a clean, hard rubber float should be used for spreading grout and fully packing joints. Excess grout should be removed by holding the float at a 90 degree angle and moving it across the panel diagonally to the joints. In order to maintain a uniform grout color and texture, colored grout should be cleaned off the panels with cheesecloth prior to grout setting. Proper cleaning of tile surfaces eliminates the need for acid cleaning. However, should some residue be left on the tile or terra cotta surface, after ten days it can be cleaned with sulfamic acid, provided the surface is thoroughly rinsed with clean water after cleaning.

Mortar stains may be removed from brick panels by thoroughly wetting the panel and scrubbing with a stiff brush and a masonry cleaning solution. A prepared cleaning compound is recommended; however, on red brick a weak solution of muriatic acid and water (not to exceed a 10% muriatic acid solution) may be used. Acid should be flushed off the panel within 5 minutes of application. Buff gray or brown brick should be cleaned in accordance with the brick manufacturers' recommendations, using proprietary cleaners, rather than acid to prevent vanadium or manganese stains. Following application of the cleaning solution, the wall should be rinsed thoroughly with clean water. High pressure water cleaning techniques, with a 1000 to 2000 psi washer, may also be used to remove mortar stains. Many precasters apply a water-based retarder to the face of the brick or tile prior to placement in the mold to facilitate mortar stain removal.

In all operations after removal from forms, clay product-faced precast concrete panels are handled, stored and shipped on the concrete edge of the panel or on their backs with the facing up. The panels must not at any time rest on the face or on any of the clay product edges or corners to avoid chipping or spalling. To minimize the effects of the sun on bowing, panels are sometimes stored on edge with the length oriented north and south.

During shipping, the panels may be placed on special rubber padded racks or other measures taken to prevent chipping of edges or damage to returns. Long returns at sills and soffits generally create handling problems, unless proper procedures are worked out ahead of time.

Should minor damage occur to the clay product face during shipping, handling or erection, field remedial work can easily be accomplished, including replacement of individual clay products.

It is important to determine the quantity of patching that will be permitted since one color will not match all of the various brick hues in use on any one job. Samples should be viewed at a distance or height consistent with actual project locations. Also, the panel must be viewed in its entirety, not just examined for individual brick deficiencies.

Full or thin brick have also been hand laid at the precast concrete plant on a panel ledge, generally created by a recess on the precast concrete panel face, to create a brick faced panel. Fig. 3.5.64 shows the use of a bullnose to provide a shadow line between the brick and precast concrete. Dovetail anchors are necessary in the precast concrete and weepholes for drainage are required. Ceramic tile may also be applied to a recess in the precast concrete using latex modified mortar.

The designer must develop a rapport with both the precaster and the clay product supplier so that all parties are aware of mutual expectations and potential problems with, for example, returns, soldier courses, joints, and window details.

3.5.11 HONED OR POLISHED

Grinding of concrete produces smooth, exposed aggregate surfaces. The grinding is called honing or polishing, depending on the degree of smoothness of the finish. Polished exposed aggregate concrete finishes compare favorably with polished natural stone facades, allowing the architect great freedom of design. Honed and polished finishes have gained acceptance because of their appearance and excellent weathering characteristics, making them ideal for high traffic areas and polluted environments. Because of a corrosive and dirt-laden atmosphere, a dense polished surface texture resembling terrazzo was used on the building in Fig 3.5.65. Maintenance since 1973 has proven to be minimal for the architectural precast concrete cladding. The panels were made with an exposed

Fig. 3.5.64

quartz aggregate, silica sand, and buff tinted white cement. After fabrication and removal from the mold, the panels were ground and polished smooth on all flat surfaces. Fig. 3.5.66 shows the excellent details.

In order to produce a good ground or polished finish it is first necessary to produce a good plain finish. The compressive strength of the concrete should be 5000 psi before the start of any honing or polishing operations. All patches and the fill material on any bug/blow holes or other surface blemishes must also be allowed to reach approximately 5000 psi. It is preferable that the mortar strength of the concrete mix approach the compressive strength of the aggregates or the surface may not grind evenly or polish smoothly, and aggregate particles may be dislodged.

Since aggregates will polish better than the matrix, it is essential to have a minimum matrix area. A continuous graded concrete mix is preferred, carefully designed to provide maximum aggregate density on the surface to be polished. In choosing aggregates, special attention should be given to maximum size and hardness. Softer aggregates such as marble or onyx are much easier to grind than either granite or quartz. This will be strongly reflected in the cost of such finishes.

Equipment may vary from a simple hand grinder to a very elaborate multihead machine which has increased the size of units which can be polished and reduced the cost.

The grinding process, which can be either wet or dry, removes approximately 1/8 in. off the form face of the pre-

Fig. 3.5.65

Fig. 3.5.66

cast concrete panels. Wet grinding is preferred, since the paste which is created aids the grinding. A high standard of craftsmanship is mandatory for this treatment, as the removal of the cement skin emphasizes any defects in either formwork or compaction. It is very important to avoid any segregation. As with other finishes, final appearance and uniformity will benefit if it is possible to match or blend matrix color with aggregate color. Fig. 3.5.67 shows a gray granite aggregate, black fleck, and white cement matrix that has been polished.

Continued mechanical abrasion with progressively finer grit, followed by a special treatment which includes filling of all surface air voids and rubbing, will produce a highly polished surface. The depth of grinding determines the extent to which the aggregate shows, but the color of the cement will be important in any case. Such panels have an attractive sheen that enhances many colors. Polished panels of pastel colors tend to appear white when viewed from a distance because of their high surface reflectance. Therefore, this type of surface is recommended for panels situated relatively close to the traffic flow or for those of medium or dark shades. Rose granite aggregates with a colored matrix were used in the panels in Fig. 3.5.68. Ground and polished to achieve the appearance of natural stone at a considerable savings over natural stone, Fig. 3.5.69 shows a closeup of the building base which has both natural stone and polished precast concrete.

A combination of light acid etching of the matrix area with polishing of the aggregates has produced excellent tex-

Fig. 3.5.67

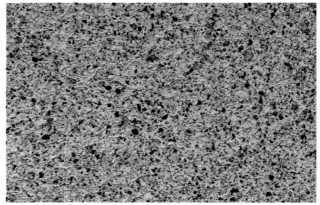

tures which weather particularly well. In Fig. 3.5.70, the precast concrete panels were ground smooth after casting and then acid etched to reveal the red granite chips. Fig. 3.5.71 shows a closeup of the surface.

Careful detailing to maximize use of automatic polishing equipment and minimize hand polishing will ensure minimum cost. For economy, only surfaces which can be ground completely by machines passing over flat areas or returns should be honed or polished. Attractive finishes may be achieved by leaving reveals or grooves in the smooth form finish, see Fig. 3.5.66, or by use of flat ribs in smooth or exposed aggregate finishes, Fig. 3.5.72 with a closeup in 3.5.73. Hand polishing of arrises and returns is expensive and should be designed out where possible.

Fig. 3.5.68

Fig. 3.5.69

Fig. 3.5.71

Fig. 3.5.70

Fig. 3.5.72

Fig. 3.5.73

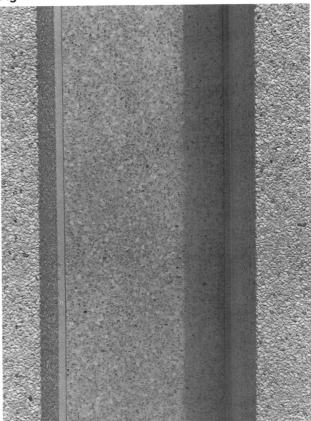

3.5.12 STONE VENEER-FACED PRECAST CONCRETE PANELS

Natural stone has been widely used in building construction for centuries due to its strength, durability, esthetic effect, availability and inherent low maintenance costs. In the 1960's, the practice of facing skeleton frame structures with large prefabricated concrete components to decrease construction time and reduce costs resulted in combining the rich beauty of natural stone veneer with the strength, versatility, and economy of precast concrete.

Some of the advantages of stone veneer-faced precast concrete panels are:

1. Veneer stock can be used in thin sections because of the short spans between anchoring points.

2. Multiplane units such as column covers, spandrels with integral soffit and sill sections, deep reveal window frames, inside and outside corners, projections and setbacks, and parapet sections can be relatively economically assembled as veneer units on precast concrete panels. Typical spandrel and column cover panels are shown in Fig. 3.5.74. Often it is desirable to use one of the veneer materials in a traditional manner around the lower portion of a building and extend a similar finish with veneered precast concrete panels up the exterior walls.

3. The erection of the precast concrete units is faster and more economical than conventional handset construc-

tion because the larger panels incorporate a number of veneer pieces.

4. The veneered concrete panels can be used to span column to column, thereby reducing floor edge loading.

GENERAL CONSIDERATIONS. As a standard practice, it is recommended that someone qualified be engaged by the purchaser of the stone to be responsible for the coordination, which includes delivery and scheduling responsibility as well as ensuring acceptable color uniformity. For larger projects, and when feasible, color control or blending for uniformity should be done in the stone fabricator's plant, since ranges of color and shade, finishes, and markings such as veining, seams and intrusions are easily seen during the finishing stages. Acceptable color of the stone should be judged for an entire building elevation rather than for an individual panel.

The responsibility for stone coordination should be written into the specifications so its cost can be bid. Frequent visits to the stone fabricator's plant may be required. With proper coordination and advance planning, fabrication and shipments will proceed smoothly. When communication is lacking, major problems in scheduling and delivery may occur.

All testing to determine the physical properties of the stone veneer with the same thickness and finish as will be used on the structure should be conducted by the owner prior to the award of the precast concrete contract. This will reduce the need for potentially costly repairs or replacement

Fig. 3.5.74 Typical spandrel and column cover panels

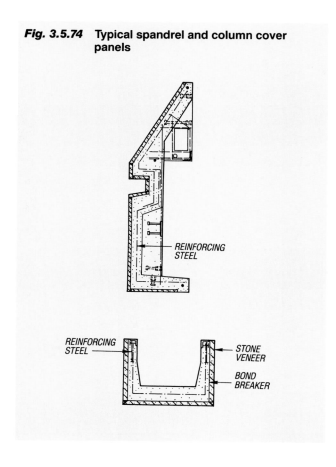

should deficiencies in the stone veneer be found after start of fabrication.

The precast concrete producer must provide the general contractor with stone quantity and sequence requirements to meet the erection sequences which are determined by mutual agreement. For reasons of production efficiency some concrete panels may be produced out of sequence relative to erection sequence. The precast concrete producer and stone fabricator should coordinate packaging requirements to minimize handling and breakage. Extra stone (approximately 2 to 5%) should be supplied to the precaster to allow immediate replacement of damaged stone pieces, particularly if the stone is not supplied from a domestic source. Deliveries should be scheduled as closely as possible to actual fabrication schedules.

Because of the difference in material properties between natural stone and concrete, veneered panels are more susceptible to bowing than all-concrete units; also, the flat surfaces of cut stone reveal bowing more prominently than all-concrete panels.

Concrete, after initial set, begins to shrink as it loses excess water to the surrounding environment. The stone veneer, especially with an impermeable bondbreaker, limits drying on the veneered side of the backup concrete. The resulting differential shrinkage of the concrete and stone veneer can cause outward bowing in a simple span panel. While all-concrete panels usually bow in response to thermal gradients through the panel thickness, stone veneered

concrete may also bow when the temperature is uniform through the panel thickness. This bowing is caused by differences in the coefficients of expansion of the stone and the concrete. Limestone has an average coefficient of expansion of 2.8×10^{-6} in./in./deg. F, while granite has 4.5×10^{-6} and marble 7.3×10^{-6}. Coefficients of 6×10^{-6} in./in./deg. F for normal weight and 5×10^{-6} for sand-lightweight concrete are frequently used. As stone becomes thinner, coefficient of expansion differentials become more important because the stone has less rigidity to resist bowing. It is desirable, therefore, to have a backup concrete with low shrinkage, and a thermal expansion coefficient that closely approximates that of the stone veneer. The coefficient of thermal expansion of concrete can be varied by changing aggregate type.

Precasters have developed design and production procedures to minimize bowing. Many manufacturers compensate by using cambered forms, e.g. 1 in. for 40 ft, to produce panels initially bowed inward and panels are sometimes stored on edge with the length of the panel oriented north and south to minimize the effects of the sun on bowing. Bowing is also a consideration in reinforcement design. If thickness is sufficient, two layers of reinforcement should be used, as this helps to reduce bowing caused by differential shrinkage or temperature changes. In some cases, reinforcing trusses are used to add stiffness; in others, concrete ribs are formed on the back of the panel, but this may require backforming and is more costly. Prestressing of panels has been effective in controlling bowing of long, flat, relatively thin panels. Minimum thickness of backup concrete on flat panels that will control bowing or warping is usually 5 to 6 in. but 4 in. has been used where the panel is small or it has adequate rigidity obtained through panel shape or thickness of natural stone. Mid-point tie back connections can also help to minimize convex bowing.

Stone Sizes. Stone veneers used for precast concrete facing are usually thinner than those used for conventionally set stone with the maximum size generally determined by the stone strength (stone breakage).

Thicknesses of marble veneer of 7/8 in. or less are not desirable as it is probable that the anchor will be reflected on the surface and the development of adequate anchor capacity is questionable. Travertine has been used in thicknesses of 3/4, 1 (2.5 cm), 1-1/4, 3cm, and 1-1/2 (4 cm) in. with the surface voids filled front and back on the thinner pieces. Thicknesses of 3/4, 1 in., 2 cm and 2.5 cm have resulted in excessive breakage and are not recommended. The 10 x 12 ft panels in Fig. 3.5.75 have 1 in. thick marble veneer on a 5 in. precast concrete backup. The concrete has 2 layers of reinforcement. Over 73,000 pieces of 1-1/4 in. travertine were anchored to 7055 precast concrete units to produce 600,000 sq ft of cladding, Fig. 3.5.76.

Granite veneer thicknesses of 3 cm or greater are recommended, unless strength of stone determined from testing indicates otherwise. The 12 x 20 ft double window panels in Fig. 3.5.77 contain as many as 39 pieces of granite veneer. The wall panels consist of 1-1/4 in. granite, 1/2 in. air space, 2-3/4 in. insulation, and 5 in. of concrete The undulating

Fig. 3.5.75

Fig. 3.5.76

shape in Fig. 3.5.78 was clad with 1-1/4 in. granite faced precast concrete panels.

Limestone thicknesses used for veneer on precast concrete have been as thin as 3 cm and as thick as 5 in. (13 cm). The performance of veneers of thicknesses less than 1-3/4 in. is questionable and their use is not recommended because of potential permeability and strength problems. The typical precast concrete panels in Fig. 3.5.79 consist of 2 in. limestone, a bondbreaker and a 4 in. concrete backup. The building in Fig. 3.5.80 has cruciform shaped precast concrete panels consisting of 2 in. limestone anchored to 5 in. of concrete with 3/8 in. stainless steel dowel pins set in epoxy at opposing angles. A bondbreaker was used between the veneer and the backup concrete.

The length and width of veneer materials should be sized to a tolerance of $+0 - 1/8$ in. since a plus tolerance can present problems on precast concrete panels. This tolerance becomes important when trying to line up the false joints on one panel with the false joints on the panel above or below, particularly when there are a large number of pieces of stone on a panel. Tolerance allowance for out-of-square is $\pm$ 1/16 in. difference in length of the two diagonal measurements. Flatness tolerances for finished surfaces depend on the type of finish. For example, the granite industry flatness tolerances vary from 3/64 in. for a polished surface to 3/16 in. for flame (thermal) finish when measured with a 4 ft straightedge. Thickness variations are less important since concrete will provide a uniform back face, except at corner butt joints. In such cases, the finished edges should be within $\pm$ 1/16 in. of specified thickness. However, large thickness variations may lead to the stone being encased with concrete and thus being unable to move. The esthetic problems that have occurred with tolerances have been the variation from a flat surface on an exposed face and stone pieces being out of square.

Fig. 3.5.77

ANCHORAGE OF STONE FACING. The responsibility for determining the type of anchorage between the stone and concrete backup varies on different projects. The stone fabricator or concrete precaster appear to have the dominant responsibility for conducting the anchor tests, with the architect or engineer of record occasionally determining the type of anchorage. However, it is preferable for the architect to determine anchor spacing so that common

Fig. 3.5.78

information can be supplied to all bidders. Contract documents should clearly define who drills the anchor holes in the stone; type, number and location of anchors; and who supplies the anchors. In most cases, the stone fabricator diamond core drills the anchor holes in the back surface for the attachment of mechanical anchors according to architectural specifications and drawings.

The precast concrete industry recommends that there be no concrete bonding between stone veneer and concrete backup in order to minimize bowing, cracking and staining of the veneer. Flexible mechanical anchors should be used to secure the veneer.

The following methods may be used to prevent bond between the veneer and concrete to allow for independent movement:

1. A liquid bondbreaker, of a thickness that allows sufficient shear displacement, applied to the veneer back surface prior to placing the concrete.

2. A one component, clear polyurethane coating, or other thin liquid bondbreaker.

3. A 6 to 10 mil polyethylene sheet.

4. A closed cell 1/8 to 1/4 in. polyethylene foam pad. The use of a compressible foam pad bondbreaker is preferred in order to have movement capability with uneven stone surfaces, either on individual pieces or between stone pieces on a panel.

Fig. 3.5.79

Fig. 3.5.80

Preformed anchors, 1/8 to 1/4 in. in diameter, fabricated from Type 304 or occasionally Type 302 stainless steel, are supplied by the stone fabricator or, in some cases, by the precaster depending on the contract document requirements. The number and location of anchors should be determined by a minimum of 5 shear and tension tests conducted on a single anchor embedded in a stone/precast concrete test sample and the anticipated loads, wind and shear to be applied to the panel. Anchor size and spacing in veneers of questionable strengths or with natural planes of weakness may require special analysis.

Four anchors are usually used per stone piece with a minimum of 2 recommended. The number of anchors has varied from 1 per 1-1/2 sq ft of stone to 1 per 6 sq ft with 1 per 2 to 3 sq ft being the most common. Anchors should be 6 to 9 in. from an edge with not over 30 in. between anchors. A typical marble veneer anchor detail with a toe-in spring clip (hairpin) anchor is shown in Fig. 3.5.81 and a typical granite veneer anchor detail is shown in Fig. 3.5.82 The toe-out anchor in granite may have as much as 50 percent more tensile capacity than a toe-in anchor depending on the stone strength.

Depth of anchor holes should be approximately one-half the thickness of the veneer (minimum depth of 3/4 in.), and are often drilled at an angle of 30 to 45 deg. to the plane of the stone. Holes which are approximately 50 percent oversize have been used to allow for differential movement between the stone and the concrete. However, in most cases, holes 1/16 to 1/8 in. larger than the anchor are com-

mon as excessive looseness in hole reduces holding power. Anchor holes should be within ± 3/16 in. of the specified hole spacing, particularly for the spring clip anchors.

Stainless steel dowels, smooth or threaded, may be installed to a depth of 2/3 the stone thickness with a maximum depth of 2 in. at 45 to 60 deg. angles to the plane of the stone. The minimum embedment in the concrete backup to develop the required bond length is shown in Fig. 3.5.83. Dowel size varies from 3/16 to 5/8 in. for most stones, except that it varies from 1/4 to 5/8 in. for soft limestone and sandstone and depends on thickness and strength of stone. The dowel hole is usually 1/16 to 1/8 in. larger in diameter than the anchor.

Limestone has traditionally been bonded and anchored to the concrete, because it has the lowest coefficient of expansion. Limestone has also traditionally been used in thicknesses of 3 to 5 in. but it is now being used as thin as 1-1/4 in. If limestone is to bonded, it is desirable to use a moisture barrier/bonding agent on the back side of the stone to eliminate the possibility of alkali salts in concrete staining the stone veneer. Moisture barrier/bonding agent materials include portland cement containing less than 0.03% water soluble alkalies; waterproof cementitious stone backing; non-staining asphaltic or bituminous dampproofing; or an epoxy bonding agent that cures in the presence of moisture. When limestone is 2 in. or thinner, it is prudent to use a bondbreaker, along with mechanical an-

Fig. 3.5.81 Typical anchor for marble veneer

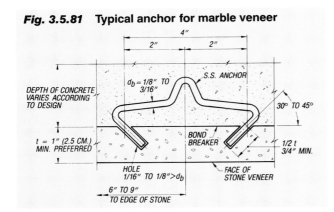

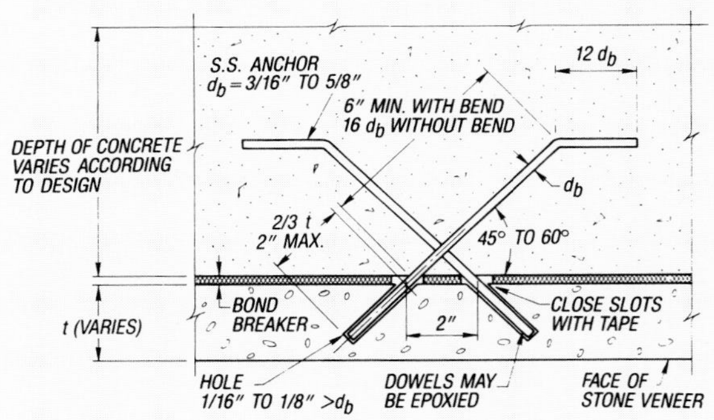

Fig. 3.5.82 Typical anchor for granite veneer

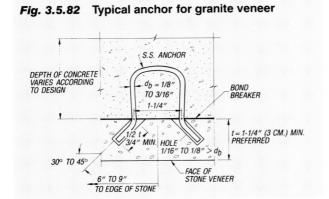

Fig. 3.5.83 Typical cross anchor dowels for stone veneer

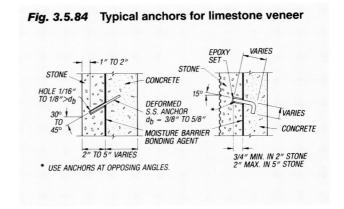

Fig. 3.5.84 Typical anchors for limestone veneer

Fig. 3.5.85

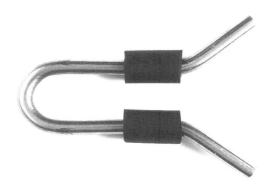

chors. Dowels and spring clip anchors have been used to anchor limestone. Typical dowel details for limestone veneers are shown in Figs. 3.5.83 and 3.5.84. The dowels in Fig. 3.5.84 should be inserted at angles alternately up and down to secure stone facing to backup concrete.

It must be emphasized that some flexibility should be introduced with all anchors of stone veneer to precast concrete panels, e.g., by keeping the diameter of the anchors to a minimum, to allow for the inevitable relative movements which occur with temperature variations and concrete shrinkage. Unaccommodated relative movements can result in excessive stress problems and eventual failure at an anchor location.

Some designers use epoxy to fill the spring clip anchor or dowel holes in order to eliminate intrusion of water into the holes and the possible dark, damp appearance of moisture on the exposed stone surface. The epoxy increases the shear capacity and rigidity of the anchors. The rigidity may be partially overcome by using 1/2 in. long compressible rubber or elastomeric grommets or sleeves on the anchor at the back surface of the stone, Fig. 3.5.85. There is also a concern that differential thermal expansion of the stone and epoxy may cause cracking of the stone veneer. However, this may be overcome by keeping the oversizing of the hole to a minimum, thereby reducing epoxy volume. It may be desirable to fill the anchor hole with an elastic fast-curing silicone which has been proven to be non-staining to light colored stones, or preferably a low modulus polyurethane sealant. The overall effect of either epoxy or sealant materials on the behavior of the entire veneer should be evaluated prior to their use. At best, the long term service of epoxy is questionable, therefore, any increase in shear value should not be used in calculating long term anchor capacity.

When using epoxy in anchor holes, the precaster needs to follow the manufacturer's recommendations as to mixing and curing temperature limitations.

Because of the expected variation in the physical properties of natural stone and the effects of weathering on the stone, recommended safety factors are larger than those used for man made building materials, such as steel and concrete. Various safety factors are recommended by the trade associations and the suppliers of different kinds of building stones. The minimum recommended design

safety factors, based on the average of the tests results, are 3 for granite, 4 for anchorage components in granite, 8 for limestone veneers, and 5 for marble veneers. If the range of test values exceeds the average by more than ± 20 percent than the safety factor should be applied to the lower bound value. The safety factors in Table 3.5.3, based on the coefficient of variation of test results, have also been recommended for different types of stones.

Fig. 3.5.86 shows typical production practices for stone veneer-faced precast concrete panels.

Table 3.5.3 Safety Factors for different types of stone

Type of Stone	Coefficient of Variation	Safety Factor (based on average values)
Igneous (Granite Serpentine)	<10 10-20 >20	4.5 6.0 8.0
Metamorphic (Marble Slate)	<10 10-20 >20	6.0 7.5 10.0
Sedimentary (Limestone Sandstone Travertine)	<10 10-20 >20	7.5 9.0 12.0

3.5.13 VARIATIONS OF FINISHES WITHIN A SINGLE UNIT

Combinations of different finishes within a single precast concrete unit offer excellent possibilities for the architectural use of tones and texture in facade treatments. Two or more finishes can be readily achieved using the same concrete mix.

Using two different facing mixes in the same panel can be an expensive procedure. The first mix is placed within an area bounded by a raised demarcation strip the thickness of the face mix. Within 1-1/2 hours, the form surface around the first cast is carefully mopped clean and the second mix is placed and vibrated. It is very important that the second mix be placed and the concrete consolidated prior to initial set of the first concrete mix. Another approach is the two-stage casting procedure discussed in Section 3.3.8. This method was used to cost effectively create precast concrete panels in three finishes, Fig. 3.5.87. The retarded rosebud quartz panel section was cast separately, set in a mold, then the remainder of the concrete was cast around it. The costs were prorated over repetitive panels.

Fig. 3.5.88 shows an effective change of texture in a panel achieved simply by using a larger size of aggregate in one of the mixes and using a recess as the demarcation feature. Concrete mixes with different colored matrices exposed at the face of the panel have casting considerations. Special care must be taken to prevent a white cement paste from leaking into a gray or vice versa.

An economical and pleasing solution to an architect's de-

Fig. 3.5.86

Typical Production Practices for Stone Veneer-Faced Precast Concrete Panels
(Sequences shown are from several projects and are used to illustrate specific points.)

1. Stone is carefully placed in the form either manually or with a vacuum lifter.

2. Compressible spacer material is placed between the stone slabs in the form.

3. Bondbreaker is placed over the back of the stone; spring clip anchors can be seen penetrating through, together with the connection hardware in place, and the prestressing strand already stressed.

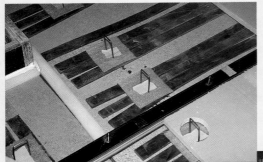

4. For an insulated sandwich panel with an airspace, rubber strips are placed to create the airspace (to be removed when panel is stripped), anchors are inserted, and a polyethylene foam pad bondbreaker is placed and taped.

5. Two layers of insulation are placed; and insulation joints are taped and caulking is used between insulation layers.

6. All reinforcement, prestressing strand, connection and lifting inserts, and additional attachments such as window washer inserts or tracks are assembled in the form prior to placing the concrete.

7. Backforms, if necessary, are fixed in the form and concrete is placed and vibrated.

8. Finished panel being lifted.

Fig. 3.5.87

sire for three different but complimentary expressions in one panel is shown in Fig. 3.5.89. One homogeneous mix, consisting of white cement, local buff sand and brown aggregates, was used throughout the panel. The ribs were left in smooth, off-the-mold finish. In portions of the flat area, aggregates were exposed by retardation, showing the brown tone of the aggregates. By creating horizontal indentations in the spandrel area under the windows, and by exposing the aggregates, this part appears to be a darker brown in color.

Combinations of various finishes on the same unit depend on the shape of the unit. For example, a retarded flat surface between ribs spaced widely apart need a groove or offset but a flat surface between close ribs may be accomplished without a demarcation feature. Proper samples should be used to assess the problems. Bushhammering and to a lesser degree sandblasting can be stopped fairly easily along specific lines. Other combinations of finishes demand demarcation features as described in Section 3.3.4. For example, a retarder applied to only a part of a mold surface may produce a ragged effect at the edge of the treated area, unless a demarcation feature is used to separate the finishes. Fig. 3.5.90 shows a panel with one concrete mix and recesses to separate the retarded and sandblasted surfaces.

The facades in Fig. 3.5.91 have half-round window head and sill profiles coupled with edge-defining reveals. They are characterized by medallions and belt courses of retarded exposed red granite aggregate, which are sur-

Fig. 3.5.88

Fig. 3.5.89

Fig. 3.5.90

Fig. 3.5.91

rounded by framing grids of light-colored, sandblasted concrete. In Fig. 3.5.92 to match the granite used on the two office towers, the banding on the spandrels for the parking structure has a retarded exposed red granite aggregate with a sandblasted center portion.

The surface textures in Fig. 3.5.93 are 1-1/2 in. deep bushhammered ribs and retarded exposed aggregate. The two textures are separated by deep reveals which reinforce the linear character of the building. In Fig. 3.5.94 precast concrete with smooth white upper and lower bands alternates with a retarded and washed center band of exposed white marble aggregate.

An acid etched finish cannot easily be applied to only a

Fig. 3.5.92

Fig. 3.5.93

portion of a unit. When only a portion of the surface is to be acid etched, a demarcation feature is necessary and costs will increase. Fig. 3.5.95 shows a combination acid etched and retarded surface where the surfaces not coated with retarder were etched and then the retarded surfaces were washed and brushed. A reveal or raised demarcation feature is necessary to keep the retarder from spreading to the area to be etched. If it is desirable to have a well-defined reveal (sharp edge), it may be necessary to keep the retarder at least 3/8 in. back from the edge of the reveal. In cases like this, the mix design must be compatible to both finishes but generally favor the gap-grading necessary for the retarded exposed aggregate finish using 1/2 to 5/8 in. maximum aggregate size. The uniformity of the acid finish may be affected by the coarser gradation necessary for the exposed aggregate. In a panel with both finishes that has a high vertical return that is acid etched, there could be a problem. The mix design in this case would have to favor the exposed aggregate and, as a result, would affect the acid etched area.

The combination of a polished or honed surface and acid etching provides a surface which exposes a very high percentage of stone. After the grinding process, the acid removes the cement matrix and fines from between the larger aggregate particles. This surface is highly resistant to weathering and self-cleaning to a high degree, see Fig. 3.5.71.

Whereas the polished/acid etching combination provides a surface finish in which the color of the aggregate predomi-

Fig. 3.5.94

Fig. 3.5.95

nates, the polished/sandblasted finish provides a dramatic contrast between the polished aggregate and the sandblasted matrix of the concrete. If this combination is desired, the architect must ensure that the overall design concept makes allowance for the change in color and texture between the two finishes and that a suitable demarcation is detailed. Since the polishing process is completed before sandblasting, it is necessary to mask the polished finish during sandblasting. The normal light hydrochloric acid wash to the sandblasted finish is applied with care to ensure that any acid that may be applied to the polished finish is removed immediately, before it can attack the matrix.

In many cases, stone veneer may be used as an accent or feature strip on precast concrete panels, Figs. 3.5.96 and 4.7.18. The major difference in the stone attachment is that a 1/2 in. space is left between the edge of the stone and the precast concrete to allow for differential movements of the materials. This space is then caulked as if it were a conventional joint. In Fig. 3.5.96, the typical spandrels are 5 x 20 ft x 4-1/2 in. thick with 12 in. high horizontal bands of 1-1/4 in. granite inset into the panels. The granite insets have 3 anchors per 5 ft length.

3.5.14 PAINTED

As precast concrete is high strength durable concrete, it does not require painting. Paints may be used for purely decorative reasons. Every paint is formulated to give cer-

tain performance under specified conditions. Since there is a vast difference in paint types, brands, prices, and performances, knowledge of composition and performance standards is necessary for obtaining satisfactory concrete paint.

The quality of paint for concrete is not solely determined by the merits of any one raw material used in its manufacture. Many low-cost paints with marginal durability are on the market. In order to select proper paints, the architect should consult with manufacturers supplying products of known durability and obtain from them, if possible, technical data explaining the chemical composition and types of paints suitable for the specific job at hand. For high-performance coatings, proprietary brand-name specifications are recommended.

Concrete is sometimes so smooth that is makes adhesion of some coatings difficult to obtain. Such surfaces should be waterblasted, sandblasted lightly, or ground with silicon carbon stones to provide a slightly roughened surface. Whenever concrete is to be painted, only mold release agents compatible with paint should be permitted or sandblasting may be necessary to assure good adhesion between the paint and the concrete.

Paint applied to exterior surfaces should be of the breathing type, that is permeable to water vapor but impermeable to liquid water. Typically, latex paints are suitable for most exterior applications. The interior surface of exterior walls should have a vapor barrier (paint or other materials) to

Fig. 3.5.96

prevent water vapor inside the building from entering the wall. Typically, interior latex paints, or epoxy, polyester or polyurethane coatings may be applied. See Section 3.5.17 for finishing procedures for interior surfaces to be painted.

The paint manufacturer's instructions regarding mixing, thinning, tinting, and application should be strictly followed.

The smooth gray finish on the spandrel panels in Fig. 3.5.97 was painted at the jobsite in tan and maroon and the ends of the large bullnoses were plated with gold flake.

3.5.15 CAST STONE

Cast stone is a type of architectural precast concrete and is manufactured to simulate natural cut stone. It is used in masonry work mostly as ornamentation and architectural trim for stone bands, sills, lintels, copings, balustrades, and door and window trimming, where natural cut stone or terra cotta may otherwise be used.

Since it is a "building stone," cast stone is specified under the masonry section (04435) rather than section 03450 of the project specifications where architectural precast concrete is normally specified.

The two most widely used casting methods in use today are the dry tamp and wet cast methods. Both require a carefully proportioned mix design consisting of graded and washed natural sand or manufactured sands of granite, marble, quartz or limestone meeting the requirements of

ASTM C33 except that gradation may vary to achieve the desired finish. A 1:3 cement/aggregate ratio mix is proportioned for maximum density and to meet the surface finish requirements. A fine grained texture similar to natural cut stone, with no bug/blow holes or air voids in the finished surface, is produced.

In the dry tamp method, a pneumatic machine rams and vibrates moist, zero-slump concrete against rigid formwork. When the concrete is densely compacted, it is removed from the form and cured overnight in a moist, warm room. The limitation of this method is that it generally requires one flat, unexposed side in the design of the basic section. L shaped, U shaped or cored out stones can be made; however, the designer should be aware that these specially shaped units must be hand molded and will result in higher unit costs. Where return legs are essential to the design, the depth of returns should be standardized at 6 or 8 in. for ease of manufacture. So long as the return leg equals the depth of the section, no additional cost is incurred, Fig. 3.5.98.

The wet cast method is similar to the production process for architectural precast concrete. Mix designs usually have a maximum 3/8 in. coarse aggregate and are comprised of an abundance of fines, typically 15 percent very fine sand. L shaped stones should be avoided as it is difficult to produce a good finish with no bug holes on L shapes.

To help strip the stone from the mold, a minimum 7 deg.

Fig. 3.5.97

Fig. 3.5.98 Returns in cast stone

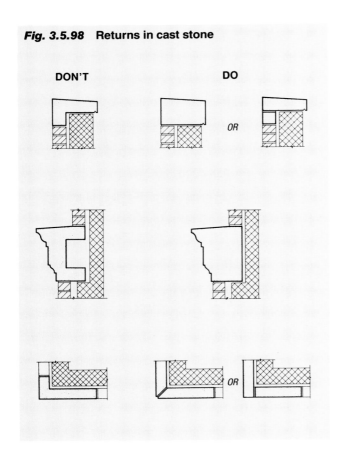

draft should be designed for all surfaces perpendicular to the face pattern. This minimizes the number of mold pieces required to create the face pattern. Negative draft profiles should be avoided whenever possible.

Cast stone should have a minimum thickness of 2-1/2 in. to reduce stripping, handling, and packaging costs. A 2-1/2 in. stone generally costs the same as a 4 in. thick stone. Dimensional tolerances are as follows:

Height and width	+ 1/16 in., – 1/8 in.
Length 2 ft or under	+ 1/16 in., – 1/8 in.
2 to 5 ft	± 1/8 in.
5 to 10 ft	+ 1/8 in., – 3/16 in.

The compressive strength of cast stone should be a minimum of 6500 psi determined by testing 2 in. cubes (equivalent to about 5000 psi for 6 in. cylinders) and the units should have a water absorption of less than 6 percent by weight.

Cast stone produced by either the wet cast or dry tamp method allows the use of air entrainment to obtain freeze-thaw durability. Also, sills, copings and projecting courses should have a slope (wash surface) of 1/2 in. in 12 in. to cause water runoff on exposed top surfaces to prevent saturation of the unit.

Cast stone is available in a variety of colors, shapes and finishes. Typical colors will match either natural stone, brick or terra cotta. The support system and setting techniques will influence the size of the cast stone pieces. Cast stone is normally anchored to masonry or concrete walls, or a steel stud grid. The size limitations of cast stone are about the same as those of natural cut stone. Similar to architectural precast concrete, the key to optimum economy is repetition of ornament. Cast stone is typically given an acid etch finish and the surface texture matches cut stone: sand texture, cleft or stippled face, cut tooled, or bushhammered. In some cases, the stone is given a rubbed finish to expose the aggregates and simulate limestone.

Curved sections should be kept shorter than 4 ft whenever possible and the major unexposed back surface should be flat. Sufficient clearance in the masonry wythe or structural wall section should be provided. A sill section generally presents no problem when designed with a radius front and rear because the major unexposed side is flat at the masonry bed joint.

Through wall flashing must be installed prior to setting of cast stone. Reglets formed into the cast stone to receive the flashing must be shown on the shop drawings. Weep holes are necessary at the level of the flashing to permit moisture to escape.

For optimum economy, standard building stone anchors should be used whenever possible. Two anchor slots are provided in the top of most trim stones to receive anchor straps. Alternatively, a continuous anchor slot may be used. This eliminates the need to locate anchors in the field. Dowel holes, continuous relieving angles and anchor straps are preferred over embedded inserts and weld plates. A continuous relieving angle is the most economical support system. The stone anchors and other accessories are all furnished by the masonry or general contractor, or stone setter. For anchors, dowels and ties, it is necessary to specify the type of material, corrosion resistance (Eraydo alloy zinc, galvanized steel, brass, or stainless steel, Type 302 or 304), and dimensions.

Since cast stone is a masonry product which is usually built into a brickwork facade, brick coursing tables must be used in determining the sizes of the units, Fig. 3.5.99. When determining the height of cast stone window sills the following should be considered:

1. The bottom of the sill to the bottom of the lintel must always equal brick coursing.

2. The lintel height (if stone) equals brick coursing minus one joint.

3. The sill height equals brick coursing to the bottom of the lintel, minus the window dimension.

4. The height of the lug must equal brick coursing minus one joint.

Slip sills have no lugs and the lengths are figured 1/2 in. less than the brick masonry opening (for a 1/4 in. mortar joint) or 3/4 in. less where sealant is desired to provide a 3/8 in. joint. Window sill joints are centered under mullions. Drips should be provided on sills where 1 in. projections occur.

When cast stone is used as part of a brick facade, consid-

eration must be given to the differential movement of the two materials by proper location of control joints.

Figs. 3.5.100 and 3.5.101 show typical applications of cast stone. These include window surrounds such as sills, mullions, jambs and headers; entry feature panels; quoins; and parapet wall cap copings. However, designers should be aware that most architectural precast concrete producers do not manufacture cast stone and local availability should be checked prior to specifying cast stone.

3.5.16 MATCHING OF PRECAST AND CAST-IN-PLACE CONCRETE

Precast concrete panels are often used in combination with architectural cast-in-place concrete. The "matching" of finishes must be planned prior to start of construction in order to consider adjustments in mix design, placement technique, methods of consolidation, and finishing procedures. Samples and full scale mockups should be prepared for both the precast concrete panel, and the architectural cast-in-place concrete, and normal differences resolved prior to finalizing either finish. Aggregate orientation cannot be controlled during placement and consolidation of the architectural cast-in-place concrete. Thus rounded or cubical coarse aggregate are best when trying to match finishes using the different production methods. An acceptable blending of the two different production techniques should acknowledge the distinction between the processes. Finishes such as chemical retardation, or sandblasting should be used to obtain the best match. In

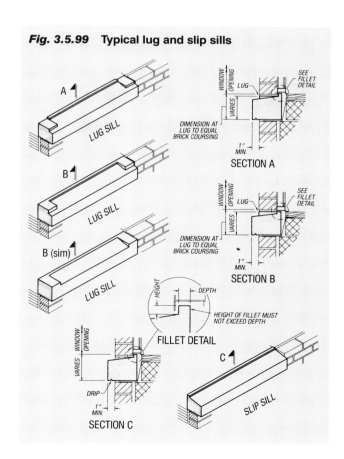

Fig. 3.5.99 Typical lug and slip sills

Fig. 3.5.100

Fig. 3.5.101

Fig. 3.5.102, the architectural cast-in-place concrete beams and columns were sandblasted to match the precast concrete panels. The bug/blow holes in the cast-in-place work are not unusual when the texture selected is a sandblast finish.

When both techniques meet in the same plane, the cast-in-place concrete tolerances must be strictly enforced. Differences in the curing methods between the two techniques, even with identical mixes, may cause color variations in the finish, particularly if the precast concrete uses accelerated high temperature curing. Even when the match-up is very good at the time of initial construction, different weathering patterns may result from dissimilarity in concrete densities.

Generally, only large projects can justify the increased cost required for mockups, mix design control, tight leakproof forms, and ready-mix concrete using special or colored cements, or uncommon aggregates. All of these features may be needed to obtain a satisfactory match-up between architectural cast-in-place concrete and precast concrete panel finishes. The architect is advised to use concrete ingredients and mix designs in the precast concrete units considered suitable for ready-mix concrete whenever a close match is required.

3.5.17 FINISHING OF INTERIOR PANEL FACES

For reasons of economy in casting and finishing, the back of a precast concrete panel (normally the face-up in the

Fig. 3.5.102

Fig. 3.5.103   Typical molds and finishes of back of panels

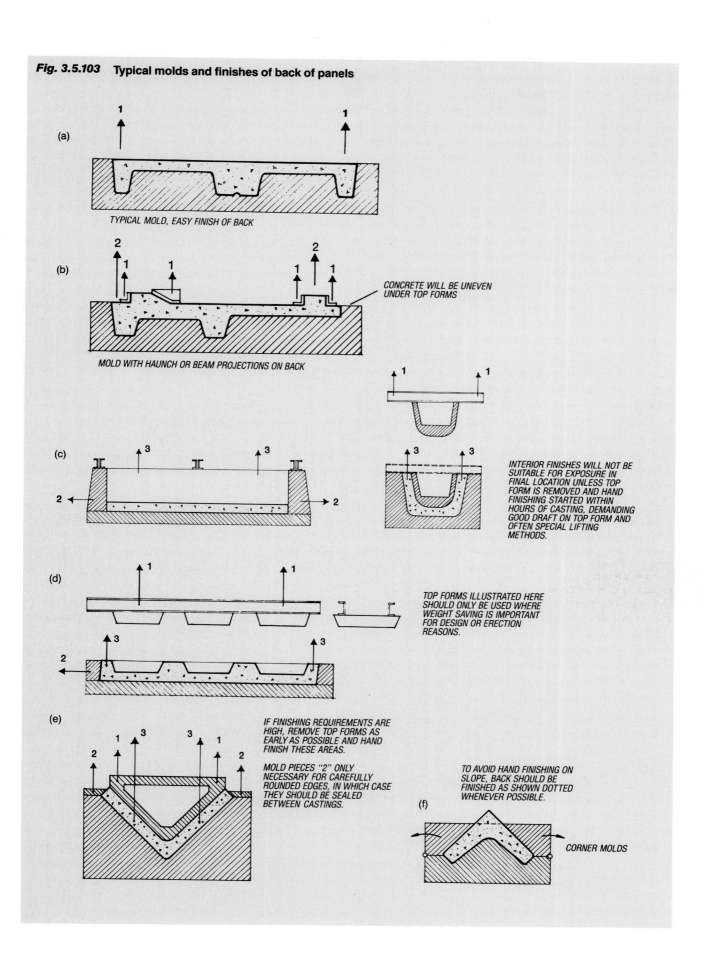

(a)

TYPICAL MOLD, EASY FINISH OF BACK

(b)

CONCRETE WILL BE UNEVEN
UNDER TOP FORMS

MOLD WITH HAUNCH OR BEAM PROJECTIONS ON BACK

(c)

INTERIOR FINISHES WILL NOT BE
SUITABLE FOR EXPOSURE IN
FINAL LOCATION UNLESS TOP
FORM IS REMOVED AND HAND
FINISHING STARTED WITHIN
HOURS OF CASTING, DEMANDING
GOOD DRAFT ON TOP FORM AND
OFTEN SPECIAL LIFTING
METHODS.

(d)

TOP FORMS ILLUSTRATED HERE
SHOULD ONLY BE USED WHERE
WEIGHT SAVING IS IMPORTANT
FOR DESIGN OR ERECTION
REASONS.

(e)

IF FINISHING REQUIREMENTS ARE
HIGH, REMOVE TOP FORMS AS
EARLY AS POSSIBLE AND HAND
FINISH THESE AREAS.

MOLD PIECES "2" ONLY
NECESSARY FOR CAREFULLY
ROUNDED EDGES, IN WHICH CASE
THEY SHOULD BE SEALED
BETWEEN CASTINGS.

TO AVOID HAND FINISHING ON
SLOPE, BACK SHOULD BE
FINISHED AS SHOWN DOTTED
WHENEVER POSSIBLE.

(f)

CORNER MOLDS

mold) should be designed to be level during the casting operation. This is particularly important if this face is exposed as part of the interior finish.

Exposed interior surfaces should have finishes that are realistic in terms of exposure, production techniques, configuration of the precast concrete units, and quality requirements. The back of a precast concrete panel may be given a screed, light broom, float, trowel, stippled, or water-washed or retarded exposed aggregate finish. If the finish is to be painted, a stippled concrete finish will normally be the most economical. This is accomplished with a roller following strike-off and wood float finishing. Although expensive, a steel trowel finish is a common interior finish. Troweling frequently darkens the surface in uneven patterns. For back surfaces of panels which receive insulation, the need for a particular degree of smoothness will depend on the insulation. The finish should normally be as smooth as a good wood float finish. If one face of a unit must be absolutely smooth and true, it should usually be the face-down surface for uniformity and economy.

The exposed aggregate procedures for finishes cast face-up are not similar to those previously described for face-down casting and may not always result in the same appearance.

Vibration (consolidation) of precast concrete usually forces large aggregate pieces to the bottom of the mold whereas finer particles are displaced to the top face. To expose aggregate on the top surface, this surface may be either water washed before the concrete sets or a retarder may be applied and the surface washed the following day. In either case, hand placing (seeding and tamping or rolling) of the larger sized aggregates should be done after concrete consolidation and before the aggregate is revealed. Interior finish requirements should be related to the configuration of the precast concrete units. This will be apparent by considering the different mold setups shown in Fig. 3.5.103. "A" is the most common condition. "B" is similar, but with bulkhead type top-forms used to form a beam (or haunch) on the back of the panel. "C" shows a mold with special top-forms for molding the interior of the returns. "D" is a typical set-up where several top-forms are used, in this case to lighten the panel. This is generally only economical where saving weight is essential, such as in earthquake zones, since the weight (or concrete) savings in most other cases, would not equal the cost of this feature. Conditions "E" and "F" are special corner panels where it is impossible to achieve level face-ups.

Where top forms are necessary, insulation or other covering of the back of panels is recommended, because finishing cost will be higher than for plain, level surfaces. If such shapes require exposed finishes, the architect should detail the interior shape with ample draft, as in "C," so that the top-forms may be removed shortly after concrete casting and the finishing done while the concrete is still plastic.

3.5.18 INTERIOR FINISHES APPLIED AFTER INSTALLATION OF PANELS

A multitude of materials may be utilized in connection with precast concrete panels, but the architect should consider the treatment of joints and connections with regard to interior finish requirements.

Ideally, load support connections at floor level can be embedded below the floor level or concealed together with mechanical services. Top connections may be above the suspended ceiling or covered by a valance. If horizontal joints are at the level of the floor slabs or edge beams, this leaves the architect with only the vertical joints to deal with visually.

If these joints are to be left exposed, they should be recessed to mask any unevenness in alignment between adjacent interior panel faces. The outside face is normally the important face for alignment, while the interior face must accommodate the tolerances in thickness and warpage.

Fig. 3.5.104 indicates interior finishes that are commonly used in connection with precast concrete panels. Many finish materials such as plaster, drywall, ceramic tile, wood paneling, etc., can be applied directly to the precast concrete panels with or without insulation. Successful installations of these finishes have been applied without stripping or furring. The architect should, however, be aware of possible movement in the panels and provide suitable control joints in the interior finish materials to accommodate such movements. This is especially true in seismic zones.

Fig. 3.5.105 shows two solutions for control joints. Maintenance crews have been known to fill the triangular space in the open joint. Therefore, the second solution is recom-

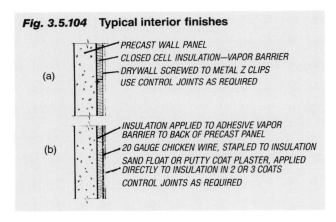

Fig. 3.5.104   Typical interior finishes

(a)
PRECAST WALL PANEL
CLOSED CELL INSULATION—VAPOR BARRIER
DRYWALL SCREWED TO METAL Z CLIPS
USE CONTROL JOINTS AS REQUIRED

(b)
INSULATION APPLIED TO ADHESIVE VAPOR BARRIER TO BACK OF PRECAST PANEL
20 GAUGE CHICKEN WIRE, STAPLED TO INSULATION
SAND FLOAT OR PUTTY COAT PLASTER, APPLIED DIRECTLY TO INSULATION IN 2 OR 3 COATS
CONTROL JOINTS AS REQUIRED

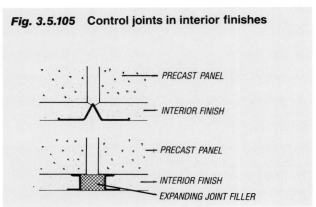

Fig. 3.5.105   Control joints in interior finishes

PRECAST PANEL
INTERIOR FINISH
PRECAST PANEL
INTERIOR FINISH
EXPANDING JOINT FILLER

mended and many suitable materials are available. See Section 5.3.7 for application of insulation.

3.5.19 MARKING OF FINISHES ON WORKING DRAWINGS

The architectural detailer should give sufficient details or descriptions on the drawings to indicate clearly the extent of all exposed surfaces of the units and their respective finishes. This is particularly important for returns and interior finishes.

The finish requirements may be given by coded numbers, by different shading behind areas representing exposed surfaces, or by any other method which remains readable after printing or possible reduction of the drawings. Simply to heavy-up the exposed surface lines has proven less than satisfactory unless carefully executed.

3.5.20 ACCEPTABILITY OF APPEARANCE

Contract documents must identify who the accepting authority will be: architect, owner, general contractor, or site inspector. One person must have final and undisputed authority in matters of color, finish, and texture, if consistent with the intent of the contract documents.

Since acceptable uniformity of color and intensity of shading are evaluated by visual examination, they are generally a matter of individual judgment and interpretation. The acceptable variations in color, texture and uniformity should be determined at the time the sample, mockup or initial production units are approved. Accordingly, it is beyond the scope of this Manual to establish precise or definitive rules of measurement of acceptability. However, a suitable criteria for acceptability requires that the finished concrete surface shall present a pleasing appearance with minimal color and texture variations, equal to the approved sample, when viewed in good typical lighting with the naked eye at a 10-ft distance, and shall show no obvious imperfections other than minimal color and texture variations at a 20-ft distance.

Units should be assessed for appearance during both wet and dry weather. The difference in tone between wet and dry panels is normally less with white concrete. In climates with intermittent dry and wet conditions, drying-out periods often produce blotchy appearances in all-gray cement units. This is particularly true of fine-textured surfaces. On the other hand, dirt (weathering) will normally be less objectionable on gray surfaces. These comparisons are based on similar water absorption or density of white and gray concrete.

It would be optimistic to imagine that every unit cast during the course of a contract will be perfect. Minor defects and blemishes will sometimes occur, and the main operations in making-good are filling bug/blow holes, removing stains and repairing physical damage.

The following is a list of finish defects and/or problems which are normally unacceptable in high quality architectural precast concrete. Design and manufacturing procedures should be developed to counteract or remedy them, unless they are specifically desired by the architect or are inherent in the design of the unit. If such "defect expressions" are specified by the architect, or are unavoidable, they should be agreed upon by the precaster and the architect in the form of approved samples and/or initial production units.

1. Ragged or irregular edges.
2. Excessive air voids, commonly called bug/blow holes, evident in the exposed surface.
3. Adjacent flat and return surfaces with a noticeable difference in exposure from the approved samples or mockup.
4. Casting lines evident from different concrete placement lifts and consolidation.
5. Visible form joints or irregular surfaces.
6. Rust stains on panel surfaces.
7. Panels not matching approved sample or non-uniformity of color within a panel or in adjacent panels due to areas of variable aggregate concentration and variations in depth of exposure.
8. Blocking stains or acid stains evident on panel surface.
9. Non-uniformity of color or texture.
10. Areas of backup concrete bleeding through the face concrete.
11. Foreign material embedded in the face.
12. Repairs visible at the 10 and 20 ft viewing distances.
13. Reinforcement shadow lines.
14. Visible cracks after wetting.

Precast concrete generally undergoes far less cracking than cast-in-place concrete. This resistance to cracking is due, in part, to the high compressive and tensile concrete strengths of the precast concrete. It should be recognized that a certain amount of cracking may occur without having any detrimental effect on the structural capacity of the member and it is impractical to impose specifications that prohibit cracking. However, in addition to being unsightly, cracks are potential locations of concrete deterioration, and should be avoided if possible. Proper reinforcement, prestressing and proper handling procedures are three of the best methods of keeping cracks to a minimum. To properly evaluate the acceptability of a crack, the cause of the cracking and the stress conditions the crack will be under with the precast concrete unit in place must be determined.

The cement film on the concrete may develop surface crazing, i.e., fine and random hairline cracks. A hairline crack is defined as a surface crack of minute width, visible to the unaided eye but not measurable by ordinary means. The primary cause of these cracks is the shrinkage of the surface with respect to the mass of the unit. Crazing has no structural or durability significance, but may become visually accentuated if dirt settles in these minute cracks. This is particularly true more with white cement than with gray or a dark color, but the effect will depend on the character of the cement film. A relatively lean, properly consolidated concrete mix will show little crazing, in contrast to a mix rich in cement and water. Where crazing occurs, a horizontal surface will be affected more than a vertical surface due to the settlement of dirt. Crazing generally will not appear where the outer cement skin has been removed. Such cracks are of little importance and should not constitute a cause for rejection.

Tension cracks are sometimes caused by temporary loads during production, transportation or erection of the products. The amount and location of reinforcing steel has a negligible effect on performance until a crack develops. As flexural tension increases above the modulus of rupture, hairline cracks will develop and extend a distance into the element. If the crack width is narrow, not over 0.010 in., the structural adequacy of the casting will remain unimpaired, as long as corrosion of the reinforcement is prevented. Accordingly, wall panels containing cracks up to 0.005 in. wide for surfaces exposed to the weather and 0.010 in. wide for surfaces not exposed to the weather should be acceptable, provided the reinforcement is galvanized or otherwise corrosion resistant. The limitation on crack size specified is for structural reasons. The esthetic limitation will depend on the texture of the surface and the appearance required. On coarse textured surfaces, such as exposed aggregate concrete, and on smooth surfaces comparable to the best cast-in-place structural concrete, the structural limitation would be esthetically acceptable. For smooth surfaces of high quality it may be desirable to limit cracking in interior panels to 0.005 in. In addition, it should be noted that cracks will become even more pronounced on surfaces receiving a sandblasted or acid etch finish.

While some of these cracks may be repaired and effectively sealed by pressure injecting a low viscosity epoxy, their acceptability should be governed by the importance and the function of the panel under consideration. The decision regarding acceptability must be made on an engineering basis as well as on visual appearance.

Every effort should be made to promptly identify the cause of any cracking, particularly when several units display similar cracking. Such cracking is often the result of a single design, manufacturing, or handling problem, which can then be rectified to prevent any recurrence.

Long-term volume changes can also cause cracking after the member is in place in the building, if the connections provide enough restraint to the member, see Section 4.5.2. Internal causes, such as corrosion of reinforcement or cement-aggregate reactivity, can also lead to long-term cracking and should be considered when materials are selected.

Erected panels not complying with the contract documents may require additional work. If the architectural precast concrete panels cannot be corrected to match the repair samples or repairs demonstrated on the mockup, they may be subject to rejection.

3.5.21 REPAIR/PATCHING

A certain amount of repair of product is to be expected as a routine procedure. Repair and patching of precast concrete is an art requiring expert craftsmanship and careful selection and mixing of materials, if the end result is to be structurally sound, durable and pleasing in appearance. Damage to sandblasted surfaces, although easier to repair than on smooth surfaces, is more difficult to repair than chemically retarded surfaces. Major repairs should not be attempted until an engineering evaluation is made to determine whether the unit will be structurally sound.

Trial mixes are essential to determine exact quantities for the repair mix. This is best determined by applying trial repairs to the project mockup or small sample panels (12 in. square). Selecting the appropriate mix should begin after the trial repairs have been allowed to cure a minimum of 7 days, (preferably 14 days), followed by normal drying to 28 days. This is important because curing and ultraviolet bleaching of the cement skin have an effect on the finished color.

It is recommended that the precaster execute all repairs or approve the methods proposed for such repairs by other qualified personnel. Repairs should be done immediately following occurrence of damage. However, the decision on the time to perform the repairs should be left up to the precaster, who should be responsible for satisfactory final appearance.

It is important that all repair of damaged precast concrete unit edges be carried out well in advance of the joint sealing operation. The repair work should be fully cured, clean, and dry prior to caulking.

Repairs should be done only when conditions exist which ensure that the repaired area will conform to the balance of the work with respect to appearance, structural adequacy, and durability. Slight color variations can be expected between the repaired area and the original surface due to the different age and curing conditions of the repair. Time (several weeks) will tend to blend the repair into the rest of the member so that it should become less noticeable. Gross variation in color and texture of repairs from the surrounding surfaces may be cause for rejection. Repairs should be evaluated when the concrete surface is dry.

The selection of techniques or materials for the repair will depend upon the following:

1. Extent of damage.

2. Function of the precast concrete unit.

3. Availability of equipment and skilled manpower.

4. Economic considerations.

5. Need for speed of repair.

6. The importance of appearance.

Since the techniques and materials for repairing precast concrete are affected by a variety of factors including mix ingredients, final finish, size and location of damaged area, temperature conditions, age of member, surface texture, etc., precise methods for repairing cannot be detailed in this Manual. (See PCI *Manual for Quality Control for Plants and Production of Architectural Precast Concrete*, MNL-117 for guidance on repair techniques and materials.)

If cracking has occurred, and if repair is required for the restoration of structural integrity or esthetics, the cracks may be repaired by the injection of a low viscosity 100 percent solids epoxy under pressure. Care should be taken to select an epoxy color (amber, white or gray) which most closely matches the concrete surface. However, this procedure may not be acceptable for high quality architectural precast concrete.

Should minor damage occur to veneer stone during shipping, handling or erection, field remedial work can successfully be accomplished. Such repairs are normally done by the precaster with repair procedures developed in consultation with the stone fabricator. Epoxy, stone dust and a coloring agent, if necessary, can be used to repair small chips or spalls. These repairs can be finished to the same surface texture as the stone facing. If it is necessary to replace a stone piece, satisfactory techniques have been developed when the back of the panel is accessible or after the panel has been erected and the back of the panel is inaccessible.

3.6 WEATHERING

3.6.1 GENERAL

A primary consideration in the architectural design of buildings should be weathering, i.e., the changes in appearance with the passage of time. Weathering affects all exposed surfaces and cannot be ignored. The action of weather may enhance or detract from the visual appearance of a building, or may have only little effect. The final measure of weathering effects is the degree to which it changes the original building appearance.

Visual changes occur when dirt or air pollutants combine with wind and rain to interact with wall materials. The run-off water may become unevenly concentrated because of facade geometry and details. The manner in which water is shed depends primarily on the sectional profiles of the vertical and horizontal discontinuities designed into the wall.

Through the years, designers acquired ways of handling traditional building materials in order to control the water flow down specific parts of a structure—copings, drip molds, gargoyles, window sills, and plinth details. However, many of these useful and relevant details have been discarded as superfluous decoration.

For architectural precast concrete (as well as all other building materials), the awareness of weathering should be reflected in the design of wall elements and the integration of windows to control water migration and collect and remove water run-off. Staining that occurs through differential surface absorption and uneven concentrations of dirt due to water run-off are considered the most common weathering problems.

Proper design and detailing for weathering of precast concrete units increase in importance whenever the atmosphere is dirty or polluted, and where one or more of the following characteristics are present:

1. Large surfaces.
2. Smooth or finely textured surfaces.
3. Lightly colored units.
4. Heavily sculptured units with either forward or backward sloping surfaces.

Weathering problems are less serious in relatively clean environments or when the precast concrete units have one or more of the following characteristics:

1. Properly designed slopes and flow patterns.
2. Rough texture and gray or dark colors.
3. Near-vertical or lightly sculptured surfaces.
4. Honed or polished surfaces.
5. Panel face broken up by several joints (real or false) or by vertical ribs (in all but the lightest colors).

Many of the effects of weathering can be predicted by studying local conditions and/or existing buildings in the area. This will often give a clear indication of the levels of pollution; the speed, principal direction and frequency of wind; and the volumes and frequency of rainfall, together with records of temperatures and relative humidity. All these factors will affect the way in which exposed concrete will get wet and dry out. With proper attention to the cause and effect of weathering, potentially detrimental results can be eliminated or at least minimized. Precast concrete will become dirty when exposed to the atmosphere, just like any other material. Fortunately, with architectural precast concrete, the designer can choose shapes, textures, and details to counteract any negative effects of weathering. Although regular cleaning of a building may make detailing a less critical factor, maintenance costs should be balanced against initial design costs.

Weathering may also affect precast concrete units during storage if they are stacked in a position different from their final orientation on the building, the panels may have to be protected to avoid streaking which may show as horizontal lines on the finished building. Protection against such weathering should be left to the precaster, but may influence the design of the units.

One of the major contributing factors to the weathering of precast concrete is dirt in the atmosphere. Atmospheric dirt or air pollutants include smoke or other gases, liquid droplets, grit, ash, soot, organic tars, and dust. Gaseous pollutants include SO_2, NO_X, H_2S, NH_3, and O_3. SO_2 can react with the lime in the concrete, and the oxygen from the air to form gypsum, see Section 3.6.11. Gypsum, due to its solubility, may be washed away, taking dirt with it. Where there is insufficient water for washing it can encapsulate dirt and hold it.

The concentration of sulfur dioxide (SO_2) and other corrosive compounds is high in some urban environments. When dissolved in rainwater, sulfur dioxide produces dilute sulfurous or sulfuric acid. These acids etch the cement-rich paste and the carbonated concrete surface, producing a gradual change in color as the fine aggregate becomes exposed.

Fig. 3.6.1 shows the pH of precipitation, taken as a weighted average, falling in the United States and Canada during the early 1980's. Each dot represents 600 metric kilotons of sulfur and nitrogen oxide emissions. Although acid rain is technically defined as precipitation with a pH level below 5.6, some researchers believe that 5.0 is a more realistic limit. Using either definition, acid rain has a far reaching impact on both the United States and Canada.

In areas with unusually high concentrations of corrosive elements (pH of rainwater lower than 5.0 and high emissions), the designer should detail the facade for water run-off, specify concrete strengths and durabilities normally associated with architectural precast concrete, provide the required cover over reinforcement, avoid soft aggregates such as limestones and marbles, and suggest more frequent washings of the building.

The flow of rainwater across the building's facade has a profound effect on weathering patterns since rain run-off redistributes the dirt and other airborne impurities that have been deposited fairly uniformly on the external wall surfaces. This deposit takes place more rapidly, however, on surfaces facing upwards and also on surfaces with a coarse texture. The designer should attempt to anticipate and plan for water flow over the wall, tracing water flow to the final drainage point or to ground level, particularly where discontinuities exist. When run-off reaches a discontinuity the water may bead and drip free. This may increase or decrease the run-off concentration, affecting both its ability to carry suspended dirt particles, and its subsequent drying behavior. Such changes of flow concentration may disfigure the building surfaces.

Rain is primarily a cleansing agent for the building surfaces. However, at some stage the water will also pick up dirt already deposited on the walls and it becomes a soiling agent. The preferred lines of water flow must be arranged through shaping of surfaces and textures so that, at the point where water is expected to become a soiling agent, it will not detract from the finishes or forms of the building elements. Dirt will drop out of the run-off water when water flow velocity is decreased; for example, when the run-off is allowed to fan out. It may be necessary to have frequent details to throw water clear of the building, collect the water, or spread the water uniformly across sloped surfaces. Such details should be continuous to prevent differential rainwashing, or must terminate at bold vertical features, or maintain a clear distinction between washed and dirty areas. These differences can then, if required, be emphasized by the use of different surface finishes.

The migration of run-off water is affected by:

1. The location and concentration of rain deposits.

2. The properties of water in contact with materials, especially surface tension.

3. The forces of wind and gravity.

4. The geometry, absorption, and texture of the building surface.

The amount of rainwater, and the velocity and angle at which it falls is markedly different on each side of a building

Fig. 3.6.1 Sources and distribution of acid rain

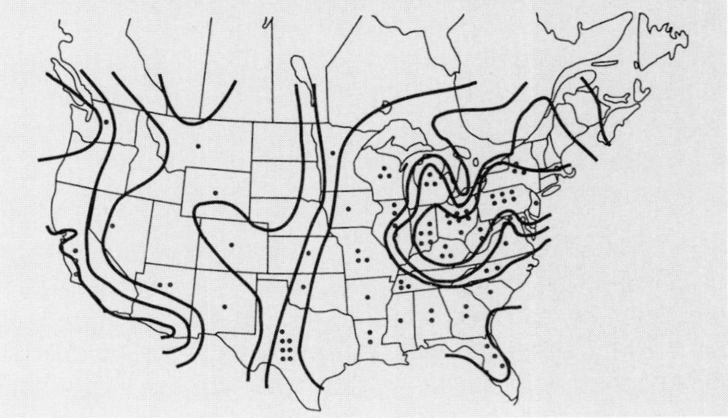

and at different heights. Therefore, it is not reasonable to expect equal weathering of all parts of the building. The influence of tall or massive buildings, projections, courts, or passages on prevailing winds can cause wind eddies to upset the natural flow of air and rain. This makes the effect of rainwater very difficult to predict.

During storms, driving rain can come from any direction, but the quantity of water available on a facade for washing is normally determined by its relationship to the prevailing wind and the intensity of rain from that direction. Wind movements around buildings are affected not only by major climatic factors but by local topography, adjacent buildings and groups of trees. All these will affect the amount and position of driving rain hitting a building and the way in which water runs down the facade. The drops of driving rain are guided for the most part by the air currents around the building and external wall components. The pattern of these air currents is independent of building height. Small obstacles give rise to sudden changes in direction of the air currents and the raindrops cannot follow these sudden changes. The mass forces carry them forward to the obstacle. On one or two story buildings, the driving rain reaches the lower parts of the walls. Dirt stain patterns do not usually occur on such low buildings.

Air currents against buildings taller than a couple of stories are, on the other hand, deflected so gently that the air has time to re-orient the raindrops. Less than half the quantity of rain which should pass through a free air cross section of the same size as the building is caught by an external wall. This applies regardless of the wind force. The rain strikes mainly the top parts of the wall. Only edge sections (corners) are reached by the driving rain and in the central sections the raindrops move almost completely parallel to the wall. As a result, water run-off very seldom reaches all the way down to the ground, except at corner areas and projecting components, unless the duration of the rain is quite long. Therefore, special care should be taken to ensure that water is not allowed to run unchecked down surfaces unless there is enough water to wash the surfaces completely. A typical weathering pattern caused by rain and prevailing wind is illustrated in Fig. 3.6.2. Parts of the building facade are clean in areas where it is washed by rain, even though the remainder of the building has become soiled.

Fig. 3.6.2

Fig. 3.6.3

Rounded or splayed corners reduce wind speed at the edges of buildings and may be useful to avoid the heavy concentrations of driving rain which are typical of these locations. Also, continuity of water flow between surfaces is improved when corners between them have rounded edges instead of sharp edges. A joint, groove or projection, near a corner with a long return, should be used to catch the rainwater and prevent partial dirt washing from water blown around the corner.

The raindrops which reach a wall surface are absorbed to different extents depending on absorption and moisture content of the wall material. Precast concrete normally has a medium to low water absorbency. Water run-off on concrete surfaces consists of a very thin layer (0.01 in. thick) and only occurs if the absorption of the concrete is lower than a certain value. The run-off flows slowly (up to 3 ft/min.) and vertically down the wall with lateral winds having an insignificant influence. When the water reaches lower sections, which have been struck by less driving rain and are thus drier, it is absorbed. The dirt accompanying the water is deposited in new places, unevenly soiling the surface. Also, a facade with high absorption normally becomes wet rapidly and remains damp for a longer period than a facade with low absorption. Airborne dirt (soiling particles) adheres easily to high absorption concrete. It is desirable to break up large areas of concrete, extending over several stories, with horizontal features which either collect or throw off the water at intermediate positions. These features will reduce the amount of water on the sur-

face, reduce the differences between panels at different levels on the facade, make the change from washed to unwashed into a gradation instead of a clearly visible line and by producing interest and shadows will make any changes less noticeable.

The water run-off on concrete surfaces has a tendency to divide into separate streams determined by microscopic irregularities or differences in absorption of the surface when the water layer thickness decreases below a critical value. This breakdown into irregular, separate streams takes place mainly on smooth or lightly textured surfaces but can also occur on surfaces with exposed aggregates. However, a uniformly distributed broken flow is more likely to occur over heavily textured materials, Fig. 3.6.3. These streams recur at the same locations during most rainfalls and are reflected in the soiling pattern. The streams of water broaden out laterally when they meet horizontal or moderately sloping obstacles. They also follow surfaces facing downwards (horizontal surfaces) in a similar manner. Consequently, the design of drips is extremely important. The path followed by the water from the lowest points of these drips should also be taken into consideration in the design.

Glass areas cause build-up of water flow. Typically on non-absorbent materials the water flows in discrete streams rather than as a continuous film due to surface tension causing the droplets of water to coalesce. Because glass is a non-absorbent material, the flow rate of water down its surface is fast, and there is little time lag in its throw-off— only a light rain is required to form rain run-off all the way down to ground level. By contrast, rain water flowing down an adjacent concrete wall surface will be slower (depending on the surface texture and absorption) and its throw-off will be less complete. As a result, there is a concentration of water at the bottom of each window—the very thing the designer must guard against if differential patterning is to be avoided. This flow must be dissipated, breaking up its concentration. Further, there is always a tendency for water flow to be in greater volume at the edges of the glass (the smallest amount of wind tends to drive rain toward the edges of the glass). Fig. 3.6.4 shows the dirt pattern caused by water run-off carrying the dirt down the mullion. An attempt to minimize staining, after the fact, resulted in grooves being cut under the mullions.

Fig. 3.6.4

Fig. 3.6.5 shows the volume of rain assumed to hit a building surface depending on the orientation of the surface. For diagrammatic purposes, the rain is assumed to be 10 deg. to the vertical. However, the variability of rain under actual conditions makes all but a general prediction difficult. Vertical or near-vertical surfaces receive insufficient rainwater to be self-cleansing. Steep forward-sloping surfaces usually weather cleaner. Large areas may begin to collect dirt at the lower end unless the angle is steep. With heavy rain, the dirt on horizontal surfaces and surfaces that have little slope may be partially washed off, streaking the surfaces below. In the case of light rain or drizzle, the dirt may collect and slowly flow down other surfaces in the general direction of the water flow, resulting in pronounced random streaking. Backward-sloping surfaces collect little or no rain but are likely to be subject to a partial, nonuniform water flow from above which may carry dirt and cause serious streaking. Often backward-sloping surfaces are seen in shadow. In this case, the accumulation of dirt is not particularly noticeable if the dirt is acquired evenly without disfiguring streaks.

The facade geometry of buildings is usually responsible for local concentrations of run-off. Such concentrations lead to the characteristic marking patterns frequently observed on building surfaces: dirt accumulation, dirt-washing, and white washing, Fig. 3.6.6. New buildings may show dirt washing at locations of concentrated run-off while their over-all surfaces are still quite clean. Later, the same areas may exhibit white washing (lighter, cleaner streaks) after

Fig. 3.6.5 Volume of rain likely to hit wall surfaces

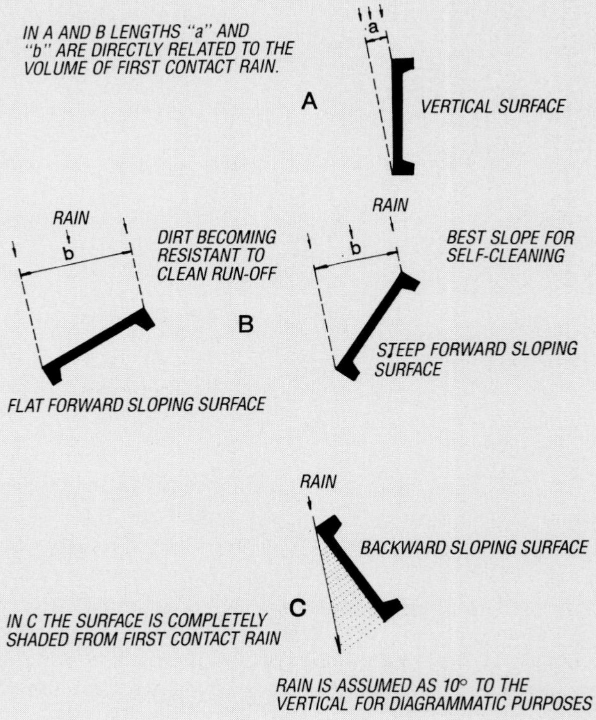

IN A AND B LENGTHS "a" AND "b" ARE DIRECTLY RELATED TO THE VOLUME OF FIRST CONTACT RAIN.

A

VERTICAL SURFACE

RAIN

DIRT BECOMING RESISTANT TO CLEAN RUN-OFF

RAIN

BEST SLOPE FOR SELF-CLEANING

B

STEEP FORWARD SLOPING SURFACE

FLAT FORWARD SLOPING SURFACE

RAIN

BACKWARD SLOPING SURFACE

IN C THE SURFACE IS COMPLETELY SHADED FROM FIRST CONTACT RAIN

C

RAIN IS ASSUMED AS 10° TO THE VERTICAL FOR DIAGRAMMATIC PURPOSES

Fig. 3.6.6 Characteristic marking patterns

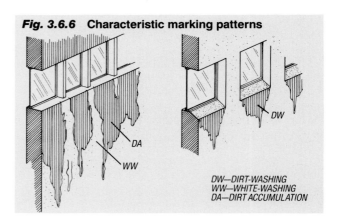

DA
WW
DW

DW—DIRT-WASHING
WW—WHITE-WASHING
DA—DIRT ACCUMULATION

adjacent surfaces have been darkened by dirt accumulation.

When mullions and other vertical elements meet a horizontal element, the configuration often causes a concentration of flow and uneven weathering, Fig. 3.6.7. Vertical surfaces can be protected and weathering minimized by providing steeply sloping overhangs with drips. These tend to reduce dirt accumulation and the washing of dirt onto the vertical surfaces below.

The intersection of horizontal and vertical projecting elements almost always creates dirt streaks. Such streaks run back from the edge of exposed columns and below the ends of horizontal elements even when they are steeply

Fig. 3.6.7

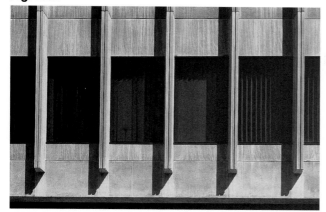

Fig. 3.6.8

sloped at the top surface. To avoid such streaks, the horizontal element should be stopped short of the column. This confines rain run-off to the horizontal element and permits unimpeded washing of the column. Channeling of the column faces also will help to prevent water from running back along the edges.

Water flowing laterally or diagonally downward on a surface will concentrate where it encounters vertical projections or recesses. The secondary airflow due to wind is also important. It concentrates run-off at the outside corners of the building, at columns, and at inside corners of vertical projections. Surface tension contributes to this effect by preventing flow back from vertical edges of small elements such as window mullions, often concentrating the flow at the corners.

In areas where nearby buildings show the undesirable effects of weathering and the atmosphere is laden with pollutants, it is recommended that consideration be given to the use of sealers (see Section 3.6.13) to increase rain run-off, reduce surface absorption of the concrete, and facilitate cleaning of the surface.

Careful attention should be given to the effects of exterior details on the weathering pattern of the building. Appropriate design details can avoid many of the more unsightly effects of dirt streaking and differential washing.

3.6.2 WATER DRIPS

Surface tension allows water flow to take place along the underside of horizontal surfaces. If no drip devices or projections are provided on a building face, run-off water can flow over the wall materials and windows for the total height of the building. Dirt may be deposited in sufficient quantities to cause disfiguring stains on wall units and, in a short period, may streak and stain the glass, Fig. 3.6.8, (see Section 3.6.11). Water always runs down a wall or window over the same preferential paths. Water drips will reduce or stop streaking due to uneven washing of a backward-sloping surface when the drips are correctly dimensioned and placed close to the forward edge. Water drips are cumbersome for the precaster to incorporate and should not be used indiscriminately. When placed under horizontal projections, drips will prevent streaking on the vertical surfaces below. These drips also prevent water

run-off (after a storm) from slowly running over the window glass, a primary cause of glass streaking. Water will leave a drip at its lowest point and it is important to follow its course thereafter. Small chips and cracks may concentrate the flow, so that water will bridge drip details and allow wetting of the surface below. If dirty water falls onto other surfaces, the problem may be merely relocated. However, if the wind tends to spread the water out on the surface below, uniformity of weathering may be obtained. To avoid streaks on the sides of window panels, the drip may be stopped about 2 in. short of the sides. Often recesses or grooves may be incorporated in the side walls to further direct the water.

Water flow may be evenly distributed Fig. 3.6.9 or deliberately directed toward the center of the panel, Fig. 3.6.10. In the later case, the panel should have a dark, rough textured surface slope below the window to break and mask the water's dirtying effect.

The drip section should be designed in relation to the slope of the concrete surface, Fig. 3.6.11, to prevent water from bridging the drip. To avoid a weakened section the drip should not be located closer than 1-1/2 in. to the edge of the precast concrete unit. Where the window is not at least 3 to 4 in. back, it is difficult to get a drip groove in the panel.

A clear sealant bead applied to precast concrete units after erection, or plastic drips glued to the concrete, are remedial drip solutions used with varying success depending on their care in application (Fig. 3.6.11). However, a drip incorporated initially into the precast concrete or window frame is the least costly and best solution. Fig. 3.6.12 shows the

Fig. 3.6.9 Straight water drip

Fig. 3.6.10 Water drip with aimed discharge

use of an extrusion (either aluminum or neoprene) across the head of the window which have either an integral gutter or extended drip lip of at least 1 in.

Fig. 3.6.13 shows a panel configuration which should prevent water dislodged by the drip from streaking the sides of fins or the fins may be terminated behind the drip to minimize streaking of the fins.

3.6.3 PARAPET AND ROOF EDGES

Parapet and roof edges should be designed to avoid run-off from flat roofs onto the building facade. A parapet of sufficient height (8 to 12 in.) will normally prevent water on the roof from blowing over the parapet onto the face of the building. The top of the parapet should slope backwards towards the roof for its full width and be narrow so that dirt accumulating on them does not cause streaking on the building face when washed off, Fig. 3.6.14. A continuous roof flashing detail suitable for a low parapet is also shown on this figure. Flashing should project at least 1 in. beyond the vertical wall surfaces and have a proper drip device to throw run-off water clear. Projections of less than 1 in. may permit water to either flow back or be blown back against wall surfaces. It is important that the flashing is installed straight and level, and without gaps to avoid streaking. The choice of flashing material and/or its treatment against corrosion should be based on preventing potential staining of the precast concrete surface, see Section 3.6.7.

Fig. 3.6.11 Design of water drip in relation to slope

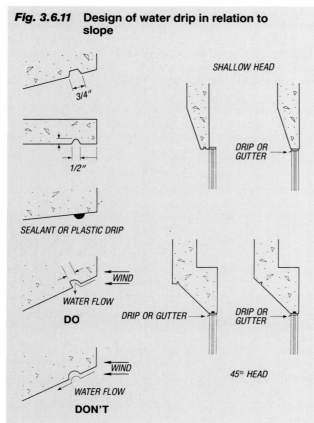

Fig. 3.6.12 Drip or gutter incorporated into head section gasket

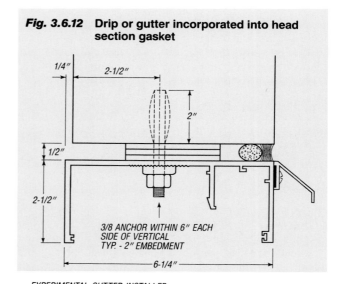

EXPERIMENTAL GUTTER INSTALLED TO CORRECT GLASS STAINING

Fig. 3.6.13 Drip details to minimize streaking of fins

Water may be directed off some low roofs by scuppers which should have sufficient projection to prevent the water from hitting lower surfaces. A good precast concrete scupper detail is shown in Fig. 3.6.15. The scupper should be placed away from panel joints to keep water from running into these joints. If a collection box is used, the top of the box should be at the same elevation or slightly lower than the bottom of the scupper to assure that tolerances allow positive drainage of the roof under all conditions. Also, consideration should be given to the clearance necessary to accommodate tolerances of the roof slab depth, roof insulation and roofing placement to allow proper drainage.

If the scupper is located within the panel face, a minimum blockout dimension of 6 in. is necessary in order to minimize the plugging of the opening.

3.6.4 SURFACE FINISH

Concrete surface finishes vary considerably in their ability to take up and release dirt under weathering conditions. They should therefore be chosen for these so-called "self-cleansing" properties. The selection of color and texture may have an esthetic significance greater than the effects of weathering. The economics of varying the surface finish from one part of the building to another should be investigated as the weathering characteristics will be different.

The surface of smooth precast concrete is hard and impervious and easily streaked by rain, unless there is enough

Fig. 3.6.14 Diagram of parapet profile

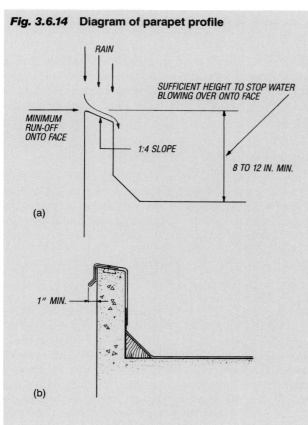

RAIN

SUFFICIENT HEIGHT TO STOP WATER BLOWING OVER ONTO FACE

MINIMUM RUN-OFF ONTO FACE

1:4 SLOPE

8 TO 12 IN. MIN.

(a)

1" MIN.

(b)

Fig. 3.6.15

water to form a complete film on the surface. Weathering patterns are determined by the shape and smoothness of the units and joints which are particularly vulnerable. Any irregularity in a smooth surface will be exaggerated by weathering patterns. Light-toned and smooth surfaces aggravate the contrast between washed and unwashed areas.

Honed or polished surfaces are alternative smooth finishes that have good weathering characteristics and are less susceptible to surface streaking or crazing.

Textured finishes accumulate more dirt, but they can maintain a satisfactory appearance. The aggregate tends to break up and distribute water run-off more evenly, thus reducing the streaking that appears on smooth surfaces, Fig. 3.6.16. It is not reasonable, however, to expect an exposed aggregate finish to deal with all problems of weathering. The way in which water moves on such a surface is different but concentrated flows or their effects will still be visible and must be controlled.

Rounded aggregates are preferred since they tend to collect less dirt than angular aggregates of rough texture. However, dirt absorption is generally confined to the matrix. For this reason, as well as for architectural appearance, the area of exposed matrix between aggregate particles should be minimized. The smooth nonporous surface of the aggregates allows less dirt to deposit and promotes more run-off to increase washing action. At the same time the slightly recessed matrix helps to absorb and mask the

Fig. 3.6.16

pollution deposition. A darker matrix will reduce the effect of atmospheric pollution.

Extreme color differences between aggregates and matrix will create uniformity problems. For example, large-size exposed aggregates of light color provide a heavily textured surface that may seem to be very dirty after a time because the matrix becomes very dark and the high spots of the aggregate are washed clean. In some cases, uniformly colored light surfaces contrasted with uniformly colored dark surfaces may be used to accentuate the depth of relief on a building face. Smooth units made with dark colored sands will slowly become darker with age when subject to weathering because the surface film of cement paste erodes away, exposing the sand. Therefore, wide differences between the color of the cement and the sand should be avoided.

The use of appropriate, dark colors and rough textured surfaces can help to mask the effect of dirt deposits. The overall darkening in tone that takes place is unlikely to be objectionable unless streaking occurs. Medium textured finishes may still allow water to run or be wind-driven into streams to cause irregular streaks. Vertical ribs or flutes which help the designer give expression to a facade will also help to control the run-off and prevent it from spreading horizontally. As dirt collects in the hollows, it emphasizes the shadow and therefore, the texture itself. The rib must not be too wide otherwise a soiled pattern may develop in the middle area of the rib's upper surface. If the ribs are terminated above the lower edges of the walls, streaking below the ribs may occur depending on the depth of projection and the wind force and direction. As water reaches the bottom edge of a vertical or inclined panel, surface tension effects cause it to slow down before dripping clear and it tends to deposit any dirt it has been carrying. Horizontal ribs or flutes spread stains rather than prevent them and can be used to protect the underlying surface by deflecting the flow of water. Concentrated water flows on diagonal ribs create a weathering pattern difficult to predict.

3.6.5 WINDOWS

Openings in walls that form the windows of a building help to break up large flat surfaces, but contribute their own weathering problems. Individual windows in a wall of architectural precast concrete must be designed with two principles in mind: (1) to contain the water flow within the window area, and (2) to disperse it at the bottom of the window in such a manner that it is spread, not concentrated.

Areas of glass cause buildup of water flows. There is always a tendency for water flow to be in greater volume at the edges of the glass than in its center, as the effect of the smallest amount of wind will be to drive the rain towards the edges. Window surfaces are impervious and do not absorb water, so a large amount of water and the pollutants which collect on the windows, are flushed over the sills, the very thing the designer must guard against if differential patterning is to be avoided. In most cases, rainwater flow should be encouraged to pass over windows, sills and spandrels as evenly and freely as possible. Window frames should be of materials that do not stain easily, and sills should be well sloped and smooth so as to minimize dirt accumulation. A drip groove should be provided under any outward-sloping sills to prevent dirty water staining on the panels.

There are many different ways of detailing windows, depending to some extent on their shape and the degree with which they interrupt water flow, Fig. 3.6.17. Many recent buildings have used precast concrete spandrel panels and windows in more or less the same plane, without sill ledges or projections, and they have weathered well without staining. This method seems to perform best when the concrete below the window is textured or profiled (vertical ribs) to break up the water flow and avoid streaking. Darker color finishes are also desirable in dirty urban areas. The path of water flow must be anticipated and provision made to collect and drain away the water in due course. Also, as mentioned in Section 3.6.1 low-rise buildings do not have the same weathering problems as high-rise buildings.

One detailing approach is to box the whole window out from the general wall face, Fig.3.6.17b. This detail can be very effective if the cut back at the base of the window is reasonably perpendicular to the wall. A reasonable slope must be maintained to ensure sufficient draft to strip the unit out of the mold. If the projecting window is surrounded by sloping surfaces, staining and streaking may occur. The window jamb should be detailed to avoid water being blown around the reveal where it can streak the wall at the sides of the window projection. A projecting window is more expensive than a simple opening, but it does avoid most unacceptable staining.

A coffered window profile also minimizes staining and streaking. In this design, the window is set in a deep coffer whose margins project slightly in front of the general wall surface, Fig. 3.6.17c. The splayed lower plane of the coffer may streak, particularly in the corners, but these streaks will not be objectionable, since they will emphasize the modeling of the feature. If the coffer is reasonably deep, there is little danger of water being carried off the edges of the window.

The easiest windows to design are vertical slot windows. These interrupt the water flow least of all. The important consideration is the water directed to the base of the win-

dow. If the water volume is considerable (a very high window) and the windows are not kept clean, a large volume of dirt will be directed to the sill. The dirt can present a problem if the under-window or spandrel panels are of light colored concrete. Fig. 3.6.17e and Fig. 3.6.18 show a solution where concrete scuppers, cast integrally with the window panels, direct all water from the windows away from the building. The walls of the building in Fig. 3.6.18 are all precast concrete with brick-faced concrete panels between the windows.

Detailing of precast concrete at the base of a glass curtain wall is important as large quantities of water and dirt will flow from tall sections of glass, see Fig. 3.6.4. Ideally, detailing should be such that water flow from these areas falls clear, by setting the precast concrete wall behind the line of the mullions and curtain wall sill. As alternatives, water may be made to flow in rustications lined up with the mullions for the glass or collected in a gutter and then drained. Shadows on rustications usually help mask any streaks, particularly when the recessed depth is equal to or greater than the recess width.

The design details at the base of a continuous horizontal strip of glazing are as important as those at the bottom of an individual window. The taller the area of glazing, the more dirty water is involved. This water should be thrown clear of the building. Figs. 3.6.17d and e and Fig. 3.6.19 show a concrete panel shaped to prevent water from running over the face of the panel beneath the window. The water is collected at sill level and directed to joint features

Fig. 3.6.18

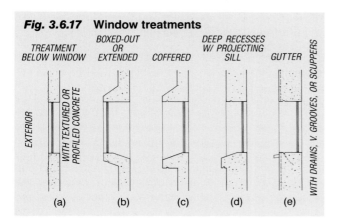

Fig. 3.6.17 **Window treatments**

TREATMENT BELOW WINDOW

WITH TEXTURED OR PROFILED CONCRETE

EXTERIOR

BOXED-OUT OR EXTENDED

COFFERED

DEEP RECESSES W/ PROJECTING SILL

GUTTER

WITH DRAINS, V. GROOVES, OR SCUPPERS

(a)　(b)　(c)　(d)　(e)

Fig. 3.6.20

Fig. 3.6.19 **Channeling of rain water hitting windows**

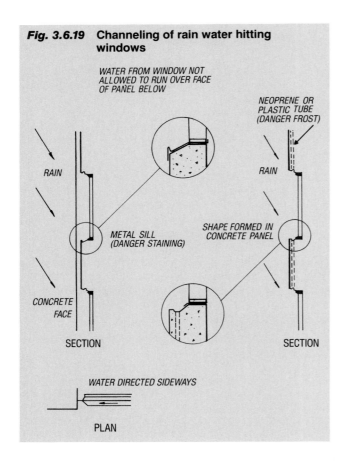

WATER FROM WINDOW NOT ALLOWED TO RUN OVER FACE OF PANEL BELOW

NEOPRENE OR PLASTIC TUBE (DANGER FROST)

RAIN

RAIN

METAL SILL (DANGER STAINING)

SHAPE FORMED IN CONCRETE PANEL

CONCRETE FACE

SECTION

SECTION

WATER DIRECTED SIDEWAYS

PLAN

along the windows, or to tubes which take it to the nearby joints (or grooves) or an internal drainage system. Tubes may create problems during freezing temperatures.

For continuous horizontal windows, a concealed gutter may be used. The gutters collect the water from the line of glazing and discharge it through a series of scuppers or drainpipes, Figs. 3.6.17e and 3.6.20. This is an expensive way of overcoming the problem. In a high building, or with high wind velocity, this solution is ineffective. An alternate in areas where rainfall is rarely heavy, is to retain the rainwater in lined gutters and allow the water to evaporate. The occasional storm in which gutters overflowed would do less harm to the appearance of the building than the slow trickle of water from frequent light rain.

3.6.6 JOINTS

Joints are important features in creating weathering patterns. In addition to leading the water down the joints (real or false), the designer should determine where the water will finally emerge.

Edges of joints should have a reasonable radius (chamfer) to reduce the possibility of chipping. Chips disrupt water flow and concentrate dirt.

Non-staining elastomeric type joint sealants should be selected to prevent the possibility of bleeding and heavy dirt accumulation, common problems with mastics. Also, care should be taken to avoid sealants that collect dirt as a

result of a very slow cure or inherent static conditions.

Two-stage joints, which must be vented and drained to the outside, should have vents located at the junction of the horizontal and vertical joints projecting at least 1/4 in. beyond the sealant, see Section 4.7.2. Therefore, if any moisture does come out of the vent tube, it will run down the face of the joint sealant and not over the face of the panels.

Vertical joints help in channeling water, provided the joint is not pointed flush with a sealant or gasket. The concentration of water at such joints requires careful detailing to prevent moisture penetration. Fig. 3.6.21 shows an elevation where some of the vertical joints, into which water is channeled, discharge this water over a vertical concrete surface with fewer joints than at higher levels causing a marked washing effect at the termination of the joint. The water should be directed until it reaches the ground or a drainage system.

Adjacent precast concrete units should have faces aligned within accepted industry tolerances, see Section 4.6.3. Any discrepancy may pass undetected on a new building, but weathering will eventually emphasize the error with uneven staining of adjacent units. It is helpful when the architect, engineer, precaster, and erector understand all requirements in this respect and cooperate in the solving of problems.

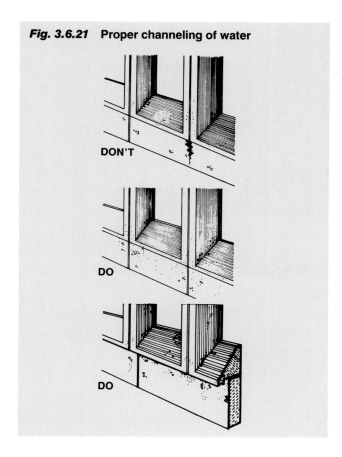

Fig. 3.6.21 Proper channeling of water

DON'T

DO

DO

3.6.7 DEPOSITS WASHED ONTO THE PRECAST CONCRETE FROM AN ADJACENT SURFACE OR MATERIAL

A partial water flow may be stopped from running over backward slopes by water drips. If dirty water, thus directed, falls partially onto other surfaces the problem may be merely relocated. The effect of dirt washed onto precast concrete surfaces may be diminished by rough textured surfaces with reasonable dark color tones to minimize the visual appearance of the dirt. Alternatively, the dirty water may be directed off the building.

An example of this is shown in Fig. 3.6.22. The three-story panels have spandrel areas below windows which project out from the building at each successive level. The same good weathering precaution may be obtained by cantilevering successive floors using story-high panels or spandrel panels as shown in Fig. 3.6.23. This type of construction may complicate erection of the panels (see Section 4.5.3) unless the floors are also precast concrete and their erection coordinated with the panels.

Water flowing over copper, bronze, weathering steels, sheet metal flashing or aluminum and which subsequently flows over concrete, may create green, rust-brown or black stains over a period of time. Consideration should be given to using flashing with a drip detail as shown in Fig. 3.6.14 and to possible protection against corrosion. All of these types of discoloring are more difficult to remove than ordi-

Fig. 3.6.22

nary climatic dirt. Also, maintenance procedures such as window cleaning, can produce dirt markings on precast concrete unless care is exercised.

3.6.8 THE SURFACE COVERING OF PRECAST CONCRETE WITH DEPOSITS EMERGING FROM THE MATERIAL ITSELF

Organic growths (algae, lichen etc.) are rather uncommon for precast concrete in North America, except in cases of frequent exposure to high ambient humidity and temperature. Lichen do not usually develop until the alkalinity of the surface has reduced to about pH 7.5 but colonization can then be quite rapid. The remedy is simply a concrete with low absorption or avoiding having the surface in a continually wet state.

In concrete at early ages, the hydrated cement contains some calcium hydroxide (soluble) as an inevitable product of the reaction between cement and water. When this calcium hydroxide is brought to the surface by water, it combines with carbon dioxide in the air to form calcium carbonate (very slightly soluble), which then appears as a whitish, crystalline deposit (efflorescence). In a second, slower reaction, calcium carbonate can react with additional carbon dioxide and water to form calcium hydrogen carbonate which is, in turn, soluble in water. By this mechanism, the efflorescence weathers away. The acid constituents of the atmosphere (e.g., sulfur dioxide) also affect a dissolution of the calcium hydroxide (lime) deposits on the concrete surface. Efflorescence disappears faster in an industrial climate than in the purer air of a marine or mountain climate. It also may be removed by washing the surface of the units with dilute hydrochloric acid followed by copious washing with water (see Section 3.5.5.)

The conditions necessary for the development of this type of efflorescence are that the concrete becomes wet and remains damp for several days; it has not had time to carbonate to any appreciable extent into the surface layer (after such carbonation, the quantity of calcium hydroxide which reaches the surface is far too small for visible efflorescence to be formed); and that a sufficient quantity of carbon dioxide (carbonic acid) reaches the same surface.

A distinction should be made between so-called primary efflorescence, which occurs during the curing of concrete, described above, and secondary efflorescence, which

Fig. 3.6.23

results from the action of external water and may occur long after production of the concrete units.

The secondary efflorescence is a white powdery surface effect caused by soluble alkalies or salts that migrate to the concrete surfaces with evaporating water. In most cases, salts that cause efflorescence come from beneath the surface; but chemicals in the materials can react with chemicals in the atmosphere to form efflorescence. Discoloration may occur, but this usually is temporary and the white coating should be left to weather naturally. Dark surfaces tend to multiply the unsightliness of efflorescence that might not otherwise be seen on a light or whitish background. It occurs most frequently during the winter or early spring months when a slower rate of evaporation due to low temperatures and high relative humidities allows excessive migration of the salts to the surface.

The formation of both types of efflorescence appears to be dependent on high concrete porosity (high water absorption). The denser the concrete, the less tendency it has toward efflorescence. Attention to mix design and curing will reduce the possibility of efflorescence. Also, the use of a concrete sealer (see Section 3.6.13) will minimize the absorption of moisture into the unit thereby reducing the migration of water to the surface.

A feature of limestone aggregates is their tendency to exude self-produced efflorescence when used for exposed aggregate finishes. On white or near-white aggregate this is of little consequence, and on white finishes it might, on occasion, even bring some slight improvement to the final result. On dark material, however, the white film will not only show but will dull down the original color of the aggregate significantly. Treatment with a clear sealer after cleaning the surface can normally be relied on to prevent a recurrence.

3.6.9 SURFACE CHANGES IN THE MATERIAL ITSELF

The cement-rich film on smooth concrete may develop surface crazing when exposed to wetting and drying cycles, see Section 3.5.2. Although crazing is a surface phenomenon and has no structural or durability significance, its presence may become visually accentuated when dirt settles in these minute cracks.

Crazing is more likely to show up in white or light concrete than with gray, because the dirt is more visible on the white, but the effect will depend on the character of the cement film. A relatively lean, properly consolidated concrete mix will show little crazing in contrast to a mix rich in cement and water. Where crazing occurs, a horizontal surface will be affected more than a vertical surface due to the settlement of the dirt on the former. Crazing generally will not appear where the outer cement skin has been removed by a surface finishing technique.

Rust stains caused by reactive iron pyrites or other contaminants will occur where such materials are found as part of some aggregates. Consequently, such aggregates should not be used for exposed concrete panels unless the possible staining is acceptable to the architect, or the rusting particles are removed prior to shipment or when they show up.

Rust stains may also be caused by particles of steel left by the aggregate crusher, pieces of tie wire from the cage assembly, or particles of steel burnt off in welding and accidentally left in the mold. These stains and the particle causing them should be removed from the surface as soon as visible.

Rust stains caused by corroding reinforcing bars are not common. When reinforcing steel does corrode, it reflects shortcomings in design, concrete quality or workmanship. When manufacturing of precast concrete is performed under plant controlled conditions such incidents should not occur if adequate concrete cover is provided. To control the location of reinforcing steel, spot checks of finished units should be made with the aid of a pachometer (or covermeter) to verify concrete cover.

Rust stains due to corrosion of hardware should not occur if the hardware is protectively coated or where it is entirely behind a weatherproofed joint, see Section 4.5.7.

If panels are prestressed, anchorages (stressing pockets) must be recessed and packed with nonshrink mortar to provide the same cover to the reinforcement as at other panel locations. It may or may not be necessary to coat the strand ends with a rust inhibitor paint such as a bitumastic, zinc rich paint or epoxy to avoid corrosion and possible rust spots which could mar the surface after years of service.

3.6.10 FOREIGN MATTER DEPOSITED ON CONCRETE SURFACES

All precast concrete units should be delivered to the jobsite in a clean and acceptable condition. Protection of the units during transportation is normally not required. The need for protective covering against road stains and weather during delivery should be determined by the precaster. This protection need should be evaluated against its cost and effectiveness enroute. If the architect desires a specific form of protective covering, this fact should be stated in the contract documents, so that the precast concrete manufacturer may make proper allowance for this expense in the bid.

Rainwater, or water from hoses used during the construction of the building, can cause discoloration of exposed precast concrete by first washing across other building materials (such as steel, concrete, or wood) and then across the units. The general contractor should provide and maintain temporary protection to prevent damage or staining of exposed precast concrete during construction operations after installation. Particular care should be taken to avoid jobsite water washing over the precast concrete. Dirt, mortar, plaster, grout, fireproofing, or debris from concrete placing should not be permitted to remain on the precast concrete and should be brushed or washed off immediately with clean water.

Precast concrete units and adjacent materials such as

glass and aluminum should be protected from damage by field welding or torch cutting operations. Non-combustible shields should be provided during these operations. To minimize staining, all loose slag and debris should be removed when welding is complete. All welds and exposed or accessible steel anchorage devices should be painted with a rust inhibitive primer, or in cases of galvanized plates, a cold galvanized coating (zinc rich paint containing 95% zinc). Such protection should be applied immediately after cutting or welding.

Apart from emphasizing care by the precaster, erector, general contractor and other trades, the architect during actual construction has little control over potential staining, but may add the following recommendation as part of the specifications:

"All staining and damage caused by other than the precaster (such as oil from cranes and compressor lines, bitumen from roofing operations, smearing by caulking or painting contractors) should be repaired by the precaster or by qualified personnel using methods approved by the precaster."

Such repairs cannot be part of the precaster's contract, and the precaster should be compensated for repairing any damage caused by others following supply of the precast concrete (or erection if part of the contract).

3.6.11 GLASS STAINING OR ETCHING

Glass, like all building materials, is subject to the effects of weathering. When damp material is in contact with or applied to glass, the glass surface may undergo subtle changes in the contact area. If the material in contact with the glass is inert and moistureproof, the glass surface will be protected by the material from changes caused by exposure to moisture. Later, if the contacting material is removed, a differential surface change may become quite visible and unattractive under some lighting and viewing conditions, even though the change is slight. Finely divided damp materials, e.g., dirt and dust, in contact with glass cause the glass constituents to dissolve slightly and be redeposited at the evaporating edge of the material resulting in staining. In addition, some silicone sealants have ingredients that may leach out and stain the glass. Also, primer or silicone sealant which overlaps the joint onto glass will leave a permanent mark.

When glass (sodium calcium silicate) is exposed to moisture, a minute amount of the glass will dissolve. (Glass will normally lose some of its sodium by dissolution in water but the calcium then stops most of the dissolution. In polluted atmospheres, the acids — SO_x or NO_x — will attack the calcium and permit further dissolution.) If the dissolved material is washed away, little change can be seen by the human eye. But when the solution remains on the glass, atmospheric carbonation of the alkali and alkaline earth silicates causes a subsequent deposit of silica gel. The gel on aging and exposure to atmospheric acids becomes difficult to remove. (Alkaline washing compounds will dissolve some of the silica gel leaving the glass again susceptible to moisture.) When this happens uniformly, the eye does not detect the differences. The silica gel deposit or the glass etch depth need not be thicker than a wavelength of light for the eye to detect it.

All materials which tend to increase the length of time moisture is in contact with glass during wet-dry cycles are likely to speed staining. Dirt accumulation on glass, for example, holds water on the glass longer causing moisture attack. Once the silica gel builds up and becomes hard, it retains moisture and also causes run-off water to flow along the same paths. The process of surface corrosion then becomes self-perpetuating. Frequent washing of the windows tends to remove the gel before it becomes hard, minimizing staining and etching.

Directed slow water run-off and the resultant dirt accumulation cause the glass to be attacked non-uniformly and eventually, the cycle of water drying, gel forming, acid atmosphere attack, and alkali washing compounds, causes in-depth glass dissolution. Then, no amount of cleaning or buffing will remove the stain or etch. Thus, architectural design and detailing must prevent this condition from occurring.

Staining will be more noticeable on tinted heat-absorbing glass because of the greater contrast between the light color of the stain or etch and the darker color of the glass. In addition, heat absorption will increase the rate of etch. There is no known difference in the composition of tinted glasses, which contributes to this staining, as compared to clear glass.

The usual explanation for the etching of glass in concrete structures is that concrete contributes alkaline materials to the run-off water. Hydration of cement results in the formation of hydrated calcium silicates $CA(OH)_2$, and aluminates with the remaining internal water becoming highly alkaline. It is well known that alkali, meaning high pH material, will attack glass. What is not well known is that atmospheric acids (NO_x, SO_x and CO_2) can quickly neutralize these alkalies from concrete to produce neutral salts of calcium, sodium and potassium. Of these salts, only the carbonates of sodium and potassium are truly alkaline. However, even these salts are quickly converted to the bicarbonates which are only very weakly alkaline.

Since, the atmosphere is usually very acid in the larger cities, Fig. 3.6.1, very little if any alkali (high pH) material will be leached more than a few millimeters away from its source except in the case of very young concrete (less than 28 days).

Rainfall can permeate concrete having high absorption and cause efflorescence, see Section 3.6.8. The efflorescencing salts are usually neutralized by the carbon dioxide in the air before they can go very far. The leaching of concrete (efflorescence) ceases in 1 to 2 years because the surface lime is mostly carbonated in place and the interior lime cannot be reached.

In addition, chemical reaction of the cement compounds with sulfur and nitrogen oxides in the air occurs with the subsequent precipitation by evaporation of solutions containing the reaction products, such as gypsum ($C_aSO_4 \cdot 2H_2O$). The transference to and deposition of these materials on the window glass by rainwater can result in surface staining and etching if they are allowed to remain on the glass for a period of time. (The gypsum acts in the same way as dirt in causing a stain.) The time period for a stain to result depends to some degree on the ambient temperatures with warmer temperatures causing the stain to occur sooner.

The plasticity of concrete lends itself to use in many shapes which may not incorporate proper water run-off in the design. In addition, the rough surface textures of exposed aggregate concrete increase water retention and results in slow water run-off. When uniform wetting of windows occurs, staining of glass generally does not occur. However, when differential wetting of the windows occurs from a slow run-off of rainwater, such as by dripping, stains will occur regardless of the construction material.

Weathering steels, bronze, limestone or aluminum curtain wall buildings as well as precast concrete buildings may experience staining of window glass, Fig. 3.6.24. Analysis of powder scraped from glass stains on both a metal clad and concrete clad building show that a good portion of the stain is composed of gypsum ($C_aSO_4 \cdot 2H_2O$) indicating that SO_x from the atmosphere also plays a role in staining. The calcium on the metal building had to come from airborne dust or from the glass.

Building details can reduce the amount of water discharged to the glass. Concrete frames at window heads should, wherever possible, be designed so that they do not splay down and back toward the glass unless drip details are incorporated into the frames. Without drip details, a direct slow wash down of the glass should be anticipated.

The introduction of edge drips and a second drip or gutter serve as an important line of defense against slow run-off. This can be accomplished by having a cast-in drip in the concrete (see Section 3.6.2) or by using extrusions (either aluminum or neoprene) across the head of the window which have either an integral gutter or extended drip lip of at least 1 in., see Fig. 3.6.12.

It is important that glass be washed, rinsed and dried with a clean squeegee following rain or other wash-off conditions, particularly during building construction. Since it is costly to ask for more than one washing during the construction phase, it might be advisable to include a provision for at least monthly examinations of the glass surfaces. Then, if dirt, dust, plaster, drywall, spackle, grout, paint splatter, or other construction refuse is found, it can be removed before permanent damage occurs. (These materials can combine with dew or condensation to form mild chemicals which may etch or stain the glass). Washing of the windows when deposits are first noticed minimizes staining and etching of the glass. Care should always be taken to prevent window cleaning compounds from being washed over the precast concrete panels.

A mild soap, detergent, or perhaps a slightly acidic cleaning solution should be used to clean the glass. Harsh cleaners and abrasives, particularly those of an alkaline character, are not recommended. If a light stain or etch remains, it may be removed with a slurry of cerium oxide and water or 4F pumice and water mixed to a paste consis-

Fig. 3.6.24

tency. As an alternative, 4F pumice plus Windex may be rubbed on the glass, applying light pressure with a clean damp cloth, using 10 to 20 strokes.

If the above procedure does not remove the stain, a motor driven lambs wool buffing head dampened lightly with a 25% solution of hydrochloric acid and sprinkled with 4F pumice should be carefully applied to the stained area. This is followed with another lambs wool head dampened with Windex plus a 5 to 10% solution of ammonium hydroxide. This solution can be sprayed on the glass or the buffing head and is used to neutralize the acid on the glass. The window is then rinsed with clean water. Household ammonia should not be used in place of ammonium hydroxide solution, as some brands have been known to contain caustic alkali which can etch the glass. Careful handling of the acid solution is required since acid can irritate skin, and damage eyes and clothing. Great skill is required in use of buffing heads to avoid creation of "bullseyes" and other non-uniform surface effects.

The hydrochloric acid cleaning process will not remove a silica gel deposit. Polishing with levigated alumina or a dilute (0.25 to 2%) solution of either hydrofluoric acid or sodium or potassium acid fluoride removes the silica deposit and etches other areas at the same time, so that the eye sees a uniform surface. However, if the etch is already deep, the acid will not remove the streak because it is already in a different plane and reflects light at a different angle. The hydrofluoric acid or fluoride treatment should be tested on several glass panels in different locations

before proceeding with the whole job. In some cases, replacing the glass may be more economical than removing the stain. Proper precautions should be taken by workmen using these acid treatments, since the acids are highly corrosive, and in the case of the fluorides, severe eye or skin damage may occur.

3.6.12 DESIGN OF CONCRETE FOR WEATHERING

The assessment of concrete mixes with respect to appearance, strength and durability is discussed in Section 3.2.6. This section deals with weathering not as it affects structural durability, but with special emphasis on the visual results, particularly staining of the concrete.

It is apparent that certain concrete surfaces weather better than others, when in the same environment.

Concrete qualities will influence the degree to which staining of concrete surfaces can be predicted and limited, and the results of later cleaning of those surfaces.

The duration of partially wet conditions and the penetration of water, dirt, acidic rainwater, carbon dioxide, and sulfur dioxide are directly related to the absorption of the concrete surface. This absorption and penetration also create difficulties in cleaning such surfaces to restore them to the original appearance.

Low absorption for the surface concrete demands a high density of concrete and is primarily influenced by the following factors:

1. **Mold Design.** Impermeable mold surfaces are necessary to provide uniformity and high density at the concrete face. When using wood or concrete molds, it is therefore advisable that they be sealed with resins or similar materials. Form joints which are necessary for mold breakdown and release of panels must also be completely sealed between each casting.

2. **Concrete Curing.** Curing of concrete should follow established quality procedures to achieve surfaces with low absorption. It is particularly important that rapid drying be prevented.

   The panel surface exposed to weather is normally the down-face in the mold, so rapid drying is not likely to occur until the unit is stripped. At this stage, precautions against rapid drying are essential for development of high quality concrete with good weathering capability. These precautions might include high stripping strength; protection from sun or drying winds, and uniform moisture retention of the panel faces until a reasonable concrete strength has been reached.

3. **Concrete Consolidation.** Consolidation of the concrete mix should accomplish full coating of the reinforcement with a cement film. Air pockets around uncoated reinforcement creates a potential corrosive environment.

   Consolidation should result in a uniform and dense concrete. The exposed concrete surface will be superior with respect to density, if it is cast face-down in the mold.

   Although density on returns might diminish with increasing depth of the return, careful placing and consolidation techniques can greatly improve density uniformity. The best techniques involve either high frequency external vibration or low frequency, high impact tables; combined with the deposit of concrete in relatively thin layers to permit free water and air to escape.

   For large returns, the two-stage casting production technique described in Section 3.3.8. including proper bonding of the dry joints, is recommended.

   If it is important to obtain high and uniform quality of concrete surfaces for weathering, vertical casting of entire wall units should usually be avoided. Horizontal casting of exterior wall units is a normal procedure even for plants with vertical molds for interior wall units.

4. **Mix Proportions.** Concrete mix design for optimum durability and weathering is dependent on the durability of the individual ingredients, as well as the density of the entire mix. Concrete mixes should be designed and/or evaluated for each individual project with respect to strength, absorption and resistance to freezing and thawing, where such environments exist.

Optimum quality of concrete for durability and weather staining should be based on the following major criteria for mix design:

Relatively Low Cement Content. A reasonable balance should be established between a minimum cement content for stripping and service strength requirements and a maximum cement content to diminish its negative qualities, such as shrinkage and a hardness lower than that of the aggregates.

Low Water/Cement Ratio. Water should be limited to a minimum as excess water will effect strength, density and absorption.

Workability. This should relate to the previous criteria, combined with assurance of proper consolidation and a full coating of the reinforcement with cement.

Solid Mass and Durable Aggregates. Aggregates should always be checked for potential alkali reactivity. It is recommended that no aggregates which are potentially alkali reactive be used for exterior concrete surfaces. This requires a petrographic examination of the aggregate by qualified personnel according to ASTM C295, *Standard Practice for Petrographic Examination of Aggregates for Concrete.*

A water absorption test of the proposed facing mixes may provide an early indication of predictable weather staining (rather than durability). For the concrete strengths normally specified for architectural precast concrete (5000 psi at 28 days), a reasonable water absorption should not be a problem unless cement-rich and/or wet (high slump) concrete mixes are used.

The architect is advised to verify the water absorption of the proposed face mix, which for average exposures and based upon normal weight concrete (150 lbs per cubic foot) should not exceed 5 to 6% by weight. As an improved weathering (staining) precaution, lower absorption between 3 and 4% (by weight) is feasible with some concrete mixes and consolidation methods. In order to establish comparable absorption figures for all materials the current trend is to specify absorption percentages by volumes. The stated limits for absorption would in volumetric terms correspond to 12 to 14% for average exposures and 8 to 10% for special conditions. To establish the water absorption for a particular mix a minimum of three absorption tests should be performed with concrete from three different batches in order to also obtain an assessment of the uniformity of the mix. A uniform water absorption is important as different absorptions may cause a blotchy or streaky appearance during wetting and drying cycles.

Special absorption tests for precast concrete are currently unavailable except in the PCI *Manual for Quality Control for Plants and Production of Architectural Precast Concrete Products* (MNL-117) or in CSA A23.2 in Canada.

A description of this water absorption test follows. All references to water absorption requirements in this Manual have been based upon this test.

Water absorption is an indication of concrete density. Dense concrete is less susceptible to the effects of wetting/drying (highly impermeable) and, therefore, will absorb less dirt in a polluted environment.

WATER ABSORPTION TEST: Three suitable test samples no less than 4 by 8 in. cylinders or 4 in cubes shall be cast from each of the mixes being tested. Test samples shall be consolidated, cured and finished similarly to the products they represent and shall be tested after 28 days. Test samples shall be clean and free from any parting or form release agent or any sealer. If possible, samples should be cast in containers made from the mold material intended for the actual production unit.

The specimens shall be dried in an oven at a temperature between 180 deg. to 225 deg. F until the loss in weight in 24 hours is less than 0.1%. Test samples shall be allowed to cool to room temperature, weighed and then submerged in water to half the height of the specimens. After 24 hours they shall be fully submerged in water flush with the top of the specimens. The water shall in both cases be maintained at a temperature between 65 deg. F and 75 deg. F. Following another 24 hours the specimens shall be removed, the surface water wiped off with a damp cloth, and specimens weighed on a scale which has an accuracy of 0.1 gram.

The percentage absorption is the difference between wet weight and oven dry weight, divided by the dry weight and multiplied by 100. This figure may be transformed to volume percentage based upon the specific weight of the concrete tested.

Laboratory freezing and thawing tests have been conducted to evaluate the ultimate durability of concrete under severe climatic conditions. These tests can be made on prismatic samples prepared from laboratory trial mixes or even from cores cut from the face of finished production units. Such tests, however, take several months to complete. The verticality of wall units seldom allows concrete to reach the saturation point on which such tests are based. However, where horizontal areas allow water or snow to accumulate, an air-entraining admixture should be specified. In addition, it is probably a prudent policy to have air entrained concrete in all precast concrete exposed to freeze-thaw cycles.

Equivalent evaluations can be obtained in a matter of days by conducting "air void studies" (amount of entrained air in cores taken from the production unit) in accordance with ASTM C457, *Recommended Practice for Microscopical Determination of Air-Void Content and Parameters of the Air-Void System in Hardened Concrete.*

3.6.13 SPECIAL PROTECTION OF CONCRETE SURFACES

Weather sealers or protective coatings may be considered for the possible improvement of weathering characteristics. (For this discussion, the term sealers will be used exclusively). A multitude of sealers have become available in recent years. The quality of concrete normally specified for architectural precast concrete, even with minimum practical thickness, does not need sealers for waterproofing.

Sealers may be applied for the following reasons:

1. The prime justification for their use is the potential improvement of weathering qualities in urban or industrial areas. A sealer may reduce attack of concrete surface by airborne industrial chemicals.

2. To facilitate cleaning of the surface if it becomes dirty.

3. To prevent change in appearance, particularly darkening of surfaces that are wetted, by making them water repellent. Also, uneven drying is caused by untreated concrete absorbing moisture at different rates over its surface. The surface will then dry unevenly, leaving a patchy appearance. Sealers keep the moisture at the surface. Drying is therefore speeded up and the surface retains its natural appearance.

4. To reduce efflorescence, particularly with a gray or buff cement matrix. The use of a concrete sealer will reduce the absorption of moisture into the surface, thereby minimizing or eliminating the wet-dry cycle and, therefore, the migration of water to the surface.

5. To reduce the incidence of concrete surface leaching which may assist etching of the glass, aluminum, and other lime susceptible construction materials. However, the use of a sealer will not replace proper design for water run-off.

6. To reduce the tendency of soiling in the yard, in transportation, and on the building. Special sealers (often with short-term effect) which will not change concrete appearance, are used by some precasters to protect concrete during yard storage, particularly where dirty atmospheric conditions exist.

7. To brighten aggregates and develop color tones that would otherwise be hidden.

The effectiveness of sealers is dependent upon the properties of specific sealers. Some problems that have occurred with certain sealers are:

1. Appearance changes vary with the age of the panel being treated. If a methyl methacrylate resin sealer of high solids content is applied, the panel will take on a "wet" look. Some sealers may create a blotchy appearance if applied before panels are fully cured and dried, while others may lose effectiveness if applied too soon after casting.

2. Certain sealers, such as silicones have been found to attract airborne hydrocarbons.

3. Sealers will often interfere with patching of the concrete surface or adhesion of joint sealants. Some methyl methacrylate resin sealers inadvertently sprayed in the joints may peel away from the concrete surface leaving a void between sealant and concrete, while silicone water repellents in the joints will prevent adhesion of joint sealant to concrete surface. Also, the sealer/sealant compatability should be checked. Application of sealers should, therefore, be delayed until these operations are completed. Sealers may also accentuate a patched area of different density. If it is necessary to remove sealers, use a solvent or wire brush, grind or lightly sandblast the surfaces.

4. The possibility of severe and permanent discoloration of the concrete surfaces to various shades of yellow, brown, or gray, or of peeling, varies considerably between types and sources of sealers. Moisture per-

Table 3.6.1 Clear Sealers

Sealer Type	Film Forming/ Penetrating	Water Absorption Resistance	Carbonation Resistance	Breathability	Non-yellowing
Acrylic	F	1	1	2	1
Butadiene-Styrene	F	1	1	2	5
Chlorinated Rubber	F	1	1	2	5
Epoxy	F	1	1	5	5
Oil	P	2	3	1	5
Polyurethane					
Aliphatic	F	1	1	5	2
Aromatic	F	1	1	5	5
Polyester	F	1	1	5	5
Siliconate	P	2	5	1	1
Silicone	P	2	5	1	1
Silane	P	1	5	1	1
Siloxane	P	1	5	1	1
Stearate	P	2	5	1	2
Vinyl	F	1	1	2	2

Rating 1 to 5: **1 Excellent, 5 Poor**
Note: This table represents average performance properties. There can be differences in performance in each generic group as many sealers are blends of various systems. Durability/longevity can also vary within each generic group. In all cases, the manufacturer should be consulted to obtain weatherometer results and field experiences.

meability (breathing) of the sealer is a requirement to prevent blistering and peeling of the sealers, Table 3.6.1.

5. Proper application, following sealer manufacturer's instructions, depends on qualified operators and possible expensive pre-treatment of precast concrete units.

6. Uncertain life expectancy.

It may be well to omit the use of sealers on precast concrete in locations having little or no air polution or in dry climates, in view of the additional cost, the bother of possible recurring periodic applications, and uncertain results of the application of sealers to precast concrete panels.

A careful evaluation should be made before deciding on the type of sealer. This should include consultation with local precasters. Suggested sealers should be tested on reasonably sized samples of varying age, and their performance verified over a suitable period of exposure or usage based on prior experience under similar exposure conditions. Any sealer should be guaranteed by the supplier or applicator not to stain, soil, darken, or discolor the precast concrete finish. Also, some sealers may cause sealants to stain the concrete. The manufacturers of both the sealant and the sealer should be consulted before application or the materials specified pretested before application.

The sealer must make the concrete surface less water absorbent while allowing the outward transmission of water vapor permitting the surface to breathe.

The type of solvent used in sealers, as well as the solids content, can affect the resulting color of the concrete sur-

face. Thus, neither the type nor source of sealer should be changed during the project. Generally, sealers having higher solid contents tend to produce darker surfaces. The amount of color change depends primarily on the type of material in the sealer and their concentration, as well as the porosity of the concrete. The active ingredient of the sealer must be chemical-resistant to the alkaline environment of the concrete. Also, the sealer should dry to a tack-free finish to prevent a buildup of airborne contaminants which results in surface staining. The sealers should be evaluated on how well they penetrate concrete surfaces that vary in absorption and texture. The new generation of penetrating sealers, generally silanes or siloxanes, develop their water repellent ability by penetrating the surface to depths in excess of 1/4 in., reacting with the cementitious materials in the concrete and making the concrete hydrophobic. The penetrating ability of the sealer system depends on the molecule size of the active ingredient, the viscosity of the system, and the solvent-carrying system.

Silane products based on monomeric alkylalkoxysilane (AS) have an extremely small molecule that provides excellent penetrating power. However, while the active substance is being formed, the relatively volatile silane can evaporate. The rate of evaporation increases as drying conditions increase and as high concentrations of active ingredient are used (typically 40% up to 100% silane).

Silane products based on oligomeric alkylalkoxysiloxane (OAS) also have an extremely small molecule, but they have practically no vapor pressure under application conditions. This means that they do not evaporate so readily and can remain in capillaries until conditions are favorable.

They can, therefore, be used at much lower levels of active ingredients.

While generally more expensive than other types of sealers, silane sealers are recommended and are typically longer lasting and less subject to deterioration from sun exposure (ultraviolet exposure). Because appearance is generally unaffected by application of these sealers, it is difficult to monitor their performance visually.

Sealers consisting mainly of the methyl methacrylate form of acrylic resins, having a low viscosity and high solids content also produce durable finishes and usually result in a glossy surface.

Surface coatings should not be applied until joints are caulked and all repairs and cleaning has been completed. In cases where the precast concrete units have been coated at the manufacturing plant, and additional cleaning is required, it may be necessary to recoat those particular panels.

Sealers should be applied in accordance with each manufacturer's written recommendations. Generally, low pressure airless spray equipment should be used to apply the sealer uniformly and prevent surface rundown. Two coats are usually required to provide a uniform coating, because the first coat may be absorbed into the concrete. The second coat does not penetrate as much and provides a more uniform surface color. Care should be taken to keep sealers off glass or metal surfaces.

3.6.14 CLEANING

If required, the final cleaning of the precast concrete units should take place only after all installation procedures, including joint treatment, are completed. This cleaning should be done a minimum of 3 to 7 days after any repairs have been completed.

It is recommended that the precast concrete manufacturer assist where possible in the final cleaning by checking the cleaner's procedures, prior to actual execution, to ensure that no permanent damage to the precast concrete work or adjacent materials is likely to occur. Before cleaning, a small (at least a square yard) inconspicuous area should be cleaned and checked to be certain there is no permanent damage to the precast concrete work or adjacent materials before proceeding with the cleaning. The effectiveness of the method on the sample area should not be judged until the surface has dried for at least one week.

Removing stains from old concrete sometimes leaves the area much lighter in color than the surrounding concrete because surface dirt has been removed along with the stain or because the surface may have become slightly bleached. If at all possible, cleaning of concrete should be done when the temperature and humidity allow rapid drying. Slow drying increases the possibility of efflorescence and discoloration.

A suggested order for testing appropriate procedures for the removal of dirt, stains and efflorescence from precast concrete (beginning with the least damaging) is:

1. Dry scrubbing with a stiff fiber brush.

2. Wetting the surface down and vigorous scrubbing of the finish with a stiff fiber brush followed by additional low pressure washing of the surface.

3. Steam cleaning or high pressure washing. Brushes and abrasive stones are usually necessary to assist in removing dirt. Steam cleaning is also done in conjunction with chemical cleaning.

4. Chemical cleaning compounds such as detergents, muriatic or phosphoric acid or other commercial cleaners used in accordance with the manufacturer's recommendation. If possible, a technical representative of the product manufacturer should be present for the initial test application to ensure its proper use. Consideration should be given to the chemical's effect on the concrete surface finish and adjacent materials, such as glass, metal or wood.

Areas to be cleaned chemically should be thoroughly dampened with clean water prior to application of the cleaning material to prevent the chemicals from being absorbed deeply into the surface of the concrete. Surfaces should also be thoroughly rinsed with clean water after application. Cleaning solutions should not be allowed to dry on the concrete finish. Residual salts can flake or spall the surface or leave difficult stains. Care should be taken to protect all adjacent corrodible materials, glass, or exposed parts of the building during acid washing. A strip-off plastic that is sprayed-on can be used to protect glass and aluminum frames.

Since the use of a dilute solution (5 to 10%) of hydrochloric (muriatic) acid may slightly change the color and texture of a panel and, thus, affect the appearance of the finish, the entire precast concrete facade should be treated to avoid discoloration or a mottled effect. With some finishes, a more dilute solution (2%) may be necessary to prevent surface etching that may reveal the aggregate and change surface color and texture. Hydrochloric (muriatic) acid may leave a yellow stain on white concrete. Therefore 3% phosphoric acid should be used to clean white concrete.

Rubber gloves, glasses, and other protective clothing must be worn by workmen using acid solutions or strong detergents. Materials used in chemical cleaning can be highly corrosive and frequently toxic. All precautions on labels should be observed because these cleaning agents can affect eyes, skin, and breathing. Materials which can produce noxious or flammable fumes should not be used in confined spaces unless adequate ventilation can be provided.

5. Sandblasting may be considered if this method was originally used in exposing the surface of the unit. An experienced subcontractor should be engaged for sandblasting. A venturi-type nozzle should be used on the gun for its solid blast pattern, rather than a straight bore nozzle that produces lighter fringe areas.

6. For information on removing specific stains from concrete, reference should be made to *Removing Stains and Cleaning Concrete Surfaces*, IS 214, published by the Portland Cement Association, Skokie, IL.

CHAPTER FOUR

DESIGN

4.1 DESIGN RESPONSIBILITY

4.1.1 GENERAL

The design and construction of a structure is a complex process involving many different activities. The scope of work and the responsibilities of each of the parties involved must be defined by the contract documents to ensure that the end result will be a safe, high quality structure, see Table 4.1.1.

Practices vary throughout North America with respect to design of structures using precast concrete members. However, it is of fundamental importance that every aspect of the design conform to the requirements of both the local building codes and the state laws governing the practice of architecture and/or engineering. This generally requires that a registered professional engineer or architect accept responsibility for the design.

The legally responsible parties involved in the design and construction of structures containing precast concrete members may include the owner, architect, construction manager, engineer, general contractor, precast concrete manufacturer, erector and site inspector.*

In general, the architect selects the cladding material for appearance, provides details for weatherproofing, selects tolerances for proper interfacing with other materials, and specifies performance characteristics. The structural engineer of record designs the structure for the weight of the cladding, designates connection points, and evaluates the effects of structural movement on the performance of the cladding. The precast concrete manufacturer designs the cladding for the specified loads and all plant/erection handling loads, details how it connects to the structure, and provides for the weatherproofing, performance and durability of the cladding itself. Full cooperation between these three professionals is absolutely essential to the success of the project. All three must fully comprehend all facets of architectural precast concrete panel design, production, and installation.

It is in the interest of all parties, architect/engineer, general contractor, and precast concrete manufacturer, to have a clear understanding of the work assignments and responsibilities of each party (in a general sense) to avoid confusion of roles and to promote a more uniform working relationship. Responsibilities should be fully expressed in the contract documents for each project and clearly understood by all parties. Where part of the design responsibility is vested with the precast concrete manufacturer, close coordination between the architect and the precaster is necessary to ensure that all statutory requirements are met.

4.1.2 RESPONSIBILITIES OF DESIGNER

The design responsibilities of the designer are subject to the contractual relationship with the owner and to the rights, duties, obligations, and responsibilities assigned in the contract documents. Exceptions to items of responsibility should be clarified in the owner/architect agreement and in a written instrument to which the owner and general contractor are parties.

The basic responsibility for both structural and esthetic design of architectural precast concrete should rest with the architect/engineer, who should:

1. Provide complete, clear, and concise drawings and specifications and, where necessary, interpretations of the contract documents. It should be made clear whether the project contract, specifications or drawings prevail in the event of conflicts. It is recommended that they prevail in the order given. All esthetic, functional, and structural requirements should be detailed.

2. Determine the part, if any, played by the precast concrete in the support of the structure as a whole.

3. Determine the load reactions necessary for the accurate design of both reinforcement and connections.

4. Design the supporting structure so that it will have the strength to carry the weight of the precast concrete and any superimposed loads and the stiffness to allow erection of the precast concrete units within the specified tolerances. The gravity supports of precast concrete panels are almost always eccentric to the centerline of the supporting steel or concrete structure.

---

*Reference should be made to *Recommendations on Responsibility for Design and Construction of Precast Concrete Structures,* PCI JOURNAL, July-August 1988.

Table 4.1.1 — DESIGN RESPONSIBILITIES

Contract Information Supplied by Designer	Responsibility of the Manufacturer of Precast Concrete
	OPTION I
Provide complete drawings and specifications detailing all esthetic, functional and structural requirements plus dimensions.	The manufacturer shall make shop drawings (erection and production drawings), as required, with details as shown by the designer. Modifications may be suggested that, in manufacturer's estimation, would improve the economics, structural soundness or performance of the precast concrete installation. The manufacturer shall obtain specific approval for such modifications. Full responsibility for the precast concrete design, including such modifications, shall remain with the designers. Alternative proposals from a manufacturer should match the required quality and remain within the parameters established for the project. It is particularly advisable to give favorable consideration to such proposals if the modifications are suggested so as to conform to the manufacturer's normal and proven procedures.
	OPTION II
Detail all esthetic and functional requirements but specify only the required structural performance of the precast concrete units. Specified performance shall include all limiting combinations of loads together with their points of application. This information shall be supplied in such a way that all details of the unit can be designed without reference to the behavior of other parts of the structure. The division of responsibility for the design shall be clearly stated in the contract.	The manufacturer has two alternatives: (a) Submit erection and shape drawings with all necessary details and design information for the approval and ultimate responsibility of the designer. (b) Submit erection and shape drawings for general approval and assume responsibility for part of the structural design, i.e., the individual units but not their effect on the building. Firms accepting this practice may either stamp (seal) drawings themselves, or commission engineering firms to perform the design and stamp the drawings. The choice between alternatives (a) and (b) shall be decided between the designer and the manufacturer prior to bidding with either approach clearly stated in the specifications for proper allocation of design responsibility. Experience has shown that divided design responsibility can create contractual problems. It is essential that the allocation of design responsibility is understood and clearly expressed in the contract documents. The second alternative is normally adopted where the architect does not engage a design engineer to assist in the design.
	OPTION III
Cover the required structural performance of the precast concrete units as in Option II and cover all or parts of the esthetic and functional requirements by performance specifications. Define all limiting factors such as minimum and maximum thickness, depths, weights and any other limiting dimensions. Give acceptable limits of any other requirements not detailed.	The manufacturer shall submit drawings with choices assuming responsibilities as in Option II. The manufacturer completes the design in accordance with the specified performance standards and submits, with the bid, drawings and design information including structural calculations. The manufacturer accepts responsibility for complying with the specified performance standards. After acceptance of the bid, the manufacturer submits shop drawings for review by the designer and for the approval of the constructor.

For this reason, the design should include provisions for deflection and possible rotation of the supporting structure during and after erection of the precast concrete.

5. Design the supporting structure to withstand the temporary loading conditions that might be encountered as a result of the sequence of loading of the structure. This responsibility is particularly important when the stability of the structure is dependent on the sequence or method of erection. However, the general contractor is responsible for the construction means, methods, techniques, sequences, or procedures. The contract documents should indicate which party is responsible for the design of the construction bracing for non self-supporting precast concrete units and how long such bracing must remain in place.

6. Make provisions in the design for the effects of differences in material properties, stiffness, temperatures, and other elements which might influence the interaction of precast concrete units with the structure.

7. Evaluate thermal movements as they might affect requirements for joints, connections, reinforcement and compatibility with adjacent materials.

8. Provide details for the interfacing of precast concrete and other materials and coordinate with general contractor.

9. Analyze the watertightness of exterior concrete wall panel systems. The analysis should include an evaluation of joint treatment, including the performance of adjacent materials for compatibility with the joint treatment, and proper sealing of windows and other openings.

10. Design for durable exterior walls with respect to weathering, corrosive environments, heat transfer, vapor diffusion, and moist air or rain penetration.

11. Design sandwich wall panels, where applicable, for proper structural and insulating performance.

12. Make selection of surface finishes giving due consideration to the limitations in materials and production techniques with regard to uniformity of color and texture, as well as performance. It is especially important that the limitations which are inherent in natural materials be recognized.

13. Make selection of interior finishes, being sure to define both the area of exposure and the interior appearance for occupancy requirements. Again, material and production limitations must be considered.

14. Specify dimensional tolerances for the precast concrete and the supporting structure, location tolerances for the contractors' hardware, and erection tolerances for the precast concrete. Any exceptions to PCI standard tolerances, Section 4.6, should be clearly identified.

15. Review and approve all of the manufacturer's shop drawings (erection and shape drawings), as required in the specifications and in keeping with Section 4.3, to

ensure that the design intent is achieved. The timeliness of this review may be critical if the production schedule is to be met. Delays in precast concrete production (and erection) resulting from slow approval of shop drawings can result in costly delays in the overall project schedule.

16. Establish the standards of acceptability for surface finish, color range and remedial procedures for defects and damage. Following the pre-final inspection of the erector's work, the architect should immediately prepare a Punch List setting forth in accurate detail any items of the work that are not found to be in accordance with the contract documents so that proper corrective action may be taken. Following the preparation of this Punch List, a meeting between the contractor, the precast concrete manufacturer, the erector, and the architect/engineer should be held promptly to discuss any questions concerning what the architect/engineer requires to be done before the work can be accepted as complete. Units found to be slightly damaged may be authorized for repair at this time. All repairs should conform to the architect's requirements (for matching the original finish) and should be structurally sound. If the repairs cannot be brought to a satisfactory level — one that matches the approved mockup — the repairs may be rejected. The architect's right to reject unacceptable work enables fulfillment of the duty to guard the owner against defects in the work which will render the building, or any portion of it, unusable for the purpose for which it was intended, or from having the intended appearance.

The contract drawings should provide a clear interpretation of the configurations and dimensions of individual units and their relationship to the structure as a whole. To do this, the contract drawings must supply the following information:

1. All sections and dimensions necessary to define the sizes and shapes of the typical units.

2. Location of all joints, both real (functional) and false (esthetic). Joints between units should be completely detailed.

3. The materials and finishes required on all surfaces, and a clear indication of which surfaces are to be exposed when in place.

4. Details for the corners of the structure.

5. Details for jointing to other materials.

6. Details for unusual conditions and fire endurance requirements.

7. Design loads and moments, which include eccentricities.

8. Deflection limitations.

9. Specified tolerances and clearance requirements for proper panel installation.

10. Support locations for gravity and lateral loads.

11. Specific mention of any required erection sequences.

12. Details of connections to the supporting structure, see Option 1, Table 4.1.1.

13. Reinforcement of units, see Option 1, Table 4.1.1.

This information must be sufficiently complete to enable the precast concrete manufacturer to produce the units and the erector to install them. In most cases, the precast concrete manufacturer will make his own erection drawings to identify the information needed to translate these details into production and erection requirements. A major reason for leaving most reinforcement and hardware details to the precasters would be that they normally have extensive experience in this field and can choose details suitable for their plants' production and erection techniques.

The above practice is sound provided all precast concrete manufacturers bidding on a project are known to have this capability or can enlist the service of someone who has this capability. Some may not be prepared to execute such designs or possess the expertise necessary. Where these precast concrete manufacturers are likely to submit bids, it is generally recommended that the architect or structural engineer show the reinforcement and the connections for each typical unit. The designer should bear in mind that an overly conservative design not only adds to costs, but may cause other difficulties. For example, reduced concrete qualities may result when placement of the concrete proves difficult because of reinforcement congestion.

Whether reinforcement and connection details are shown by the designer or left to the precast concrete manufacturer is the prerogative of the architect. This decision may also be controlled by local practices or determined by the project specifications. If reinforcement and connections are not detailed, the performance requirements of the design must be clearly identified. The space allowed for connections should also be indicated. It is generally recommended that the architect, or the structural engineer working with the architect, design the reinforcement and the connection for each typical unit. This approach provides a design compatible with structural capability, concrete cover, hardware protection, appearance, and clearance for mechanical services. In addition, it establishes parameters for reasonable structural and functional modifications which may be suggested by the precast concrete manufacturer. The precast concrete manufacturer is responsible for recommending to the designer for approval such additional reinforcement and/or connection hardware as may be necessary for fabrication, transportation, or erection stresses. Design deviations requested by the precast concrete manufacturer or erector should match the required quality and remain within the parameters established for the project. The design changes should not be implemented until the architect has approved the proposed change. Once again, timely approval is extremely important to maintaining the project schedule. To give favorable consideration to alternate proposals is particularly advisable if the suggested modifications are made to conform to the precaster's normal and proven production and/or erection procedures.

In general, there is considerable variety in allocation of design responsibility for architectural precast concrete, from total design by the engineer of record — following the architect's requirements for dimensions, finishes, and profiles — to a performance specification addressed to the precaster based on the architect's drawing and some criteria laid down by the engineer of record, see Table 4.1.1.

4.1.3 RESPONSIBILITIES OF GENERAL CONTRACTOR/CONSTRUCTION MANAGER

When the construction of a building that was independently designed for the owner is undertaken, the general contractor's responsibility is to build the structure in accordance with the contract documents. When design responsibility is accepted by the general contractor, the professional nature of this function must be recognized and allowed to be achieved in a manner that does not compromise integrity or impair quality.

The general contractor is responsible for administration of the project schedule; coordination with all other construction trades; and developing adequate construction means, methods, techniques, sequences, and procedures of construction not specially addressed in the contract documents. In addition, the general contractor has responsibility for the safety precautions and programs in connection with the project. The general contractor should:

1. Be responsible for coordinating all information necessary for the precast concrete manufacturer to produce the precast concrete erection drawings.

2. Review and approve or obtain approval for all precast concrete erection drawings which include the scheme of handling, transporting, and erecting the precast concrete, as well as the plan for temporary bracing of the structure. (Handling and transporting responsibilities depend on whether precast concrete is sold F.O.B. plant, F.O.B. jobsite or erected).

3. Be responsible for the coordination of dimensional interfacing of precast concrete with other materials and construction trades.

4. See that proper tolerances are maintained to guarantee accurate fit and overall conformity with precast concrete erection drawings.

It is the responsibility of the general contractor to establish and maintain, at convenient locations, control points and benchmarks in an undisturbed condition for use of the erector until final completion and acceptance of a project. If the building frame is not precast concrete, then the benchmarks and building lines should be provided by the general contractor at each floor level.

The general contractor must be responsible for coordinating the precast concrete erection drawings with shop drawings from other trades so that related items can be transmitted to the design architect in one package. As part of this responsibility, the general contractor must immediately notify the precast concrete manufacturer and erector of any deviations found in dimensions due to plan or construction errors or changes to the structure.

After erection of the precast concrete panels, the general contractor should notify the architect for the pre-final inspection of the work. Representatives of the precast concrete manufacturer and the erector should be prepared to participate in this inspection tour and answer any questions posed by the architect.

The responsibilities of the construction manager (CM), who is engaged by the owner to manage and administer the construction, may be different than those of the general contractor depending on the CM's agreement with the owner and local practice. The responsibilities of the construction manager should be clearly defined in the contract documents.

4.1.4 RESPONSIBILITIES OF PRECAST CONCRETE MANUFACTURER

The precast concrete manufacturer is either a material supplier or subcontractor to the general contractor. Many precast concrete manufacturers will perform component design of the members that they produce when this task is required by contract and sufficient information is contained in the contract documents. The precast concrete manufacturer may also accept responsibility for design of the connections when the forces acting on the connections are defined by the engineer of record. At the time of bidding, if there is insufficient information in the contract documents to fully cover all of the reinforcement requirements, the precaster customarily assumes that industry minimum standards are acceptable. The fact that a registered professional engineer, retained by the precast concrete manufacturer, has sealed the documents that depict the component design of the precast concrete members or the connections does not relieve the architect or the engineer of record of the responsibility to approve or accept the design as meeting the requirements of state laws and local authorities. The precaster should not proceed with the fabrication of any products prior to receiving this approval.

The precast concrete manufacturer should be responsible for furnishing samples; shop drawings (erection/production drawings); and tooling, manufacturing, and installation procedures. All drawings and specifications that convey the full requirements for the precast concrete must be provided. Pertinent drawings might include architectural, structural, electrical, and mechanical drawings depending on the size and scope of the project; approved shop drawings from other trades; and site plans showing available storage areas. The precast concrete manufacturer cannot be held responsible for problems arising from the use of obsolete or outdated contract documents. The basic aim is to ensure that the finished job will match the specified quality level.

The precast concrete manufacturer should review the product design of the architectural precast concrete for structural soundness and feasibility with respect to finishes, connections, handling stresses, material quality, joint treatment, tolerances for both manufacturing and installation, and unusual structural conditions. The performance of the product should be evaluated for manufacturing, erection, and service conditions. Precast concrete erection and shape drawings should be submitted to the architect for approval or acceptance. Since the precast concrete manufacturer is generally a subcontractor, this submittal is often done through the general contractor. The precaster should request clarification in writing of any discrepancies found in the contract documents, preferably prior to bidding. This review by the precast concrete manufacturer does not relieve the designer of the responsibilities described in Section 4.1.2.

The precast concrete manufacturer should analyze and design all precast concrete units for handling (stripping, storage, shipping and erection) stresses or temporary loadings imposed on them prior to and during final incorporation into the finished structure. The precaster may suggest modifications that would improve the economics, structural soundness, or performance of the precast concrete products. Alternate proposals from a precast concrete manufacturer should match the required quality and remain within the parameters established for the project. In the case of alternates, the precast concrete manufacturer should design and provide reinforcement or temporary strengthening of the units to ensure that no stresses are introduced which will exceed the requirements of the codes or standards governing the project or have an adverse effect on the units' performance or safety after installation.

From the time the erection drawings and finishes are approved for production, the precast concrete manufacturer should be given responsibility and authority for properly implementing the design drawings, furnishing materials as well as workmanship, and maintaining the specified fabrication and erection tolerances. In some cases, others will assume responsibility for erection of the precast concrete products.

The architect, in the specifications, may leave the responsibility for erection of architectural precast concrete open for decision by the general contractor, or may require that the erection be part of the precast concrete manufacturer's work, to be done by the precaster's own forces or subcontracted to specialized erection firms. The inclusion of erection in the precast concrete contract should be governed by local practice.

The precast concrete erector should perform a continuing survey of the building as constructed and check cast-in-place or contractor's hardware (miscellaneous steel), and set out joint locations prior to actual product installation. This will keep the differential variation in joint width to a minimum. For multistory buildings, a joint layout and building elevation check should be made every third or fourth floor. This survey should identify any problems caused by building frame, columns or beams which are incorrectly dimensioned or out of alignment. Any discrepancies between site conditions and the architectural, structural and erection drawings, which may cause problems during erection, such as structural steel out of alignment, hardware improperly installed, errors in bearing elevation or location, and obstructions caused by other trades, should be noted in writing and sent to general contractor/construction manager. Erection should not proceed until discrepancies are corrected by the GC/CM, or until erection requirements are modified.

After the precast concrete units have been installed in the structure in conformance with plans and specifications and accepted by the architect, subsequent responsibility and liability should rest with the general contractor and the architect. Provisions for any construction loads which are in excess of stated design requirements, and which may occur after installation of the units should not be the responsibility of the precaster or erector. They should be the responsibility of the general contractor.

4.1.5 ADDITIONAL DESIGN RESPONSIBILITIES

Any additional design responsibility vested with the precast concrete manufacturer should be clearly defined by the architect. Most precast concrete work today is covered in Option I, Table 4.1.1. Additional design responsibility for the precast concrete manufacturer may occur when the architect uses the methods of communication listed in Options II and III. Option II (b) may occur when no engineer is involved with the architect for panel design and where doubtful areas of responsibility therefore may exist. Under this option, the precast concrete manufacturer should employ or retain a structural engineer experienced in the design of precast concrete. This engineer would ensure the adequacy of those structural aspects of the erection drawings, manufacture and installation for which the precaster is responsible. Option III is not a common practice, but might be used for design of systems buildings.

4.2 STRUCTURAL DESIGN

4.2.1 GENERAL CONSIDERATIONS

This section constitutes a checklist of design considerations over and above those listed in Chapters 2, 3 and elsewhere in this chapter. The sequence of these considerations is not of importance, but the architect should make decisions based on their ultimate effects on the quality and economy of the final product. Design approaches, determination of loads, dimensioning of precast panels, reinforcement and contract drawings are among the subjects treated in detail in this chapter.

The factors typically encountered in the structural design of precast concrete units include:

1. Shape and its impact on mold design, see Section 3.3.

2. Properties of the concrete.

3. Inserts for handling and connection, see Sections 4.5.4 and 4.5.6.

4. Loads for plant handling, transportation and erection, see Section 4.2.9.

5. In-service loads, see Sections 4.2.2 to 4.2.7.

6. Reinforcement, see Section 4.4.

7. Connections, see Section 4.5.

8. Tolerances, see Section 4.6.

9. Joints, see Section 4.7.

Design Objective of structural integrity of the completed structure is primary. Deflections must be limited to acceptable levels, and distress that could result in instability of an individual element or of the complete structure must be prevented. The inherent stiffness of architectural precast concrete panels will significantly stiffen a structure, thereby reducing deflections and improving stability.

Economy is an important design objective. The total cost of a completed structure is the determining factor. What seems to be a relatively expensive precast concrete element may result in the most economical building because of the reduction or elimination of other costs. In some cases, precast concrete elements are not economical due to improper application or the necessity for complex on-site construction procedures. The designer should attempt to optimize the structure by using precast concrete panels to serve several functions, and take advantage of the economy gained by standardization.

Standardization reduces costs because fewer molds are required. Also, productivity in all phases of manufacture and erection is improved through repetition of familiar tasks. There is also less chance of error.

A very strict discipline is required of the designer to avoid a large number of non-standard units. Preliminary planning and budgeting should recognize the probability that the number of different units will increase as the design progresses. If non-standard units are unavoidable, costs can be minimized if the units can be cast from a "master mold" with simple modifications for the special pieces, rather than requiring special molds.

The esthetic design objectives for the structure should be a matter of concern to the structural engineer. A precast concrete element or system may achieve all other design objectives but fall short of esthetic objectives through treatment of structural features only.

To limit damage to architectural precast concrete panels during an earthquake, three paths are open to the designer. First, the readily damaged units may be uncoupled from the structural system, so that these units are not forced to undergo as much deformation as the supporting structure. This separation between the precast concrete panels and the building frame can minimize contact during an earthquake. Second, the deflection of the supports could be reduced in order to minimize deformations in the architectural precast concrete units. Third, the connections between individual units and the supporting frame could be designed to sustain large deformations and rotations without fracture (exhibit connection ductility). Generally, the first or third approach is adopted for non-loadbearing wall panels.

Design Criteria selection is one of the many important choices that must be made in the design process. In some cases, criteria are used in lieu of analysis. An example would be the use of a 1/4 in. provision for differential movement between two adjacent stories in a multistory building as a criterion for the design of the connections. By adopting this criterion, the designer is making a judgment instead of calculating the amount of movement for the specific structure under applied loads. Whenever criteria are adopted, the designer should consider the limitations involved in the application of those criteria.

Building codes contain design criteria. In some cases, these criteria are general and it thus becomes necessary for the designer to develop specific criteria. The Precast/Prestressed Concrete Institute, the American Concrete Institute, and other organizations have, as a part of their purpose, the development of useful design criteria and methods.

The designer of a precast concrete building can choose to transfer loads through architectural precast concrete elements or intentionally avoid significant load transfers through the precast concrete elements. In the preliminary design phase, the structural engineer should therefore recognize that he is able to choose the structural characteristics rather than simply analyze a predetermined set of criteria.

Design Analysis, in some cases, of a single precast concrete element can be completed with very little consideration of other materials and elements in the structure. The weight of the element and the superimposed loads are simply transferred to supports, and the element can be considered independently of the structure. Occasionally, however, it is necessary to consider the characteristics of other materials and elements within the structure. For example, neglecting the relative movement between the precast concrete cladding and the supporting structure may lead to inaccurate estimates of connection forces. These movements result from differential volume changes between the panel and the supporting structure and load deformation of the supporting structure. Forces induced by restrained differential movements between the panel and the supporting structure are best avoided by allowing sufficient movement of the connections.

In other cases, architectural precast concrete elements may interact with other parts of the structure in the transfer of loads. The design of the interacting system and the individual precast concrete elements must be based on analysis of the whole system.

The potential for element volume change requires consideration. Volume changes may be a result of:

1. Expansion and contraction resulting from temperature change.

2. Shrinkage.

3. Elastic and inelastic (creep) deformations.

Potential movement in the support system (the beam, wall, column, foundation or other part of the structure which provides vertical and horizontal support for the element) must also be considered. Movement can result from:

1. Elastic and inelastic deformation from gravity loads.

2. Horizontal displacement resulting from wind and earthquake.

3. Foundation movement.

4. Temperature change.

5. Shrinkage of concrete.

In most cases, potential movements may be estimated by analysis, and provisions made to accommodate these movements.

The structural design of architectural precast concrete requires that, after all loads are determined, the following be considered:

1. Forces and strains caused by handling, transportation and erection.

2. Acceptable crack width and location.

3. Strain gradients and restraint forces from thermal and moisture differentials through the panel.

4. Localized wind forces and the response of a precast concrete element to these transient loads.

5. Forces transferred to the element and connections due to distortion of the structural frame.

6. Differential deflections between the precast concrete element and the supporting structure.

7. Accommodation of tolerances allowed in the supporting structure.

8. Experience with various types of connections.

The designer must recognize that loads and behavior cannot be established precisely, particularly with respect to members continuously subjected to the environment. The imprecise nature of the actual loading will generally not affect the safety of the precast concrete member, provided that reasonable design loads have been established and the above factors considered. The structural engineer is referred to *PCI Design Handbook—Precast and Prestressed Concrete,* and *Design and Typical Details of Connections for Precast and Prestressed Concrete* for design procedures.

4.2.2 DETERMINATION OF LOADS

The structural design of the architectural precast concrete elements and the system of which they are a part involves load transfer and consideration of stability. Achievement of structural design objectives also requires the consideration of movement or the potential for movement within the system of which the precast concrete element is a part.

The forces which must be considered in the design of architectural precast concrete structures can be classified as follows:

1. Those caused by the precast concrete member itself, e.g., the self weight and earthquake forces.

2. Loads during handling and erection (prior to installation on the building).

3. Those structural loads that are externally applied to the unit or transferred to the unit by the behavior of the supporting structure. These loads can include wind, snow, floor, or construction loads.

4. Forces that develop as a result of restrained volume

change or support system movement. These forces are generally concentrated at the connections.

Design for the second condition is normally the responsibility of the precaster. There should be a structural engineer on the staff, or the precast concrete company should retain an engineer to evaluate the handling loads. As stresses during handling and erection often govern the design of the precast concrete units, the danger of developing doubtful areas of design responsibility may occur.

In the interest of both the architect and precaster, the design responsibilities of each party should be clearly defined. It is recommended that this be done in the contract documents. Responsibilities are discussed in Section 4.1.

The forces or stresses imposed during the manufacturing-erection process occur primarily as a result of the member being in an orientation differing from that of its final position in the structure, and the limitations of these stresses are controlled by the fact that the concrete strength at time of stripping may only be a fraction of its final design strength. Thus, for maximum economy of material, the production process should be given consideration as part of the design process. In particular, architectural precast concrete panels should be shaped such that they will be as stiff as possible in the direction of handling-induced stresses. Limitations on product dimensions imposed by transportation must also be considered during the design process. The designer should be familiar with legal load limitations, and with the cost premiums that are associated with transporting members which are overheight, overwidth or overlength.

In-service loads will generally not be as critical as those loads imposed during the manufacturing-erection process, except for panels in zones of high seismicity, or load-bearing panels in tall structures. Generally, the in-service loads including, loads on the structure, are the responsibility of the designer.

4.2.3 VOLUME CHANGES

Volume changes of precast concrete are caused by variations in temperature, shrinkage due to air-drying, and by creep caused by sustained stress. If precast concrete is free to move or deform, volume changes are of little consequence. For example, unrestrained thermal bowing of a panel induces no stress in the panel. If these members are restrained by foundations, connections, steel reinforcement, or connecting members, significant stresses may develop over time.

The volume changes due to temperature variations can be positive (expansion) or negative (contraction), while volume changes from shrinkage and creep are only negative.

Since architectural precast concrete members are generally not subjected to large sustained stresses, creep volume changes are usually small. Thus, the major volume changes in architectural concrete which must be recognized and accounted for are temperature and drying shrinkage. The consideration of temperature and shrinkage in design may be critical for some architectural elements. They can affect performance, appearance, extent of cracking, effective tensile strength, and warping. The amount of temperature change and drying shrinkage movement that is tolerable depends on jointing and design details of the structure.

Architectural precast concrete members are usually 30 days or older when they are erected. Thus, much of the drying shrinkage will have taken place during yard storage, particularly if the elements are thin sections. However, connection details and joints must be designed to accommodate the long-term drying shrinkage and creep, if any, and temperature change movements which will occur after the precast concrete member is erected and connected to the structure. In most cases, the shortening that takes place prior to making the final connections will reduce the shrinkage and creep strains to manageable proportions.

TEMPERATURE EFFECTS. The coefficient of thermal expansion of concrete varies with the aggregate. Ranges for normal weight concrete are 5 to 7×10^{-6} in./in./deg. F when made with siliceous aggregates and 3.5 to 5×10^{-6} when made with calcareous aggregates. The approximate values for structural lightweight concretes are 3.6 to 6×10^{-6} in./in./deg. F, depending on the type of aggregate and amount of natural sand. Coefficients of 6×10^{-6} in./in./deg. F for normal weight and 5×10^{-6} for sand-lightweight concrete are frequently used. If greater accuracy is needed, tests should be made on the specific concrete.

Since the thermal coefficient for steel is also about 6×10^{-6} in./in./deg. F, the addition of steel reinforcement does not significantly affect the concrete coefficient.

SHRINKAGE. Precast concrete members are subject to air-drying as soon as they are removed from the molds. During this exposure to the atmosphere, the concrete slowly loses some of its original water causing shrinkage volume change to occur. About 40% of the drying shrinkage occurs by an age of 30 days and about 60% occurs by an age of 90 days.

Total unit length change at age one year due to drying shrinkage of normal weight concrete typically ranges from about 400 to 650 millionths in./in. when exposed to air at 50% R.H. Lightweight concrete containing all natural sand fines has one year shrinkage values that range from 550 to 900 millionths in./in. Concrete with a unit shrinkage of 600 millionths shortens about 0.72 in. per 100 ft while drying from a moist condition to a state of moisture equilibrium in air at 50% R.H. In comparison, this equals approximately the thermal contraction caused by a decrease in temperature of 100 deg. F.

Sufficient reinforcement must be used in each unit to control the distribution of any shrinkage cracking. Where units have complex shapes and particularly where they have unbalanced volumes, unsymmetrical reinforcement, large protrusions or changes of section, the risk of shrinkage cracking is increased. Distortion (bowing or warping) of the shape of the panel can also occur from these causes.

CREEP. When concrete is subjected to a sustained load, the deformation may be divided into two parts: (1) an elastic deformation which occurs immediately, and (2) a time-dependent deformation which begins immediately and continues over time. This long-term deformation is called creep.

For design, it is convenient to refer to specific creep, which is defined as the creep strain per unit of sustained stress. The specific creep of architectural precast concrete panels made with normal weight aggregates per unit stress (psi) can range from 0.5 to 1.0 millionths in./in./psi. About 40% of the creep occurs 30 days after load application, and about 60% occurs 90 days after load application.

4.2.4 DESIGN CONSIDERATIONS FOR NON-LOADBEARING WALL PANELS

Non-loadbearing (cladding) panels are those precast concrete units which transfer negligible load from other units of the structure. Generally, they are closure panels only, and are designed to resist wind, seismic forces generated from the self weight and forces required to transfer the weight of the panel to the support. It is rare that these externally applied loads will produce the maximum stresses; the forces imposed during manufacturing and erection will usually govern the design, except for the connections. On occasion, wind imposed on slender precast concrete panels may control.

All non-loadbearing panels should be designed to accommodate movement freely and, whenever possible, with no redundant supports, except where necessary to restrain bowing.

The relationship of the deformations of the panel and the supporting structure must be evaluated, and care taken to prevent unintended restraints from imposing additional loads. Such deformation of the supporting structure may be caused by the weight of the panel, volume changes in concrete frames, or rotation of supporting beams. To avoid imposing loads on the panel, the connections must be designed and installed to permit such deformations to freely occur.

Deflection of a panel support is a function of the stiffness of the support. Thus, where adjacent panels are supported on different portions of the building frame with differing stiffnesses, relative deflections between adjacent panels may occur. This is often the case at building corners, where the structural arrangement may result in significantly different stiffnesses. It is also a concern where the structural frame cantilevers. If panels are attached in a manner which will tend to prevent relative displacement, panel stresses developed must be evaluated.

Behavior of a series of small panels supported on a long span flexible beam is shown in Fig. 4.2.1. The support beam will deflect in increments as each panel is erected, resulting in an in-plane rotation of the panels previously erected. This rotation can result in joint widths as illustrated in the figure. See also discussion in Section 2.5 on wall-supporting panels.

Loading from other sources may also cause a problem. For example, if precast concrete is erected prior to floor slab construction, the weight of the floor may deflect the support beam and cause a problem similar to that shown in Fig. 4.2.1. The connections should be designed to allow the

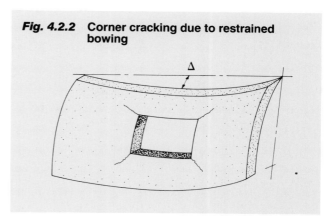

Fig. 4.2.1 **Deformation of panels on flexible beam**

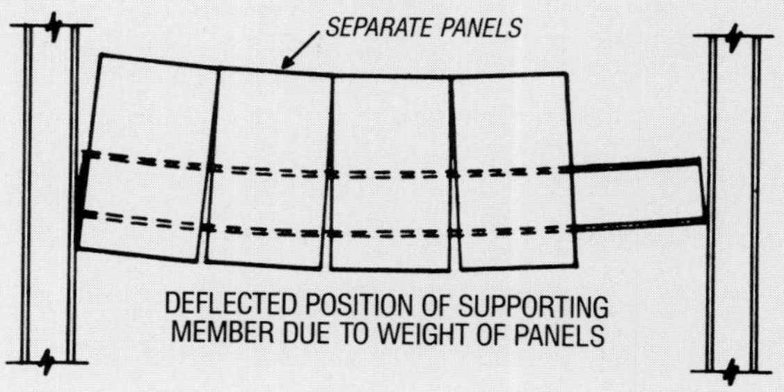

SEPARATE PANELS

DEFLECTED POSITION OF SUPPORTING MEMBER DUE TO WEIGHT OF PANELS

Fig. 4.2.2 **Corner cracking due to restrained bowing**

Δ

supporting beam to deflect, but the beam should be stiff enough that panel joint widths remain within specified tolerances.

Bowing due to thermal gradients is the most prevalent cause of panel deformation after placement of the precast concrete panel in the structure. If supported in a manner that will permit bowing, the panel will not be subjected to stress. However, if the panel is restrained laterally between supports, stresses in the panel and forces in the structure will occur.

Non-loadbearing panels which contain openings may develop stress concentrations at these openings, resulting from unintended loading or restrained bowing. Fig 4.2.2 illustrates the profile of a panel which tends to deflect outward due to warming of the exterior surface. Experience indicates that if the panel is restrained on all four edges, hairline cracking may develop, radiating from corners. While these stress concentrations may be partially resisted by reinforcement, the designer should consider methods of eliminating imposed restraints. Good design practice requires that areas with abrupt changes in cross section be reinforced.

Non-loadbearing panels should be designed and installed so that they do not restrain frames from lateral translation. If such restraint occurs, the panels may tend to act as shear walls (significant diagonal compression may occur) and become overstressed, Fig. 4.2.3. Panels which are installed on a frame should be connected in a manner to allow frame distortion. In some cases, especially in high

Fig. 4.2.3 Panel forces induced by frame distortion

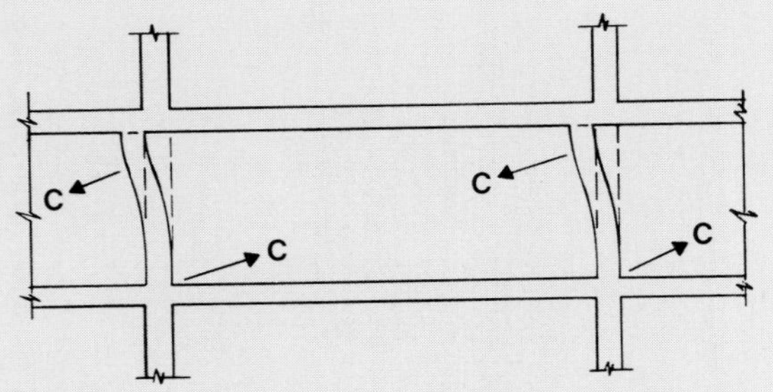

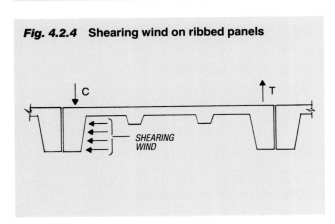

Fig. 4.2.5 Forces on a panel subjected to wind suction and eccentric loading

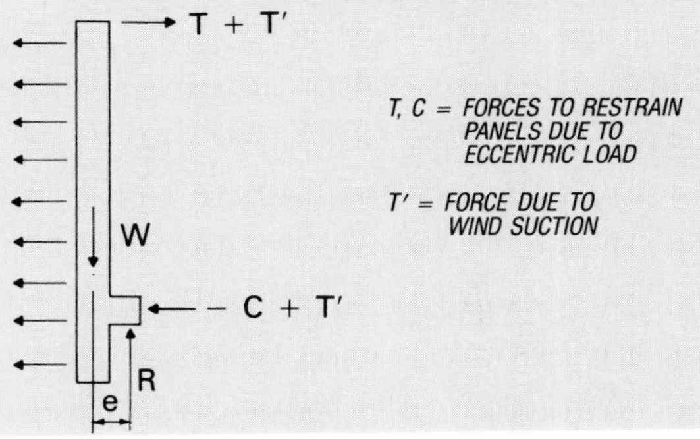

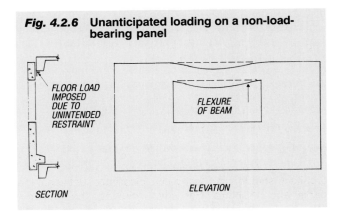

T, C = FORCES TO RESTRAIN PANELS DUE TO ECCENTRIC LOAD

T' = FORCE DUE TO WIND SUCTION

Fig. 4.2.4 Shearing wind on ribbed panels

SHEARING WIND

Fig. 4.2.6 Unanticipated loading on a non-load-bearing panel

FLOOR LOAD IMPOSED DUE TO UNINTENDED RESTRAINT

FLEXURE OF BEAM

SECTION   *ELEVATION*

seismic regions, special connections which allow movement may be required.

Wind Loads are the most common lateral force. Building codes specify the minimum wind pressure for which a building of any height should be designed. However, these loads may not be adequate for localized portions of a tall structure which may be subject to gusting or funnel effects produced by adjacent structures. The lateral deflection of thin panels when subjected to wind should be determined, particularly if they are attached to, or include, windows. Panels with deep protruding ribs may require analysis for shearing winds, as indicated in Fig. 4.2.4, and the connections may have to be designed for the twist produced. Wind suction must also be considered in design with the magnitude dependent upon the panel shape or building configuration. Although the design of the panel itself will generally not be critical for wind, the design of the connections may be. This is particularly true for tension connections which resist eccentric gravity loads, as indicated in Fig. 4.2.5.

Frame Shortening on buildings and how it impacts precast concrete panels is only of concern if it produces a differential movement between the building cladding and the structure. There is little cause for concern if the building and the precast concrete cladding move approximately the same amount. However, this seldom occurs.

A 40-story cast-in-place concrete building can shorten as much as 5 to 6 in. during and after construction, assuming that it is founded on an unyielding base. Each of the floors exhibits a proportional shortening. These shortenings are cummulative for the height of the structure. At each level, the differential shortening between two adjacent floors is a small amount which can be accommodated by the cladding panels. At the lowest level, if the panel is rigidly supported at the base (such as foundation or transfer girder), and the panels are stacked to support the wall above and tied into the structure, the gradual shortening of the structure above may induce unintended loading on the panel. In such cases, the panel connection should be designed to permit the calculated deformation. The full shortening of concrete columns may take several years, although a major part occurs within the first few months after construction. Frame shortening must be considered to determine true fabrication elevations. Unless this is done, panel heights may not match the structural frame and connections will not line up. Panel connections must line up, at the time of erection, and there must be sufficient space at the joints to compensate for future movement.

A similar design situation will occur when two adjacent columns have significantly different loads. For example, the corner column of a structure will usually be subjected to a smaller load than the adjacent columns. If both columns are the same size (as is often the situation for architectural reasons) and reinforced approximately the same, they will undergo different shortening.

Loads from adjacent floors can be imposed on non-load-bearing panels by methods of joinery. These loads can cause excessive stresses at the "beam" portion of an opening, Fig. 4.2.6. This condition can be prevented by locating connections away from critical sections. Unless a method of preventing load transfer can be developed and permanently maintained, the "beam" should be designed for some loads from the floor. Determining the magnitude of such loads requires engineering judgment.

Column covers and mullions are normally supported by the structural column or the floor, and are usually designed to transfer no vertical load other than their own weight. The vertical load of each length of column cover or mullion section is usually supported at one elevation, and tied back top and bottom to the floors for lateral load transfer and stability. In order to minimize erection costs and horizontal joints, it is desirable to make the covers or mullions as long as possible, subject to limitations imposed by weight and handling. With thin flexible members, such as mullions, consideration should be given to prestressing, to prevent cracking. Since mullions generally resist only wind loads applied from the adjacent glass, they must be stiff enough to maintain deflections within the limitations imposed by the window manufacturer.

Since column covers and mullions are often isolated elements forming a long vertical line, any variation from a vertical plane is readily observable. This variation is usually the result of the tolerances allowed in the structural frames. To some degree, these variations can be handled by precast concrete connections with adjustability. The architect should plan adequate clearance between the panel and structure. For steel columns, the architect should consider the clearances around splice plates and projecting bolts.

Column covers and mullions which project from the facade will be subjected to shearing wind loads. Connections must be provided to resist these forces.

Due to vertical loads and the effects of creep and shrinkage, cast-in-place concrete columns will tend to shorten. The horizontal joint between abuting column covers and mullions should be wide enough to permit this shortening to occur freely as well as handle rotation from temperature gradients.

The architect must clearly envision the erection process. Column cover and mullion connections are often difficult to reach and, once made, difficult to adjust. This problem of access is compounded when all four sides of a column are covered for a height exceeding the length of a precast concrete cover. Sometimes this problem can be solved by welding the lower piece to the column and anchoring the upper piece to the lower with dowels set in front, or by a mechanical device that does not require access.

Insulation on the interior face of the column cover will reduce heat loss at these locations. Such insulation will also minimize temperature differentials between exterior columns and the interior of the structure.

4.2.5 DESIGN CONSIDERATIONS FOR LOADBEARING WALL PANELS

Most of the items for non-bearing wall panels also must be considered in the analysis of loadbearing wall panels.

The design and structural behavior of exterior architectural precast concrete bearing wall panels is dependent upon the panel shape and configuration, and should consider the following:

1. Gravity loads and the transfer of these loads to the foundation.

2. Magnitude and distribution of lateral loads and the means for resisting these loads using shear walls and floor diaphragms.

3. Location of joints to control volume change deformations due to concrete creep, shrinkage and temperature movements; influence upon design for gravity and lateral loads, and effect upon non-structural components. The volume change effects will only be significant in high-rise buildings, and then only differential movements between elements will significantly affect performance of the structure. This can occur, for example, at the corner of a building where loadbearing and non-loadbearing panels meet.

4. Connection concepts and types of connections required to resist the various applied loads. For example, local practice may suggest the use of bolts rather than welds.

5. Tolerances required for the structure being designed in regard to production and erection for both precast concrete units and connections, including the interfacing tolerances of different materials.

6. Specific design requirements during the construction stage which may control designs, such as site accessibility.

The design of exterior walls using loadbearing architectural panels follows usual engineering procedures. Perhaps the only design consideration difference is recognizing the role precast concrete panel production and erection play in the overall design process. Similarly, usual accepted procedures and code requirements apply to the design of an individual precast concrete panel and its components.

Architectural precast concrete panels used in exterior loadbearing walls can be classified as either flat panels or ribbed panels. Both flat and ribbed panels may have window, door or other openings.

Panels may be designed to span horizontally or vertically. Whether or not the architectural panel of the exterior wall is placed horizontal or vertical depends primarily upon handling and erection requirements and the methods or details selected for making connections. When spanning horizontally, they are designed as beams, or, if they have frequent, regularly spaced window openings as shown in Fig. 4.2.7(a), as Vierendeel trusses.

A horizontal Vierendeel truss type panel lends itself to simple handling since it is shipped in its erected position, requires vertical load transfer connections at each story level, allows for minimum erection handling and, because of its projected height during erection, demands more sophisticated erection bracing procedures.

When the panels are placed vertically, they are usually

Fig. 4.2.7 Horizontal and vertical rib panels

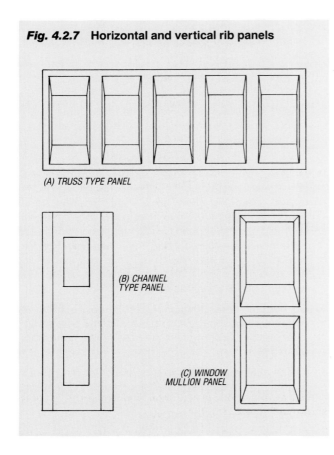

(A) TRUSS TYPE PANEL

(B) CHANNEL TYPE PANEL

(C) WINDOW MULLION PANEL

designed as columns, and slenderness should be considered. If a large portion of the panel is window opening, as in Fig. 4.2.7(c), it may be necessary to analyze it as a rigid frame.

Fig. 4.2.7 shows architectural wall panels, generally used with relatively short vertical spans, although they will sometimes span continuously over two or more floors.

When stemmed floor members are used, the width of load-bearing walls or spandrels should module with the double tee width. In other words, for 8-ft double tees, walls should be 8, 16, 24 or 32 ft wide. Local precast concrete producers should be contacted for their particular module.

4.2.6 DESIGN CONSIDERATIONS FOR NON-LOADBEARING SPANDRELS

Non-loadbearing spandrels are precast concrete elements which are less than story height, made up either as a series of individual units or as one unit extending between columns. Support for spandrel weight may be provided by either the floor or the column, and stability against eccentric loading is achieved by connections to the underside of the floor or to the column (see Fig 4.2.8).

Spandrels are usually part of a window wall, so consideration should be given to the effect of deflections and rotations of the spandrel on the windows, particularly if the floor beams are structural steel.

While structural steel frames tend to have good dimen-

Fig. 4.2.8 Typical spandrel connections

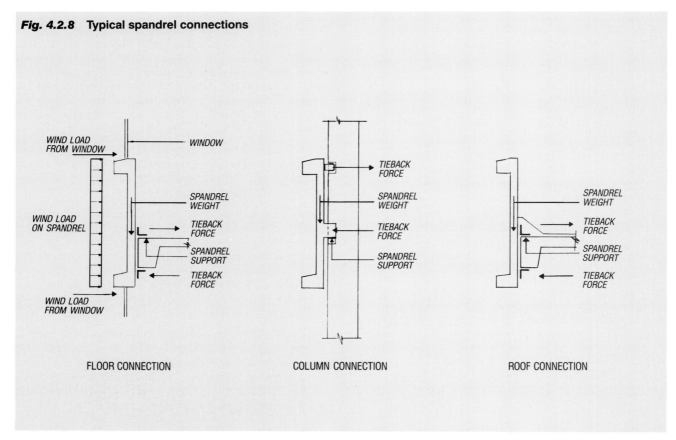

FLOOR CONNECTION

COLUMN CONNECTION

ROOF CONNECTION

sional control, rotation and deflection of the steel support beams is much more common than in concrete framing. The design engineer should account for this deflection and rotation in the steel design. Connections for cladding should generally be designed to transmit the vertical loads to the centerline of the steel beam because of potentially large eccentricities. Provisions should be made for torsional stresses by specifying heavier members, braces or gussets.

For elements which extend in one piece between columns, it is preferable that the connections which provide vertical support be located close to the column centerlines. This arrangement will minimize interaction and load transfer between floor and spandrel.

The weight of a series of wall panels supported on a flexible beam will cause deflection (or rotation) of the edge beam. To prevent imposing loads on the panel, the connections must be designed and installed to permit these deformations to freely occur. The advantages of wall panels spanning between columns (see Section 2.5) becomes apparent in steel structures, where these units may be supported directly off the columns. They may also help to balance the load on the columns.

Consideration should also be given to spandrels which are supported at the ends of long cantilevers. The designer must determine the effect of deflection and rotation of the support, including the effects of creep, and arrange the details of all attachments to accommodate this condition,

Fig. 4.2.9. A particularly critical condition can occur at the corners of a building, especially when there is a cantilever on both faces adjacent to the corner.

When panels supported on cantilevers are adjacent to panels supported in a different manner, this may result in joint tapers and jogs in alignment.

Connections might allow for final adjustment at a later date. However, normal erection procedures assume a panel can be set and aligned without returning for later adjustment. Often the best way to solve this problem is to use a support scheme that does not rely on cantilever action, such as is shown in Fig. 4.2.10. The auxiliary strut supports in Fig. 4.2.10 are normally supplied and erected by the structural steel or miscellaneous steel subcontractor.

4.2.7 DESIGN CONSIDERATIONS FOR LOADBEARING SPANDRELS

Loadbearing spandrels are panels which support floor or roof loads. Except for the magnitude and location of these additional loads, the design is the same as for non-loadbearing spandrels.

Loadbearing spandrels support structural loads which are usually applied eccentrically with respect to the support. A typical arrangement of spandrel and supported floor is shown in Fig. 4.2.11.

Torsion due to eccentricity must be resisted by the span-

Fig. 4.2.9 Effect of cantilever supports

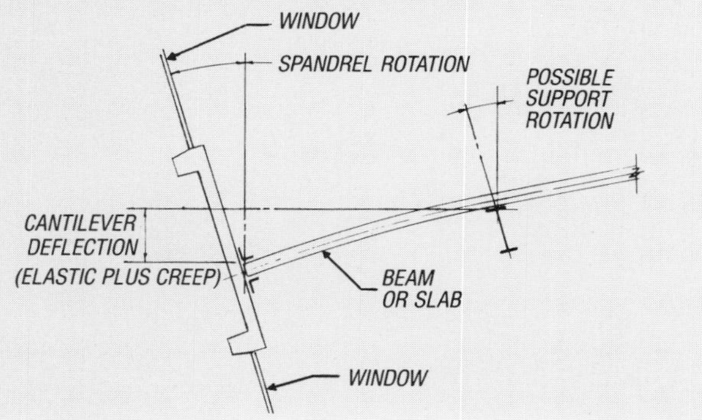

WINDOW
SPANDREL ROTATION
POSSIBLE SUPPORT ROTATION
CANTILEVER DEFLECTION (ELASTIC PLUS CREEP)
BEAM OR SLAB
WINDOW

Fig. 4.2.10 Auxiliary strut support

PRECAST CONCRETE PANEL
SUPPORT MEMBER
AUXILIARY STRUT SUPPORT

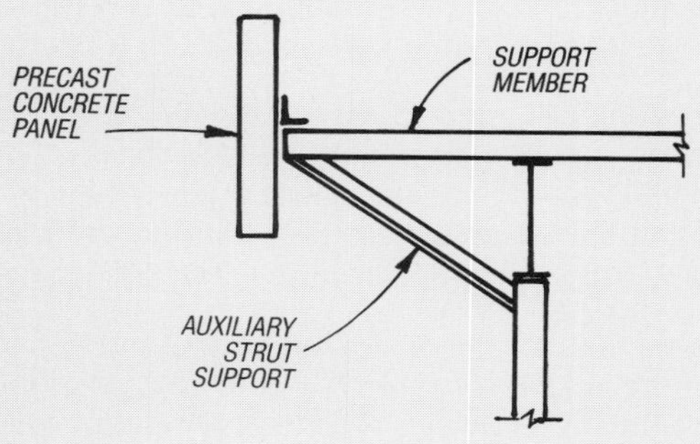

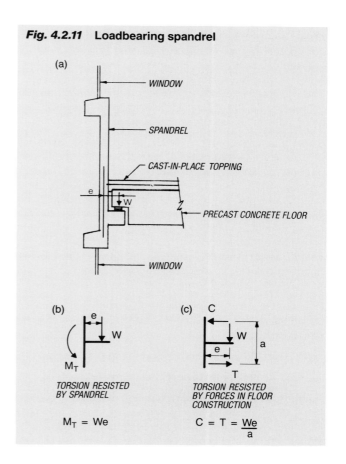

Fig. 4.2.11 Loadbearing spandrel

(a)

WINDOW
SPANDREL
CAST-IN-PLACE TOPPING
e
W
PRECAST CONCRETE FLOOR
WINDOW

(b)

e
W
M_T

TORSION RESISTED BY SPANDREL

$M_T = We$

(c)

C
W
e
a
T

TORSION RESISTED BY FORCES IN FLOOR CONSTRUCTION

$C = T = \dfrac{We}{a}$

Torsion due to eccentricity must be resisted by the spandrel, or resisted by a horizontal couple developed in the floor construction. In order to take torsion in the floor construction, the details must provide for a compressive force transfer at the top of the floor, and a tensile force transfer at the bearing of the precast concrete floor element. The load path of these floor forces must be followed through the structure, and considered in the design of other members in the building. Even when torsion is resisted in this manner in the completed structure, twisting on the spandrel during erection must be considered.

If torsion cannot be removed by floor connections, the spandrel panel should be designed for induced stresses.

4.2.8 DIMENSIONING OF PRECAST CONCRETE UNITS

As a general rule, the precast concrete panels should be made as large as possible without creating special handling requirements or losing repetition. Flat panels should not be made any thicker than necessary for obvious economical and/or weight reasons. Neither should they be made so thin that structural or performance requirements cannot be fulfilled. Because of the multitude of combinations of sizes, functions, applications and finishes, no chart for sizing has been attempted. Instead some considerations for optimum dimensioning follow.

The smaller the unit, the greater the number of pieces required for enclosure. More pieces usually means more handling, more fastening points and higher costs. Therefore, large units are preferable unless they lack adequate repetition, require high tooling costs or incur significant cost premiums for transportation. The maximum size of individual units requires consideration of production repetition, handling ease, shipping equipment required, erection crane capacity, and loads imposed on support system. When desired, the scale of large panels may be reduced by using false joints, as shown in Fig 4.2.12.

Structural considerations such as in-service loads will rarely govern by themselves, unless the panels are load-bearing or special conditions require a large spacing between connections. For most precast concrete exterior bearing wall structures, except for tall slender panels, the gravity dead and live load condition will control dimensions rather than load combinations which include lateral loads. Minimum dimensions for grouting panels together at horizontal joints, space for placing reinforcement, locating handling devices or accommodating a variety of connection conditions can determine the minimum dimensions.

Reasonable slenderness ratios — minimum thickness over unsupported length (when in final position) — for flat panels should be:

1/20 to 1/50 for panels which are not prestressed.

1/30 to 1/60 for panels which are prestressed.

Higher ratios are feasible but must be modified to account for panel end fixity, deflection limitations, or applied loads. However, all but the lowest ratios should be subject to a careful structural and performance analysis. For example,

Fig. 4.2.12

Fig. 4.2.13

a more thorough analysis is required in those buildings without effective shear walls for lateral loads or when walls are subjected to large bending moments. Other types of walls such as window walls should be individually designed.

For large, slender flat panels, the possibility of improving structural capacities and overall performance by prestressing should be considered. The 4-in. thick panels in Fig. 4.2.13 were prestressed to accommodate handling and in-service stresses.

Handling loads at the plant and during erection may be minimized by special handling techniques, such as tilt tables, vacuum lifting, cradles, or special lifting frames.

The precaster will normally endeavor to use such methods rather than add materially to the thickness of panels.

Practical considerations may govern thickness in order to provide concrete cover and accommodate aggregate sizes. A thickness of 3 in. is required to achieve a minimum cover of conventional reinforcement for normal climatic exposures. Many surface treatments such as retarded finishes will reduce the cover and influence the minimum thickness. Attention must also be given to scoring or false joints in flat panels, where the required minimum dimension should be measured from the back of the panel to the bottom of the groove. Panel thicknesses less than 4 in. may create problems for the installation and proper concrete cover of lifting and connection anchoring devices.

Precast concrete flat units should have a minimum thickness of approximately 3 times the largest aggregate used. For example, panels with 1 in. aggregate will require a minimum thickness of 3 in.

An example of the logical determination in detail dimensions appears in Fig. 4.4.3, pg. 181. This illustration shows a typical rib for which a minimum dimension is required.

A 3/8 in. reinforcing bar is assumed as the practical minimum of reinforcing steel. Keep in mind that a deformed 3/8 in. bar is really 1/2 in. thick because of the deformations. Allowing minimum tolerance of 1/4 in. in the placement of reinforcement, the rib width should be 2-1/2 in., assuming that a 3/4 in. concrete cover is needed. (See 'a' in Fig. 4.4.3.) This minimum rib dimension presumes that the reinforcing cage can be made as shown in 'b' or 'c', both of which require properly welded cages. If normal reinforcing ties are required for structural reasons such as loadbearing or wall-supporting units, the minimum rib dimension will correspond to that shown in 'e.'

In both cases the minimum rib dimensions will increase where the concrete surface is treated, since the 3/4 in. concrete cover should be measured from the reveal of the treated surface rather than the surface itself. In addition, the 3/4 in. cover is realistic only if the maximum aggregate size does not exceed 1/2 in.

Performance requirements such as sound attenuation, fire-rating, and the desired tolerances for planeness of facades may govern minimum thicknesses of flat panels. Sound insulation requires a sufficient mass of concrete, which may govern thickness unless additional wall features help to dampen sound.

Rather than increasing the overall thickness of panels, consideration may also be given to ribbing the panels. Ribs may be part of the architectural expression or where flat exposed surfaces are required, ribs may be added to the back of panels for additional stiffness.

Sculpturing of precast concrete units may increase their structural strength and thus simplify handling. Such sculpturing may increase the depth-to-span ratio by providing ribs or projections in either direction of a unit. Contrary to common belief, sculpturing of a precast concrete cladding unit will not constitute a cost premium where sufficient repetition of the unit will keep mold costs within reason and

where the sculpturing will aid the unit's structural capacity. For further details refer to Section 3.3.5.

4.2.9 HANDLING AND ERECTION CONSIDERATIONS

Design consideration must be given to installation conditions for the project with respect to handling, transportation and hoisting. This must be done before sizes, shapes and other features are finalized as erection equipment will frequently influence panel size.

Transportation limitations have already been described in Section 3.3.9 together with handling in the plant.

Re-handling and turning of units between stripping and final installation adds to the cost and increases the danger of accidental damage. Consequently handling should be reduced to a minimum. The precast concrete manufacturer is generally responsible for designing the panels for handling stresses.

The optimum solution for economical handling is the ability to strip a unit from the mold and tilt it into a vertical position similar to the position of the unit in its final location on the building. Fig. 4.2.14 illustrates this point. This procedure has the added advantage that during prolonged storage the panel will weather much as it would following final installation.

The preceding solution is obviously not possible when more important structural or connection considerations

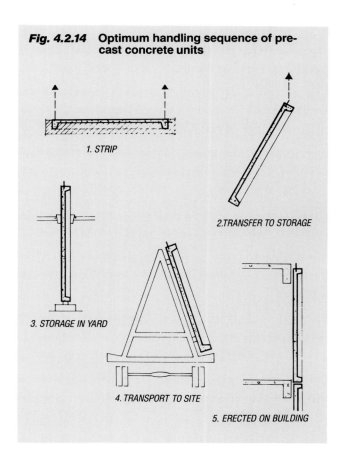

Fig. 4.2.14 Optimum handling sequence of precast concrete units

1. STRIP

2. TRANSFER TO STORAGE

3. STORAGE IN YARD

4. TRANSPORT TO SITE

5. ERECTED ON BUILDING

dictate units several stories high. (Sections 2.5, 2.6 and 2.7.)

Such units may be stacked on their long side and handled at the jobsite by turning in the air as shown in Fig. 4.2.15. In most cases this will be a more economical solution than single story units.

As mentioned in Section 3.6, precast concrete units stored in a position differing from their final orientation may have to be protected against weathering. The need for protection will depend on the configuration of the units, the length of storage time and local environmental conditions. This point may influence the precaster's preference for sizes and shapes of units.

An important consideration in designing towards optimum erection costs — a significant portion of total installed cost — is to provide proper access for trucks and mobile equipment at the jobsite and sufficient room and proximity to the structure so as to allow erection to proceed as normally contemplated. At the pre-job conference, arrangements for jobsite access, on-site storage areas, and other items affecting transportation should be made with the general contractor or construction manager. To avoid excessive erection costs, efforts should be made to allow a unit to be handled in one motion from unloading to positioning. Configuration of the building, or adjacent buildings, may prevent erection equipment from operating this efficiently and may in some cases require double handling of the units. It should be noted that cranes are rated by the safe capacity they will lift with the shortest boom and at the steepest boom-up angle. Maximum lifts will reduce rapidly as boom length increases and angle changes, and the designer should, if possible, consider this during the planning stage. For example, it is not advisable to design heavy units for a high-rise building extending from a parking or shopping plaza unless truck and equipment access to the tower is assured, at least until the units have been hoisted, see Fig. 4.2.16. The entire job could be penalized by these few difficult lifts which could have been avoided had the erection methods been considered. Early coodination of design and realistic erection are essential for overall economy of precast concrete.

To avoid unrealistic or impossible erection conditions, a review of the project by the designer with the precast concrete manufacturer and erector is beneficial. The erector should state his needs for handling and erecting the members in the safest and most economical manner. The designer should give careful consideration to all the factors that affect the methods that can be used in the construction of the project and prepare his design to best suit jobsite conditions.

The precasters, in preparing their design, must first determine where the lifting devices should be placed. One set is required for handling and erection, and another may be needed for stripping. Lifting devices should be designed for actual loads plus an impact factor which depends on configurations and finishes. The architect should insist that all lifting devices be shown on the shop drawings in order to check for interference with other functions or with finishing requirements.

Fig. 4.2.15   Hoisting and turning multi-story units

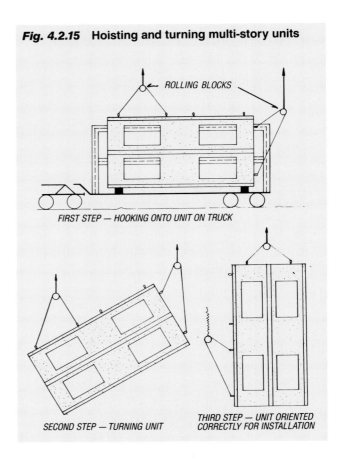

ROLLING BLOCKS

FIRST STEP — HOOKING ONTO UNIT ON TRUCK

SECOND STEP — TURNING UNIT

THIRD STEP — UNIT ORIENTED CORRECTLY FOR INSTALLATION

Fig. 4.2.16   Access for trucks and hoisting equipment

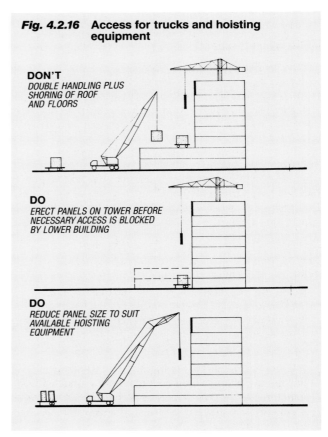

DON'T
DOUBLE HANDLING PLUS SHORING OF ROOF AND FLOORS

DO
ERECT PANELS ON TOWER BEFORE NECESSARY ACCESS IS BLOCKED BY LOWER BUILDING

DO
REDUCE PANEL SIZE TO SUIT AVAILABLE HOISTING EQUIPMENT

Erection inserts cast in precast concrete members vary depending on the member's use in the structure, the size and shape of the member, and the precast concrete manufacturer's or erector's preference. The most common lifting devices are prestressing strand or cable loops projecting from the concrete, threaded inserts, or special proprietary devices.

The type and location of lifting devices for precast concrete units should be planned with the precaster. Ideally the unit should be moved into position on the building without having to be pulled back at the top or the bottom of the unit. On occasion, the building configuration, or the precast concrete unit, may be shaped so that a special hoisting or setting jig must be constructed to hold or "cradle" the unit for setting. These devices have been employed with success and economy where conditions justify their cost.

The architect should approve the final disposal of the lifting devices, which are normally left in the unit, unless they interfere with the work of other trades or are exposed to possible corrosion. Protruding lifting devices (strand or wire) are normally removed after panel installation by burning.

All temporary lifting and handling devices cast into the members should be dealt with as specified on the erection drawings based on one or more of the following conditions:

1. Removed where they interfere with other trades.

2. Removed and surfaces patched where demanded by appearance requirements.

3. Removed or protected from possible corrosion and marring of the finished product where not exposed to view.

4. No action required.

If possible, the placing of temporary lifting and handling devices should be planned so that little or no patching will be required after use. However, when the temporary lifting and handling devices are located in finished edges or exposed surfaces, bolt or insert holes require filling and patching. Often, these devices may be recessed, filled, and finished at a later date with prior designer approval. However, specialized lifting equipment may be necessary to eliminate the patching of exposed lifting and handling devices. A spreader bar with C-hooks can be used to erect architectural precast concrete panels without using exposed inserts, Fig. 4.2.17.

Depending on local practice, the assignment of responsibility for the erection of precast concrete generally varies in the following manner:

1. The precast concrete manufacturer is contractually responsible to the general contractor and/or owner for the supply and erection of the precast concrete products. Erection may be performed by in-house labor or by an independent erector hired by the manufacturer.

2. The precast concrete manufacturer is responsible only for the supply of the precast concrete product, F.O.B. plant or jobsite. Erection is performed by an independent erector under a separate agreement with the general contractor and/or owner.

Fig. 4.2.17

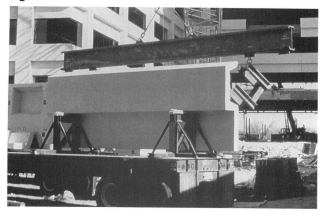

3. The precast concrete manufacturer is responsible for the supply of the product to an independent erector who has a contractural obligation with the general contractor and/or owner for the complete precast concrete package.

On projects where the precast concrete manufacturer does not have the contractural obligation for the erection, it is desirable that the manufacturer assign a representative to coordinate or report on planned erection methods. Regardless of contractural obligations, it is recommended that the precast concrete manufacturer maintain adequate contact with the firm(s) responsible for both transportation and erection to ensure that the precast concrete units are properly handled and erected according to the design and project specifications. Erection of architectural precast concrete should only be performed by competent erectors employing workmen who are properly trained to handle and safely erect the product. Safety procedures for the erection of precast concrete members are the responsibility of the erector, and must be in accordance with all rules and regulations of local, state (province), or Federal agencies which have jurisdiction in the area where the work is to be performed.

Proper planning of the construction process is essential for efficient and safe erection. The sequence of erection must be established early and the effects accounted for in the bracing analysis and the preparation of shop drawings.

After erection, each panel must be stable and offer resistance to wind, accidental impact, and loads which may be imposed due to other construction operations. Provision should be made both in the architectural panel and the support system to permit immediate bracing of the panels. The arrangement of such temporary bracing should be such as not to interfere with adjacent erection and other construction processes, and the bracing must be maintained until permanent connections are accomplished. The single story loadbearing panels shown in Fig. 4.2.18 are temporarily braced until the connections are welded and floor and roof structural elements are installed. It is desirable that each precast concrete element be braced independently of other elements, so that a panel may be moved without effect upon the adjacent panel. Permanent connections which safely sustain loads imposed during erection may be used instead of temporary bracing.

Fig. 4.2.18

Fig. 4.2.19

ring units to block and tackle equipment or setting them temporarily for final placement or alignment during daylight hours. Stiff-legs or elevators, combined with power buggies for transport across floors, and many other handling innovations have been developed for specific jobs.

When wall panels are to be set back into the face of the building, under an overhanging slab, the panels cannot be picked up from above as the structure interferes with the crane lines. These erection difficulties may often be overcome by proper planning or the use of specialized equipment. Temporary openings in floors above, or suitable scheduling of other trades can alleviate such difficulties. Wherever possible, contract documents should facilitate such provisions. Special equipment, such as a balance beam with a movable or fixed counterweight, may be used to erect the panels without defacing the exposed surface, Fig. 4.2.19. While some types of specialized equipment have become a standard item with a number of erection contractors, availability of the specialized equipment to all bidders may very well affect final job costs, unless the projects are large enough to offset the initial capital investment required to secure the necessary equipment.

Hoisting and setting the precast concrete pieces are usually the most expensive and time-critical process of erection. The smaller the unit, the greater the number of units required for enclosure. This leads to more handling, more fastening points and consequently to higher costs. Large units are therefore considerably more economical than small units, but the maximum size of individual units should be decided in relation to the capacity of available hoisting equipment. Speed of erection is directly related to the type of connections selected and the arrangement of the building frame. It is highly desirable that the connection allow for initial setting of the panel and immediate release of crane, with final alignment completed independent of crane support.

Structural limitations governing the erection and/or bracing sequence should be stated on the drawings or in the specifications. Limitations may state, for example, that loading of the structure shall be balanced, requiring that no elevation be erected more than a stated number of floors ahead of the remaining elevations; or limitations may involve the rigidity of the structure, requiring that walls should not be erected prior to completion of floors designed to carry horizontal loads. In steel frames, it

The type of jobsite handling equipment selected may influence the erection sequence, and hence affect the temporary bracing requirements. Several types of erection equipment are available, including truck-mounted and crawler mobile cranes, hydraulic cranes, tower cranes, monorail systems, derricks and others. The PCI *Recommended Practice for Erection of Precast Concrete* provides more information on the uses of each.

Mobile cranes are most commonly used where access is adequate and reasonable. Monorail systems, fed by booms and hoist, or by cranes, are favored by many erectors for buildings above 16 or 20 stories and where relocation of rails can be made in jumps of at least 10 to 15 stories. The use of either of these hoisting systems is independent of other trades requiring lifting.

The use of any type of crane requires planning. Tower cranes, however, require more than the usual planning because their structures, foundations, and presence on the site are generally permanent for as long as heavy construction phases are on-going. The use of tower cranes may have a significant effect on the planning of the structural frame and the sequencing of construction, which should be considered during the planning stages of the structural frame. When tower cranes are used for setting units, their reach and capacity may determine panel weight. Scheduling problems may be encountered unless firm time allocations for separate trades can be maintained by the general contractor. Some erectors occasionally make use of night-time hoisting with tower cranes, transfer-

Fig. 4.2.20 (a)　　　　　(b)　　　　　　　(d)

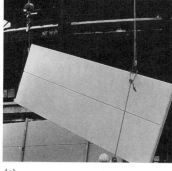

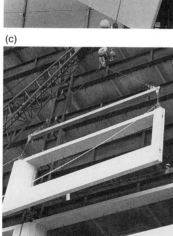

(c)

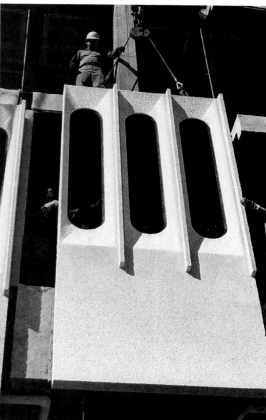

should be determined how far ahead final connections of the frame must be completed prior to panel erection. In concrete frames, it must be determined what strength of concrete is required prior to imposing loads of the precast concrete panels. The frame designer must also recognize that connections between panel and frame impose concentrated loads on the frame and that these loads may require supplementary local reinforcement. In the case of multi-story concrete frames, consideration should be given to the effects of frame shortening due to shrinkage and creep. Delay in erecting precast concrete panels to permit a portion of the shrinkage and creep to occur may be beneficial. For panels which are supported on a continuous bed of grout (such as in loadbearing wall construction) the maximum number of floor levels which can be erected using only shims should be determined and, if critical, indicated on the contract drawings. Where an erection analysis is not performed by the prime consultant, or when it is not the responsibility of the consultant, the contract documents should name the responsible party.

Erection procedures for precast concrete members will vary depending on the size, shape and design of the members, the structural elements which will receive or support them, and the overall complexity of the structure. Typical rigging arrangements are shown in Fig. 4.2.20 for architectural precast concrete panels. Fig. 4.2.20(a) shows the use of one rolling block, while (b) shows the use of two rolling blocks. A spreader beam is shown in (c) and the use of an equalizing rolling block is shown in (d).

WALL PANELS. Panels should be shipped to the jobsite in the vertical position, whenever possible, so that turning is not required. Wall panels should be rigged and hoisted onto the structure near final location and held in place until safely secured. Final alignment and final connections may be made at once or later by another follow up crew, depending on type of connection.

The connections should allow for easy adjustment in all directions. Panels should be installed on a floor by floor basis, where practical, to keep loading equal on the structure, although this approach may not be the most economical. It may be more economical to minimize crane movement and finish an elevation, as much as possible, prior to moving the crane. Panels should be aligned to predetermined offset lines, established for each floor level. This is important due to the drift of a high-rise structure caused by sun, wind and eccentric loads and forces due to construction activity being performed by others.

SPANDRELS. Precast concrete spandrels usually extend from column line to column line at the perimeter beams. They connect either at the columns or directly to the perimeter beams. Precautions should be taken against torsional rotation, due to eccentric loadings on perimeter beams, until final connections are made. Shoring may be required to assist in erection until connections are made.

Spandrels should be shipped to the jobsite in the vertical position. On long spandrels, a spreader beam can be used when hoisting to minimize rigging requirements and reduce the forces which could cause bowing or buckling of the member.

Alignment and all connections should be made, or the spandrels should be safely secured, prior to releasing the load and disconnecting the rigging. Long spandrels may be tied back at the center, if required by the design of the member, to minimize panel bowing. Spandrels should be installed on a floor by floor basis, where practical, to keep equal loading on the structure, although this approach may not be the most economical. It may be more economical to minimize crane movement and finish an elevation, as much as possible, prior to moving the crane.

Spandrels should be aligned to predetermined offset lines established for each floor level. Vertical dimensions between spandrels should be checked to ensure that opening size is within allowable tolerance.

COLUMN COVERS AND MULLIONS. Column covers are usually manufactured in single story units and extend either from floor to floor or between spandrels. However, units two or more stories in height may be used and care should be taken in hoisting and rotating them to prevent cracking. Two lines may have to be used to rotate the member.

Column covers and mullions are usually shipped in the horizontal position. Slings are attached to inserts at the top and the units are rotated to the vertical position. The unit is then hoisted and placed onto the structure to near final position. Alignment and all connections should be made, or the column covers should be safely secured, prior to releasing the load and disconnecting the rigging.

Column covers and spandrels are often erected at the same time. Caution should be taken so that one member does not stress the other. Temporary braces may be required until the spandrel is erected above the column cover and permanent attachments made.

When spandrel panels or column covers and mullions are interspaced with strips of windows to create a layered effect of glazing, precast, and glazing, the general contractor should work closely with the designer to arrive at interface details which are buildable within the sequence of construction and embody all the elements required of exterior wall design.

SOFFITS. Soffit units are normally erected under perimeter beams to form an architectural closure. To achieve this, special erection equipment is often required.

Precast concrete soffits are normally shipped to the jobsite in the horizontal position. Slings are attached to inserts and the soffit hoisted onto cantilever shoring and disconnected. The soffit unit is then rolled under the beam and jacked or hoisted into final position. Connections are made and shoring removed. Other methods of placement are cantilever speader bar, scissor lift and genie hoist. All of these methods are costly and very time consuming and require a great deal of pre-planning and preparation. Consideration should be given to combining soffit and spandrel in a single unit.

4.3 CONTRACT DRAWINGS

The architect's contract drawings should define the scope and intent of the work required. These documents should supply the information described in Section 4.1.2.

These documents should enable the precaster to deal with the forces that the architectural precast concrete units must resist. This and other sections of Chapter 4 should be thoroughly studied so that those parties preparing the contract drawings will obtain the maximum benefits from this text.

Translation of the design concepts into contract drawings is relatively simple for typical units. Difficulties arise primarily where non-typical conditions have to be solved. These conditions may include outside and inside corners, intermediate roof levels, non-typical floors (such as ground level or mechanical floors), and entrances. Such details should not be overlooked. Contract drawings should not be left open to different interpretations at the shop drawing stage. Confusion may increase costs or lead to compromises during production or erection which may be unsatisfactory to both the architect and the precaster.

In detailing architectural precast concrete for a specific design, it is important to continue applying the criteria which govern the economy and performance of precast concrete. If design modules and master mold concepts are not maintained, additional cost factors may be introduced. The detailer is advised to maintain liaison with local precast concrete manufacturers, since their services may be particularly valuable during the development of working drawings. The final and exact dimensions of precast concrete units should be copied from the preliminary design, subject to an assessment of feasibility of all salient details.

Information from the detailer should be sufficient to enable the precaster to produce the units and for the erector to install them. The precast concrete manufacturer will make shop drawings to identify the information for precast concrete and translate these details into production and erection requirements.

The architectural detailer should give sufficient details or descriptions on the drawings to indicate clearly all exposed surfaces of the units and their respective finishes. This is particularly important for returns and interior finishes. The finish requirements may be given by coded numbers, by different shading of the areas representing exposed surfaces, or by any other method which remains readable after printing or possible reduction of the drawings. Simply to heavy-up the exposed surface lines has proven less than satisfactory unless carefully executed.

The use of isometric sketches often makes it simpler to finalize details, particularly in the case of non-typical conditions. By making such isometric sketches part of the working drawings, the detailer greatly facilitates the interpretation of the project requirements. Working drawings, Fig. 4.3.1 (a), for a precast concrete structure show fairly simple architectural shapes having a single finish (exposed aggregate) for eave panels, columns and wall panels. Fig. 4.3.1 (b) shows how this was interpreted on the shop drawings. Fig. 4.3.2 shows the resulting poor appear-

Fig. 4.3.1 Detailing example

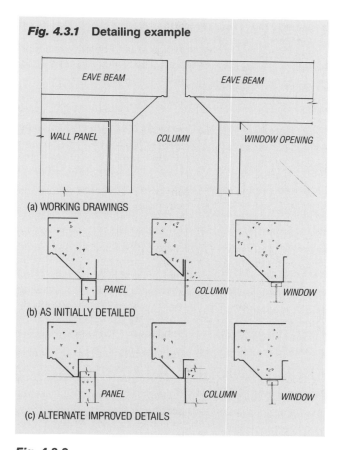

(a) WORKING DRAWINGS

EAVE BEAM

WALL PANEL

COLUMN

EAVE BEAM

WINDOW OPENING

(b) AS INITIALLY DETAILED

PANEL

COLUMN

WINDOW

(c) ALTERNATE IMPROVED DETAILS

PANEL

COLUMN

WINDOW

Fig. 4.3.2

ance of the corner detail. The 45 deg. edge without a quirk, the misalignment of horizontal lines (caused by the necessity of a space between beam and columns which did not occur along the wall panel), the mitered corner, and the difficult production might all have been avoided by better development of project details.

The more logical solution shown in Fig. 4.3.1 (c) has the lower edge of the eave panel level with a return, which is either part of a proper quirk or the window head. The detail also creates a better shiplap joint at the wall panel.

Had the detailer utilized an isometric view as shown in Fig. 4.3.3, the alternative improved solution for this corner would have been obvious. It would have avoided a mitered corner for the eave panels and created a rounded corner matching that of the column. Had the precaster seen or made this isometric sketch, it might have suggested to him to produce this eave panel corner in a combined, one-piece column and corner mold.

Manufacturer's Shop Drawings (erection and production drawings) translate project contract drawings into usable information for accurate and efficient manufacture, handling and erection of the precast concrete units. They should be prepared in general conformance to the *PCI Drafting Handbook—Precast and Prestressed Concrete*.

The first step in preparing the shop drawings necessary for a project is a thorough review and restudy of the contract documents (plans and specifications) to determine all the factors that can influence decisions regarding the precast

Fig. 4.3.3 Detailing example — isometric view

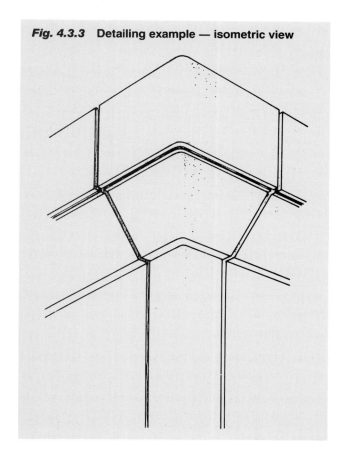

concrete. The goal of this analysis is to produce standardization of precast concrete units, modifications required of precast concrete units, connections, shop production techniques, handling methods, and erection. Aside from the general architectural shape requirements, the main factor in establishing standardization of the units is the building frame and its relationship to the architectural units (i.e. connection locations and clearances).

Erection drawings should include all precast concrete member piece marks, completely dimensioned size and shape of each member, the location of each member with respect to building lines and/or column lines and finished floors, and the details and locations of all connections from member to member or member to structure. Joints and openings between precast concrete members and any other portion of the structure should be identified. These drawings are not necessarily intended to show or describe procedures for building stability during erection. When temporary bracing is required, additional bracing drawings are recommended. These may show such items as sequence of erection, bracing hardware and procedures, and instructions on removal.

Some precasters prepare in-house shop tickets from shop drawings listing schedules of precast concrete units. Others produce separate drawings of each individual and different unit by means of sepia prints from typical master mold units. All of these need not be submitted for approval, although the architect may request record prints except of shop tickets. In this case, the number required should be stated in the specifications.

Only the different structural units and the typical units to be produced in each type of mold need be submitted for approval.

Specifications should state the number of copies required for approval or whether sepia prints are required. Current requirements usually call for two prints and one sepia of shop drawing be submitted for approval.

Generally shop (erection) drawings are submitted to the general contractor who, after checking them and making notations, submits them to the architect/engineer for checking and review. Final approved shop drawings are returned to the precaster by way of the general contractor. Timely review and approval of shop drawings and other pertinent information submitted by the precaster is essential. Fabrication should not commence until final approval or ''approved-as-noted'' has been received. If shape drawings are submitted separately, approval would allow fabrication of molds and tooling. Alternatively, shop drawings may be approved initially for mold production and subsequently for panel production.

Since mold production requires the greatest amount of production lead time, the common goal of both the architect and the precaster at the shop drawing stage is to determine all the details regarding the size and shape of the precast concrete panels for the most economical and efficient production sequence.

Because of the need for close coordination between the precast concrete manufacturer and stone veneer supplier, shop drawing preparation and submissions for stone veneer-faced precast concrete panels may vary from procedures established for non-veneered panels. It is suggested that the precaster detail all precast concrete units to the point where the fabricator of the veneer is able to incorporate details, sizes and anchor holes for the individual stone pieces.

Checking and approval of these details and shop drawings will be simplified and expedited if they can be combined and/or submitted simultaneously. Separate subcontracts and advance awards often occur in projects with stone veneered panels. While these procedures may affect normal submission routines, it is not intended that responsibilities for accuracy be transferred, or reassigned. In other words, the precast concrete manufacturer is responsible for precast concrete details and dimensions and the stone veneer fabricator is responsible for stone details and dimensions.

The manufacture of stone veneer panels requires adequate lead time in order to avoid construction delays. Therefore, it is important shop drawing approvals be obtained expeditiously. Furthermore, it is recommended that the designer allow the submission of shop drawings in predetermined stages to allow the start of manufacturing as soon as possible and ensure a steady flow of approved information to allow for uninterrupted fabrication.

The architect reviews the precast concrete manufacturer's erection drawings primarily for conformance to the specifications, then passes them along to the structural engineer for conformance to the specified loads and connection locations. This allows the engineer to confirm their own understanding of the forces for the structure at the connection points and the precaster's understanding of the project requirements. Design details, connection locations, and specified loads cannot be left to the discretion of the precaster. It is especially important in cladding panels, where a spandrel might weigh 20 tons and its torsion and shear must be taken into the frame.

Eccentricity of weight can cause deformation of the structure, see Fig. 4.2.10. To install the cladding within the specified tolerance, the structure must be stiff enough for both weight and bracing forces. Location of bracing for horizontal forces may be unfavorable.

"Approved" or "Approved-As-Noted" shop drawings means that the general contractor and architect/engineer have verified dimensions to be correct and final for the following: overall building dimensions, column centerlines, floor elevations, floor thicknesses, column and beam sizes, foundation elevations, the location of mechanical openings and other items pertinent to architectural precast concrete. It also means that the architect/engineer has reviewed the precast concrete connections and reinforcement as detailed by the precast concrete manufacturer, and that the architect/engineer takes responsibility for the strength and long term behavior of the precast concrete connections and reinforcement in service. The precaster should be responsible for the precast concrete units fitting the building dimensions and conditions as given on the final approved drawings.

Good shop drawings will provide the architect with a means of checking interfacing with adjacent building materials. They reduce plant costs and speed production by providing effective communications between the architect and the production/erection departments of the precast concrete manufacturer. The correct transfer of information from the contract documents is the responsibility of the precaster.

4.4 REINFORCEMENT

4.4.1 GENERAL

In designing architectural precast concrete panels, it is desirable that there not be any discernible cracking. In some cases, cracking may be permitted, but the crack width must be limited, see Section 3.5.20. When a reinforced concrete element is subjected to flexural tension, the amount and location of reinforcing steel has a negligible effect on member performance until a crack has developed. As stresses increase, hairline cracks may develop and extend a distance into the element. If cracks are narrow, the structural adequacy of the element will remain unimpaired.

In members in which concrete stresses during service are less than the allowable flexural tension, distributed reinforcement is needed to control cracking that may unintentionally occur during fabrication, handling or erection and also to provide ductility in the event of an unexpected overloading. In members in which the stresses are expected to be greater than the allowable flexural tension, conventional or prestressed reinforcement is required for satisfactory service load performance, adequate safety and meeting esthetic requirements. Reinforcement may serve either or both of these purposes in architectural precast concrete.

The types of reinforcement used in architectural precast concrete wall panels include welded wire fabric, bar mats, deformed steel bars, prestressing tendons and post-tensioning tendons. Non-prestressed reinforcement is normally tied or tack welded together into cages by the precast concrete manufacturer, using a template or jig when appropriate, unless the precast concrete unit is a simple flat panel. The cage, whether made for the entire casting or consisting of several sub-assemblies, must have sufficient three dimensional stability so that it can be lifted from the jig and placed into the mold without permanent distortion. Also, the reinforcing cages must be sufficiently rigid to prevent dislocation during consolidation in order to maintain the required cover over the reinforcement. The rigidity will normally improve with the tack welding and hence weldable grades of reinforcing steel are recommended. However, a designer should work with the grade of steel which is reasonably available to the precasters likely to bid on the project. Fig. 4.4.1(a) shows a typical, well-made reinforcing cage fabricated primarily of reinforcing bars for a loadbearing panel, while the cage in Fig. 4.4.1(b) consists mainly of wire fabric. The reinforcement cage must be carefully and accurately located and secured within the mold.

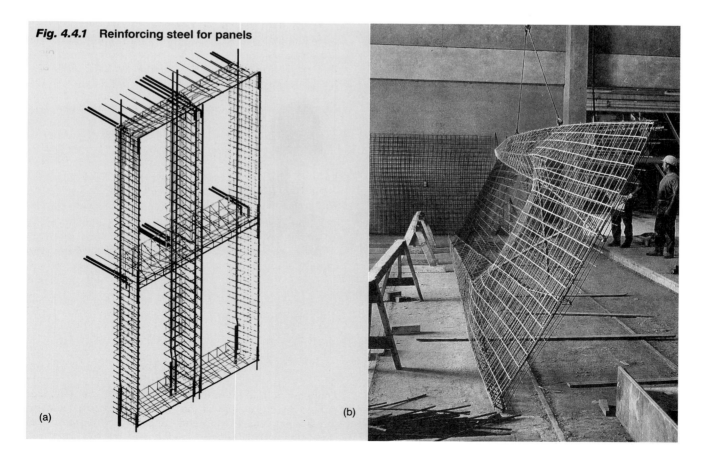

Fig. 4.4.1   Reinforcing steel for panels

(a)

(b)

The size of reinforcing bars is often governed by concrete dimensions and the required concrete cover over steel.

Reinforcement, in addition to that needed for structural reasons, is required to control thermal and moisture movement and the cracking that these might otherwise cause. As a general rule, bar sizes should be kept reasonably small even where this will reduce the spacing of the bars. Bars of small diameter, closely spaced, will control crack width and improve distribution of temperature stresses better than fewer, larger bars at a wider spacing. The use of additional reinforcing bars as compared with fewer heavier bars becomes more important in thin concrete sections. Since the sum of the widths of potential cracks in concrete are more or less constant for a given set of conditions, the more cracks there are, the smaller and less visible they will be.

The recommended maximum spacing of reinforcement in wythes exposed to the environment is 6 in. for welded wire fabric, or 3 times the wythe thickness (18 in. maximum) for reinforcing bars.

Large reinforcing bars create other problems as well. They often require anchorage lengths and hook sizes that may be impractical, making supplemental mechanical anchorage necessary. Since reinforcing bar termination points act as stress raisers and cause shrinkage cracks, it is often better to use welded cross bars or other types of mechanical anchorage when the bars are large relative to the concrete thickness, rather than rely on bond alone. Connection details with reinforcing bars crossing each other require careful dimensional checking to ensure sufficient cover.

Good bond between the reinforcing bar and the concrete is essential if the bar is to perform its functions of resisting tension and of keeping cracks small. Therefore, the reinforcing bar must be free of materials injurious to bond, including loose rust. Mill scale that withstands hard wire brushing or a coating of tight rust is not detrimental to bond.

The minimum reinforcement in each wythe should be 0.001 of the cross sectional area of the wythe in each direction, except as otherwise required by analysis or experience. Panels exposed to the environment which are less than 2 ft wide in one direction need not be reinforced in that direction — 4 ft. for panels not exposed to the environment.

The reinforcing steel, Fig. 4.4.1, should be placed as symmetrically as possible about the panel's cross sectional centroid. This is particularly true for flat and delicately shaped panels. Non-symmetrical placement may cause panel warpage due to restraint of drying shrinkage or temperature movements. Panels having a concrete thickness less than 6 in. may have reinforcement placed in one layer; however, two layers of equivalent weight are recommended to control concrete cracking. The cage should be checked to be certain that no reinforcing steel is touching the mold. Strict quality control is necessary to ensure adequate minimum cover is provided for all reinforcement, particularly in thin slab and delicately shaped units. Suitable production techniques should be utilized to maintain this

Fig. 4.4.2   Proper anchorage of reinforcing bars

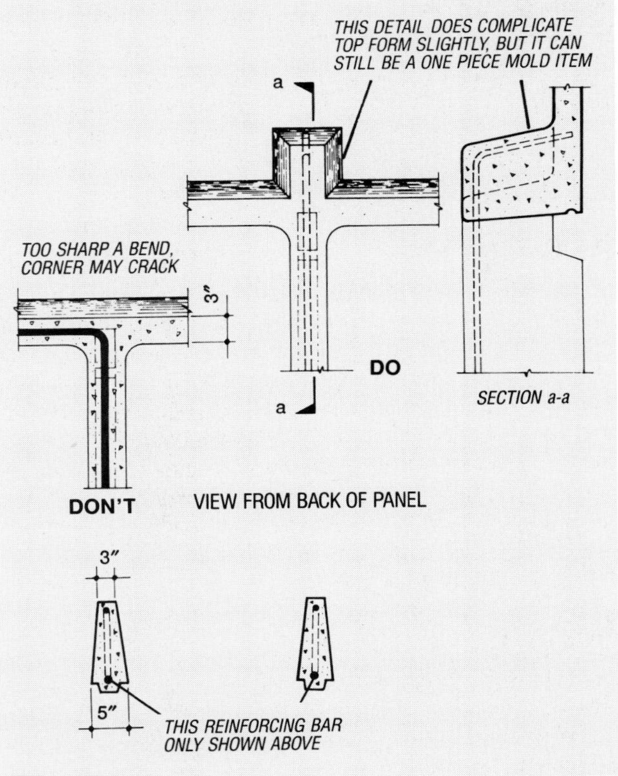

location during placement of the concrete. Reinforcing bars should not be bent sharply around corners especially in slender sections. Reinforcing bars should preferably be run straight through an intersection of bars to ensure proper anchorage, as illustrated in Fig. 4.4.2.

Normal stirrups are not recommended for slender sections. They cannot always be bent accurately or to a radius small enough to properly locate the main bars. Reinforcement in slender concrete sections may be detailed as shown in (b) and (c), Fig. 4.4.3, or consist of light wire fabric as shown in (d). Bending difficulties and tight clearance tolerances make the construction shown in (e) very difficult to achieve. Precast concrete units may contain attachment and lifting devices as well as prestress anchorages, along with their associated reinforcement. The congestion which these, in conjunction with the normal reinforcement, may cause should be given careful consideration.

If possible, reinforcing cages should be securely suspended from the back of the molds because spacers of any kind are likely to mar the finished surface of the panel. Metal chairs, with or without coating should not be used in a finished face. For smooth cast facing, stainless steel chairs are tolerable, if specifically permitted in the specifications. Chairs should preferably be of stainless steel, plastic tipped or all-plastic to ensure absence of surface rust staining.

An alternative to suspending reinforcement, or the use of chairs in flat panels, is to place the reinforcement on a layer

Fig. 4.4.3 Typical rib reinforcement

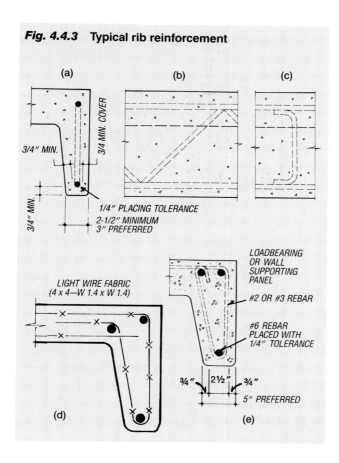

(a)   (b)   (c)

3/4" MIN.

3/4 MIN. COVER

3/4" MIN.

1/4" PLACING TOLERANCE
2-1/2" MINIMUM
3" PREFERRED

LIGHT WIRE FABRIC
(4 x 4—W 1.4 x W 1.4)

LOADBEARING
OR WALL
SUPPORTING
PANEL

#2 OR #3 REBAR

#6 REBAR
PLACED WITH
1/4" TOLERANCE

3/4" 2½" 3/4"

5" PREFERRED

(d)   (e)

of consolidated fresh concrete of the correct depth before placing the upper layer of the panel.

4.4.2 WELDED WIRE FABRIC

Welded wire fabric is the most common type of reinforcement used in architectural precast concrete. The fabric may be used as the main reinforcement, or strengthened with reinforcing bars in ribs and where otherwise required. Fabric used in architectural concrete will generally be required to comply with requirements of the ACI Building Code (ACI 318). Welded wire fabric is available in a wide range of sizes and spacings, making it possible to furnish almost exactly the cross sectional steel area required. The premium for deformed wire is about 3 percent of the price of fabric made from smooth wire. Special order sizes or types may increase costs and cause production delays. Two sizes, a heavy and a light wire fabric, will usually suffice for most architectural precast concrete. 4x4-W4xW4 (4x4-4/4) to 4x4-W1.4xW1.4 (4x4-10/10) are the most common sizes of welded wire fabric used by precast manufacturers. Specific geographical areas may have standardized on other sizes. The architect is advised to consult local precast concrete manufacturers for preference in type and size of welded wire fabric.

Welded wire fabric for architectural precast concrete is supplied in flat sheets. Fabric from rolls, if used in thin precast concrete sections, must be flattened to the required tolerances. Stock sizes of flat sheets vary, but often are 8 ft x 12 ft or 15 ft and sometimes 10 ft x 20 ft for special orders. Many precast concrete plants have the capability of accurately bending wire fabric to desired shapes, increasing its usefulness in large members. Because the wire fabric is closely and uniformly spaced, it is well suited to control cracking. Furthermore, the welded intersections ensure that the reinforcement will be effective close to the edge of the member, resisting cracking that may be caused by handling, and making it easier to repair damaged edges.

4.4.3 REINFORCING BARS

Deformed reinforcing bars are also used extensively in architectural precast concrete. However fabrication of mats and cages with reinforcing bars is often tedious and time consuming. Deformed reinforcing bars are hot-rolled from steels with varying carbon content. Deformed bars conforming to ASTM A615, *Deformed and Plain Billet-Steel Bars for Concrete Reinforcement*, are generally available in #3 through #11 in Grade 60. Selection of grades of reinforcing steel are determined by the structural design of the precast concrete units. For bars that are to be welded, ASTM A706, *Low-Alloy Steel Deformed Bars for Concrete Reinforcement*, specifies a bar with controlled chemistry that is weldable. Availability should be determined before ASTM A706 bars are specified.

Bars are classified by bar numbers, the number for any bar being its diameter in eighths of an inch. For example, a #4 bar has a nominal diameter of 4 eighths or 1/2 in. In detailing reinforcing in congested areas, it is important to note that the diameter of a bar, with deformations, is slightly greater than its nominal diameter.

4.4.4 PRESTRESSING STEEL

Prestressing may be used to minimize cracking of members by applying a precompression in the concrete that is greater than the internal tensile stresses. Prestressing may be either pretensioning or post-tensioning. In either case, the prestressing force should generally be concentric with the effective cross section. It is recommended that prestressing in a panel, after all losses, be limited to the range of 150 psi to 800 psi.

Prestressing of concrete panels may be used as partial reinforcement for the following reasons:

1. Structural requirements for in-service loads.

2. Units are to be suspended and it is desirable to maintain the concrete in compression. Units which are suspended should usually have light wire fabric reinforcing in both faces to counteract the effect of tensile stresses. Alternatively these units may be prestressed to provide a nominal residual compressive force.

3. Units are slender, and prestressing is the method chosen to facilitate handling without undue stresses.

4. For general crack control.

The belief that prestressing may reduce or control bowing or warping is, in some cases, a misconception. Unless tendons are located accurately (concentric with the effective cross section) and securely maintained in that location during casting and curing, prestressing may actually increase the occurrence of bowing. Some manufacturers prefer to have a slight inward bow in the in-place position to counteract thermal bow.

If panels are prestressed, anchorages (stressing pockets) must be recessed and packed with nonshrink mortar to provide the same cover to the reinforcement as at other panel locations. It may or may not be necessary to coat the strand ends with a rust inhibitor paint, such as a bitumastic, zinc rich paint, or epoxy, to avoid corrosion and possible rust spots which could mar the surface after years of service.

Welding in the proximity of prestressing steel should be avoided within a distance through which the spatter, direct heat, or short-circuited current may flow. The temperature rise due to the direct effect of welding heat or the indirect effect of current flow through the high tensile prestressing steel can cause a sudden strand rupture.

Numerous wire, strand, and bar prestressing materials are available that are suitable for architectural precast concrete. Strands are commonly used in both pretensioned and post-tensioned applications. They are manufactured in two grades to conform to ASTM A416, *Uncoated Seven-Wire Stress-Relieved Strand for Prestressed Concrete.* High-strength specially treated bars used as prestressing tendons are covered by ASTM A722, *Uncoated High-Strength Steel Bars for Prestressing Concrete.* Uncoated stress-relieved wires meeting the requirements of ASTM

A421 can be used in post-tensioned tendons. Unless prestressing is required for in-service load or performance, the choice of using this form of reinforcement should be left to the precast concrete manufacturer.

In order to minimize the possibility of splitting cracks in thin pretensioned members the strand diameter should preferably not exceed:

Panel thickness, in.	Strand diameter, in.
2 1/2	3/8
2 1/2 to 3 1/2	7/16
3 1/2 and thicker	1/2

The strand (tendon) most frequently used in architectural precast post-tensioned concrete is called the monostrand. Although monostrands can be fabricated to be grouted, they are usually coated with grease and covered with paper or plastic. Thus, they are typically used in the unbonded condition. If friction is exceptionally high due to length or curvature of the tendon, a strand coated with teflon and encased in a plastic tube is available. These monostrands have a very low coefficient of friction. When panels are post-tensioned, care must be taken to assure proper transfer of force at the anchorages. Provision for anchor plate protection against long term corrosion is essential. The anchorage area should be sealed immediately after the tendons or strand are post-tensioned. Straight post-tensioning cables or bars can be incorporated into the product, and this would generally require anchorages at both ends of the tendon. One method used to minimize the number of anchorages is illustrated in Fig.

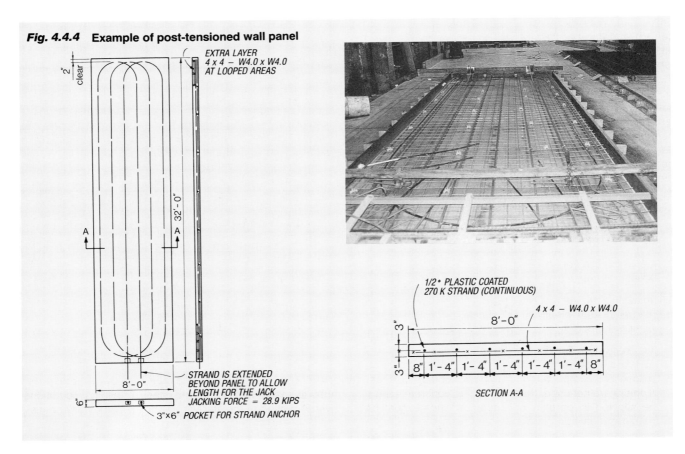

Fig. 4.4.4 Example of post-tensioned wall panel

EXTRA LAYER
4 x 4 — W4.0 x W4.0
AT LOOPED AREAS

2" clear

32'-0"

A — A

8'-0"

6"

STRAND IS EXTENDED BEYOND PANEL TO ALLOW LENGTH FOR THE JACK
JACKING FORCE = 28.9 KIPS

3"x6" POCKET FOR STRAND ANCHOR

1/2" PLASTIC COATED 270 K STRAND (CONTINUOUS)

4 x 4 — W4.0 x W4.0

8'-0"

3"

3"

8" | 1'-4" | 1'-4" | 1'-4" | 1'-4" | 1'-4" | 8"

SECTION A-A

4.4.4. Plastic coated unbonded strand with a low coefficient of angular friction ($\mu = 0.03$ to 0.05) are looped within the 8 ft by 32 ft panels. Anchorages are installed at one panel end only. Many architectural panels which do not lend themselves to being pretensioned because of difficulties with long line casting can be easily post-tensioned. The flat panels in Fig.4.4.4, the largest of which measured 8 ft. by 29 ft., were post-tensioned in the precaster's plant to provide optimum structural integrity for panel lifting, hauling and erection.

A significant design consideration is the evaluation of possible future unplanned openings. Cutting of an unbonded tendon will remove the effect of prestressing for that particular tendon. While it is unlikely that unplanned openings will be required, the designer must still be cognizant of this.

4.4.5 SHADOW LINES

Shadow lines (steel reflection) show up on some finishes depending upon mix design, placement of reinforcing steel, and consolidation of the concrete. Lack of concrete compaction in areas where steel or embed concentrations prevent full vibration of the concrete cover, is one type of shadowing. Another cause of shadow lines occurs when a minimal cover is used over the welded wire fabric or a rigid welded steel reinforcing cage is used. The fabric or cage will vibrate as an assembly in phase (resonates) during the intense vibration used to compact the low slump concrete used in architectural precast concrete panel production. During vibration of the concrete, the agitation is much more vigorous near the surface of the form under the steel. This agitation tends to cause a greater concentration of cement and fine particles of aggregate at the form interface under the steel. These fines scatter light better and tend to give the area a light color. This type of shadowing is, therefore, only a "skin deep" change in the concrete. The concrete beneath the steel surface at the ghosting is the same as the balance of the concrete. This ghosting problem may be aggravated by excess form-release agent being folded into the surface of the concrete during vibration. Mockup samples and early production panels should be checked carefully for indications of shadow lines. In most cases, modification of the manufacturing procedures will eliminate shadow lines. The following are approaches to reduce or alleviate shadowing:

1. Cages should be tied and not tack welded to reduce rigidity.

2. Galvanized steel should be treated with a chromate wash or chromic oxide added to mixing water.

3. To avoid settlement of the steel reinforcement with consequent loss of steel cover, it is often expedient to use plastic chairs or to support the steel from above.

4. Alternatively, a 2-in. layer of low slump concrete (slump as low as can be reasonably consolidated) should be placed and compacted prior to placement of the cage. If chairs are not used, the 2 in. layer of concrete should be allowed to partially set (adequate bond must be obtained between concretes) so that the initial lift is capable of keeping the steel grid from settling. When vibrating the top layer of concrete, care must be taken to avoid vibration of the steel caused by touching the cage with the vibrators.

5. If external vibration is used, steel supports should be isolated from the mold to prevent vibrations being transmitted into the cage.

If shadowing does occur, sandblasting with the smallest available gun and nozzle while using fine grit can reduce the reinforcement outline to a reasonably uniform tone. However, portions of the surface immediately over the reinforcement may be less dense than areas away from the reinforcement, making it extremely difficult not to over-erode the surface at the reinforcing steel.

4.4.6 TACK WELDING

Tack welding, unless done in conformance with AWS D1.4, may produce crystallization (embrittlement or metallurgical notch) of the reinforcing steel in the area of the tack weld. Tack welding seems to be particularly detrimental to ductility, impact resistance, and fatigue resistance. Tack welding affects static yield strength and ultimate strength to a lesser extent. Where a small bar is tack welded to a larger bar, the detrimental "notch" effect is exaggerated in the large bar. Fast cooling under cold weather conditions is likely to aggravate these effects. Tack welds which do not become a part of permanent welds of reinforcing steel are prohibited by AWS D1.4, unless approved by the engineer. However, tack welding of reinforcement at locations where neither bar has a structural function should be allowed. For example, welding the ends of the outside bars (within 10 bar diameters from the free end of the bar) may be an aid in fabrication of reinforcing cages. Reinforcing bars should not be welded within 2 bar diameters or a minimum of 2 in. with 3 in. preferred from a cold bend, Fig. 4.4.5, as this can result in crystallization and unpredictable behavior of the reinforcing bar at the bend. Tack welding must be carried out without significantly diminishing the effective steel area or bar area should be one-third larger than required. A low heat setting should be used to reduce the undercutting of the effective steel area of the reinforcing bar.

4.4.7 CORROSION PROTECTION

Protection of reinforcing steel from corrosion is obtained by providing proper embedment. A protective iron oxide film forms on the surface of the bar as a result of the high alkalinity of the cement paste. As long as this alkalinity is maintained, the film is effective in preventing corrosion. The protective high alkalinity of the cement paste is usually lost only by leaching or carbonation. Therefore, concrete of sufficiently low permeability with the required cover over the steel will provide adequate protection. Low permeability is characteristic of well-consolidated concrete with a low water-cement ratio and high cement content. This composition is typical of architectural precast concrete.

Hairline or structural cracking may allow oxygen and moisture to reach the embedded steel, providing conditions where rusting of the steel and staining of the surface may occur. A sufficient amount of closely spaced reinforcement limits the width of cracking and the intrusion of water, help-

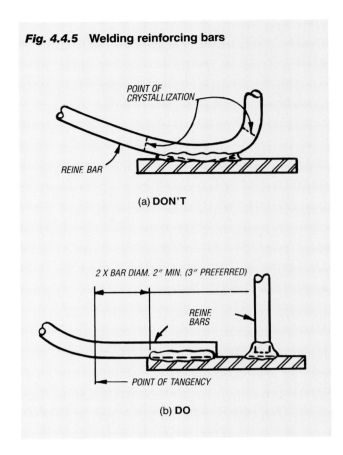

Fig. 4.4.5 Welding reinforcing bars

POINT OF
CRYSTALLIZATION

REINF. BAR

(a) **DON'T**

2 X BAR DIAM. 2" MIN. (3" PREFERRED)

REINF.
BARS

POINT OF TANGENCY

(b) **DO**

ing to maintain the protection of the reinforcement. Prestressing may also be used to limit cracking.

For concrete surfaces exposed to weather, nonprestressed reinforcement should be protected by concrete cover equal to nominal diameter of bars but not less than 3/4 in. for panel thicknesses equal to or greater than 3 in. Prestressed reinforcement should be protected by a 1 in. concrete cover.

Cover requirements over reinforcement should be increased to 1-1/2 in. for non-galvanized reinforcement or 3/4 in. with galvanized reinforcement when the precast concrete elements are acid treated or exposed to a corrosive environment or to severe exposure conditions. In addition, the 3/4-in. cover is realistic only if the maximum aggregate size does not exceed 1/2 in. and the reinforcing cage is not complex.

The 3/8 in. placing tolerance for reduction in cover of reinforcing steel may require a reduction in magnitude when the panel is exposed to weather. Where possible, excess cover of 3/8 to 5/8 in., depending on the degree of complexity of the cage, should be specified instead of a reduced tolerance because of the inaccuracy of locating reinforcing steel utilizing standard fabrication accessories and placing procedures.

For concrete surfaces not exposed to weather, ground or water, nonprestressed reinforcement should be protected by concrete cover equal to nominal diameter of bars but not less than 5/8 in. for panel thicknesses equal to or

greater than 3 in. Prestressed reinforcement should be protected by a 3/4 in. concrete cover.

A panel thickness of 3 in. is required to achieve a minimum cover of conventional reinforcement for normal climatic exposures. Precast concrete flat units should have a minimum thickness of approximately 3 times the largest aggregate used. For example, panels with 1 in. aggregate will require a minimum thickness of 3 in.

To provide too much cover is a mistake, although the ACI Building Code (ACI 318), Section 7.7, does not specifically say so, and 2 in. should be regarded as a maximum because concrete outside this limit lacks the restraint of the steel and is liable in consequence to have wide cracks.

Concrete cover refers to the minimum clear distance from the reinforcement to the face of the concrete. For exposed aggregate surfaces, the concrete cover to the surface of the steel should not be measured from the original surface. The depth of mortar removed from between the pieces of coarse aggregate must be subtracted to give a realistic measurement. Attention must also be given to scoring, false joints or rustications, and drips, as these reduce the cover. The required minimum cover should be measured from the thinnest location to the reinforcement.

Increasing the concrete cover increases the protection provided to the reinforcement. In determining cover, consideration should be given to the following:

1. Structural or nonstructural use of precast concrete member.

2. Maximum aggregate size (cover should be greater than the maximum aggregate size, particularly if a face mix is used).

3. The means of securing the steel in a controlled position, and maintaining this control during placement of concrete, and consolidation of the concrete mix.

4. Accessibility for placement of concrete, and consolidation of the concrete mix.

5. The type of finish treatment of the concrete surface.

6. The service environment (interior, or exposed to weather, ocean atmosphere, or aggressive industrial fumes).

7. Fire code requirements.

Galvanized Wire Fabric and Reinforcing Steel is often considered for architectural precast concrete panel construction. It must be recognized that galvanizing cannot take the place of quality control in mix proportioning and steel placement. Experienced and capable designers and precasters consider galvanized reinforcement a poor substitute for proper control of concrete durability. Sound concrete having strengths of 5000 to 6000 psi in 28 days and proper cover will typically provide all of the corrosion protection necessary for the reinforcing steel.

The term galvanized steel refers to the electrolytic method which bonds zinc to the steel surface. Today, zinc coating is usually applied by dipping the steel in a hot bath of zinc. The time or extent of dipping determines the thickness of the zinc coating and, thus, its durability in use. Galvanized wire fabric is generally manufactured from galvanized wire and is a stock item in some sizes. The premium usually is 15 to 20 percent over non-galvanized wire. Deformed wires are seldom galvanized. Some precast concrete manufacturers prefer galvanized wire fabric because rusting will be minimized during the prolonged storage which is often required by the economics of bulk purchasing. Other precasters avoid long storage by standardizing on two sizes only (heavy and light). The architect is advised to consult local precasters for preference in type and sizes of wire fabric. There is no ASTM specification for galvanized welded wire fabric. The amount of zinc coating on wire fabric is rarely specified for galvanized wire fabric reinforcement. Galvanized wire can be produced with thicknesses of zinc coating ranging from 0.30 oz. per sq ft to 2.0 oz. per sq ft for different grades and wire sizes.

Using zinc-coated reinforcement other than standard welded wire fabric in architectural precast concrete is a difficult and expensive process. Zinc-coating should be specified only after careful analysis. From a strictly technical point of view, there is seldom any real need for it. With proper detailing and specifications, galvanizing appears to be superfluous provided such specifications are upheld.

Galvanized welded wire fabric or reinforcing bars are often considered when minimim cover requirements cannot be achieved or when the concrete is exposed to a severe environment. However, a detrimental chemical reaction can take place when the concrete is damp and chlorides are present. Therefore, the benefit obtained by galvanizing is questionable for members subjected to de-icing salts and similar treatments.

When galvanized reinforcing steel is placed close to non-galvanized metal forms, the concrete may have a tendency to stick to the forms. This may also happen if non-galvanized reinforcement is used close to galvanized forms or form liners. Sticking may be prevented by passivating the surface of the galvanized steel reinforcement using a chromate treatment. This inhibits reaction between zinc and the alkalies of the concrete, (hydrogen gas development around the galvanized steel or on a steel mold face). Chromate coating of galvanized surfaces can be readily done in most galvanizing plants and consists of either dipping the galvanized elements for 10 to 20 seconds in an aqueous solution having a 2 percent solution of potassium or sodium dichromate or a 5 percent solution of chromic acid. Spraying may be used to dispense these solutions. The addition of chromic oxide (CrO_3) to the concrete mix (150 parts per million based on weight of mixing water or about 8 oz. of a 10 percent solution per cu. yd. of concrete) may also be effective in eliminating sticking and steel reflection (shadowing). The surface shadows shown by some finishes when galvanized reinforcement is used are unsightly. Early production pieces should be carefully examined for signs of this problem and the mix adjusted when necessary. Precautions must be taken when handling chromic

oxide to avoid dermatitis. Note: Connections, hardware fittings and other embedded items require similar consideration, if galvanized.

The fabrication of zinc-coated reinforcing bar cages may be approached in two ways. Both have disadvantages. Zinc coated reinforcing bars may be fabricated into cages by welding the bars together with special rods. However, the damage done to the zinc coating must be repaired with a cold zinc-rich paint (92 to 95 percent pure zinc). Alternatively, the cage may be fabricated using normal reinforcement and the entire cage hot-dipped after fabrication. Unfortunately, a large bath is required to zinc coat a complete reinforcing cage (some panels measure 10 ft x 45 ft). The process is quite costly. In both cases, transportation and multiple handling of the cage requires extra time and cost. Time delays may disturb the normal production sequence and impact the job completion schedule. Hot-dip galvanizing involves high temperatures, about 850 deg. F. This temperature may warp a rigid, asymmetric reinforcing cage and require cage adjustment afterwards. Furthermore, low grade steel becomes brittle at high temperatures. Zinc-coated reinforcing bars must conform to ASTM A767 *Zinc-Coated (Galvanized) Steel Bars for Concrete Reinforcement*. Class I (3.5 oz./sq ft) is normally specified.

When galvanized reinforcement is used in concrete it should not be coupled directly to ungalvanized steel reinforcement, copper, or other dissimilar metals. Polyethylene and similar tapes can be used to provide local insulation between any dissimilar metals. Galvanized reinforcement should be fastened with ties of soft stainless steel, zinc-coated or nonmetallic coated tie wire.

Epoxy-Coated reinforcing bars and welded wire fabric are also available for use in members which require special corrosion protection. Epoxy-coated reinforcing steel bars should conform to ASTM A775 and epoxy-coated welded wire should conform to ASTM A884. The epoxy provides excellent protection from corrosion if the bar is uniformly coated. Bars generally are coated when straight, and subsequent bending should have no adverse effect on the integrity of the coating. If the coating is damaged or removed during handling, the reinforcement should be touched up with commercially available epoxy compounds. Epoxy coating has a minimal effect on bond strength, but does increase the necessary development length somewhat. Supplemental items must also be protected to retain the full advantage of protecting the main reinforcement. For example, bar supports should be solid plastic and bar ties should be nylon-, epoxy-, or plastic-coated wire rather than black iron wire. Epoxy-coated reinforcing bars or fabric should be handled with nylon slings. The primary disadvantage of epoxy-coated reinforcement is that it adds significantly to the cost of the precast concrete products as the cost for the epoxy-coated welded wire fabric is 2 to 2-1/2 times plain fabric, and depends on size of the wires. The cost of epoxy-coated reinforcing bars ranges from 15 to 25% more per pound than plain bars.

4.5 CONNECTIONS

4.5.1 GENERAL

In addition to the design of the precast concrete units, the connections must be considered a major design factor influencing safety, performance, and economy. Many different connection details will result from the combination of the multitude of sizes and shapes of precast concrete and the variety of possible panel support conditions. A systematic analysis of the forces and movements of the panels and the supporting system is required to reduce these connection details to a manageable number.

Individual precasters have developed connections over the years which they favor, because they suit their particular production and/or erection techniques. Connections should comply with local codes. They may also be subject to functional requirements such as recessing for flush floors and/or exposed ceilings. These conditions lead to a variety of connections, nearly always custom-designed for each project and often defying any grouping or classification as "standard" connections. However, some basic concepts governing the design, performance and material requirements of connections can be formulated. These concepts have been divided into groups which may vary in importance and overlap, or may differ from one project to another. The designer should assess the relative importance of these concepts for the specific project and discuss the options with the precast concrete manufacturer or erector before finalizing the connections. Concepts governing the design and detailing of connections are considered under the following headings:

1. Design considerations, Section 4.5.2.

2. Handling and erection considerations, Section 4.5.3.

3. Handling and lifting devices, Section 4.5.4.

4. Manufacturing considerations, Section 4.5.5.

5. Connection hardware and materials, Section 4.5.6.

6. Protection of connections, Section 4.5.7.

7. Fastening methods, Section 4.5.8.

8. Supply of hardware for connections, Section 4.5.9.

9. Connection details, Section 4.5.10.

4.5.2 DESIGN CONSIDERATIONS

The primary purposes of a connection are to transfer load to the supporting structure and provide stability. Precast concrete connections must also meet design and performance criteria. However, all connections are not required to meet precisely the same criteria. These criteria include:

1. **Strength:** A connection must have the strength to safely transfer the forces to which it will be subjected during its lifetime. In addition to gravity loads, the forces to be considered include:

   a. Wind and seismic forces.

   b. Forces from restraint of volume change strains.

   c. Forces induced into wall panels by restrained differential movements between the panel and the structure.

   d. Forces required for stability and equilibrium.

2. **Ductility:** This is the ability to accommodate relatively large deformations without failure. In connections, ductility is achieved by designing so that steel devices yield prior to concrete failure.

3. **Volume change accommodation:** Restraint of creep, shrinkage and temperature change strains can cause severe stresses on precast concrete members and their supports. These stresses must be considered in the design, but it is usually far better if the connection allows some movement to take place, thus relieving the stresses.

4. **Durability:** When exposed to weather, or used in a corrosive atmosphere, steel elements should be adequately covered by concrete, painted, galvanized, or epoxy coated. Stainless steel is sometimes used, but with a substantial increase in cost. All exposed connections should be periodically inspected and maintained.

5. **Fire resistance:** Connections, which could jeopardize the structure's stability if weakened by fire, should be protected to the same degree as that required for the members that they connect.

6. **Constructability:** The following items should be kept in mind when designing connections:

   a. Standardize connection types.

   b. Avoid reinforcement and hardware congestion.

   c. Avoid penetration of forms, where possible.

   d. Reduce post-stripping work.

   e. Be aware of material sizes and limitations.

   f. Consider clearances and tolerances.

   g. Avoid non-standard production and erection tolerances.

   h. Standardize hardware items and use as few sizes as possible.

   i. Use repetitious details.

   j. Plan for the shortest possible hoist or crane hook-up time.

   k. Provide for field adjustment.

   l. Provide accessibility.

   m. Use connections that are not susceptible to damage in handling.

Regardless of whether an architectural precast concrete unit is used in a loadbearing or a non-loadbearing function, various forces must be considered in its design. For example, a cladding panel must resist its own self-weight, earthquake forces, when required, forces due to restraint of volume change or support system movement, and forces due to wind, snow and construction loads. If the panel is loadbearing, it must also resist and transfer the dead and live loads imposed on it by the supported structural members. These forces are transferred to the supporting structure through the architectural precast concrete panel's connections.

Bearing pads are sometimes used to distribute loads over the bearing area and to accommodate construction, fabrication and erection irregularities. These pads reduce the concentration of forces at the connection by deforming readily within their thickness or allowing slippage. The physical characteristics of bearing pad material necessary to satisfy this function are:

1. Permanence and stability.

2. Ability to equalize uneven surfaces and avoid point pressure.

3. Ability to accommodate movements.

The pad supplier or precast concrete manufacturer should be consulted when selecting bearing pads. The type and material required will depend on the imposed loads and the expected relative movements of the cladding and support structure. The two most satisfactory materials are:

1. Elastomerics with known compression, shear, and friction strength and known ability to deform with movements.

2. Plastics with low friction coefficients along with high compression and shear strength.

If significant movements are expected, soft pads or low friction rigid pads should be used. However, if relative movement is not expected, a bed of rigid material such as grout or drypack can be used to make a bearing connection.

A designer should always remember that simple statically determinant design concepts are preferred. Simple connections will usually perform best. One of the advantages of working with precast concrete is that connections may be designed for specific purposes, and, when properly designed, can be expected to perform accordingly.

The principles for the design of connections are relatively easy to follow where precast concrete units are supported on one level, at two points, hereafter referred to as **Load Support or Bearing Connections,** and held in with some degree of flexibility at other points, hereafter referred to as **Tie-Back or Lateral Connections.**

A common solution for floor to floor panels is to install load support connections near the bottom of the panel and place tie-back (lateral) connections at the top. Some designers prefer to have load support connections at the top and the lateral connections at the bottom. This is common for spandrels. Lateral support at an intermediate level for tall, thin panels such as column covers is possible. In all cases, the basic connection concepts are similar.

It is best to support the entire weight of the panel at one level. This is due to possible deflection of the supporting element. If supported by more than one floor, the varying deflections of supporting building frame members may cause the weight distribution to be indeterminate. Fig. 4.5.1 illustrates ten basic design principles.

Fig. 4.5.1 Design principles for cladding panel connections

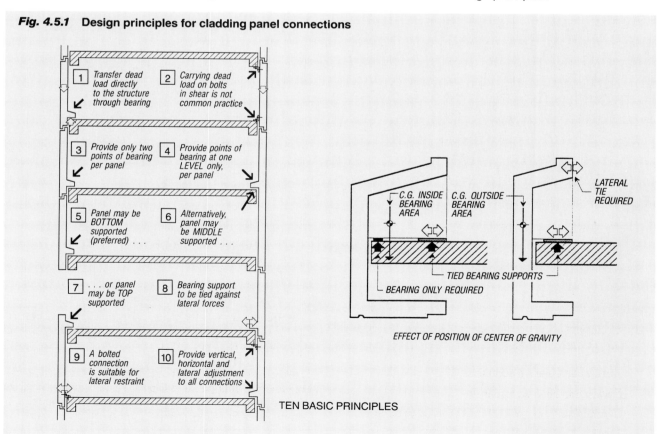

TEN BASIC PRINCIPLES

1. Transfer dead load directly to the structure through bearing
2. Carrying dead load on bolts in shear is not common practice
3. Provide only two points of bearing per panel
4. Provide points of bearing at one LEVEL only, per panel
5. Panel may be BOTTOM supported (preferred) . . .
6. Alternatively, panel may be MIDDLE supported . . .
7. . . . or panel may be TOP supported
8. Bearing support to be tied against lateral forces
9. A bolted connection is suitable for lateral restraint
10. Provide vertical, horizontal and lateral adjustment to all connections

C.G. INSIDE BEARING AREA
C.G. OUTSIDE BEARING AREA
LATERAL TIE REQUIRED
TIED BEARING SUPPORTS
BEARING ONLY REQUIRED

EFFECT OF POSITION OF CENTER OF GRAVITY

ARCHITECTURAL PRECAST CONCRETE   **187**

The arrangement and size of cladding elements with reference to the grid of the support system can vary. Since the panel size and the number and spacing of the connection points all influence the design, an optimum solution is desirable. In general, the largest possible size of panel with a minimum number of connections is the most economical, subject to limits imposed by handling, shipping, crane capacity and loads on the support system.

Fig. 4.5.2 illustrates schematically solutions for different configurations of precast concrete units. (a) represents a typical (floor to floor) wall unit. (b) is a unit with a width less than six to eight feet, or narrow enough to disregard the horizontal restraint of the load supports. (c) shows a unit of such width that two intermediate lateral connections have been utilized.

The designer should provide simple and direct load transfer paths through the connections and ductility within the connections. This will reduce the sensitivity of the connection and the necessity to precisely calculate loads and forces from, for example, volume changes and building frame distortions. The number of load transfer points should be kept to a practical minimum. It is highly desirable that no more than two connections per panel be used to transfer gravity loads, unless all are designed to carry substantially greater but indeterminate loads. Regardless, the bearing points should all occur at the same level. Load transfer should always be as direct as possible.

The impact loads associated with handling and setting pre-cast concrete units may double the dead load used in the design of a connection. The magnitude of the impact loading is dependent upon the methods and controls of hoisting and the vulnerability of the connection (or its anchorage) to damage from impact loads. Where connections are designed for loads equal to or exceeding the impact loads, the requirements for impact have been automatically satisfied.

In **High Seismic Areas,** the most common application of architectural precast concrete is as non-loadbearing cladding. The Uniform Building Code requires that "Precast or prefabricated non-bearing, non-shear wall panels or similar elements which are attached to or enclose the exterior shall be designed to resist the (inertial) forces and shall accommodate movements of the structure resulting from lateral forces or temperature changes." The force requirements often overshadow the importance of allowing for moisture and thermal movement. Panels typically have two rigid loadbearing connections with volume change relief provided only by the ductility of the connections, and two or more tie-back connections with full freedom of movement in the plane of the panels.

Ductility may be described as the ability of a material in the connection to stretch or "give a little" when overloaded, without failing and causing resultant additional overstresses within the supporting structure. Connections should be designed such that if they were to yield, they would do so in a ductile manner, without loss of load-carrying capacity.

Connections and joints between panels should be designed to accommodate the movement of the structure under seismic action. Connections which permit movement in the plane of the panel for story drift by bending of steel, properly designed sliding connections using slotted or oversized holes, or other methods providing equivalent movement and ductility are also permissible. Story drift is defined as the relative movement of one story with respect to the stories immediately above or below. Between points of connection, non-loadbearing panels should be separated from the building frame to avoid contact under seismic action. Story drift must be considered when determining panel joint locations and sizes, as well as connection locations and types.

The Uniform Building Code requires allowance for "story drift." This required allowance can be 2 in. or more from one floor to the next and may present a greater challenge to the designer than the forces. This (UBC) requirement is in anticipation of frame yielding to absorb energy. The isolation is achieved using slots or (more often) long rods which flex. The rods must be designed to carry tension and compression in addition to the induced flexural stresses. In the case of floor to floor wall panels, the panel is usually rigidly fixed to and moves with the floor beam nearest the panel bottom, see Fig. 4.5.3a. In this case, the upper attachments become isolation connections to prevent the building movement forces from being transmitted to the panel, thus the panel translates with the load supporting beam. Some designers prefer to support the panel at the top and put the isolation connections at the bottom.

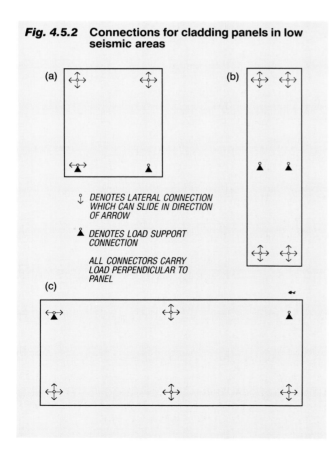

Fig. 4.5.2   Connections for cladding panels in low seismic areas

(a)   (b)

↕ DENOTES LATERAL CONNECTION WHICH CAN SLIDE IN DIRECTION OF ARROW

▲ DENOTES LOAD SUPPORT CONNECTION

ALL CONNECTORS CARRY LOAD PERPENDICULAR TO PANEL

(c)

Fig. 4.5.3 Panel connection concepts

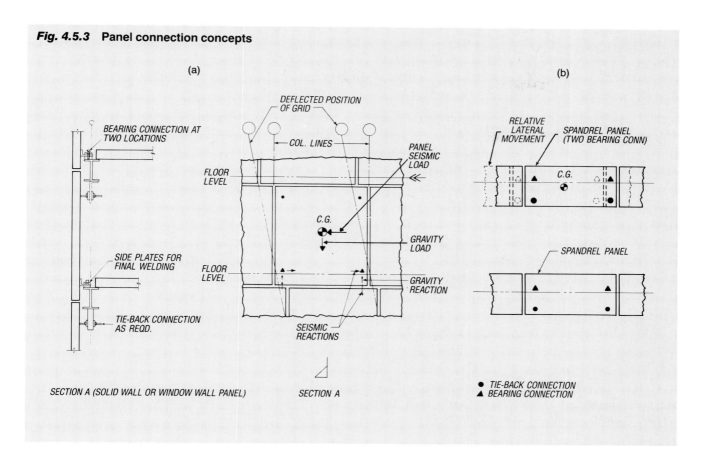

(a)

(b)

BEARING CONNECTION AT
TWO LOCATIONS

SIDE PLATES FOR
FINAL WELDING

TIE-BACK CONNECTION
AS REQD.

SECTION A (SOLID WALL OR WINDOW WALL PANEL)

DEFLECTED POSITION
OF GRID

COL. LINES

FLOOR
LEVEL

C.G.

FLOOR
LEVEL

SEISMIC
REACTIONS

SECTION A

PANEL
SEISMIC
LOAD

GRAVITY
LOAD

GRAVITY
REACTION

RELATIVE
LATERAL
MOVEMENT

SPANDREL PANEL
(TWO BEARING CONN)

C.G.

SPANDREL PANEL

● TIE-BACK CONNECTION
▲ BEARING CONNECTION

Spandrel panels usually have the loadbearing connections at the top of the floor beam with the tie-back (also known as push-pull or lateral or stay) connections located and attached to bottom of the same floor beam, see Fig. 4.5.3b. In this instance the tie-backs are not affected by story drift since the top and bottom of the floor beam move together.

If the panel or column cover is narrow, the connection system is sometimes chosen to have both the top and bottom of the panel move with their respective floors and force the panel to rotate or rock up on one of the two loadbearing connections, Fig. 4.5.4a. Since the movement occurs in both directions, each loadbearing connection must have the capacity to carry the full weight of the element without becoming tied down. Vertical movement, such as allowed with slots, must not be restricted as the panel rocks back and forth.

The connection system determines panel movement. In Fig. 4.5.4a, seismic reactions at top together with "lift off" allowance of bottom connections allow panel to rotate with its entire weight being carried on one lower connection. In Fig. 4.5.4b, all vertical and inplane horizontal loads are carried near panel c.g. with connectors that keep it plumb and make it translate with the connected floor. The upper and lower tie-backs must tolerate the drift.

These movement capabilities must not be compromised with the need for adequate production and erection tolerances. If tolerances were ± 1/2 in. and drift allowance was ± 1 in., a slot length of 3 in. plus the bolt diameter would be required.

It is essential that the types of movement (e.g., translation or rotation) be studied and coordinated not only with the connection system but with the wall's joint locations and joint widths. For example, if a rotating column cover occurs between translating spandrel panels, the joint width must accommodate the amount of rotation that would occur in their common height. Such considerations may govern the connection system or the wall's joint locations.

For seismic forces, the Uniform Building Code requires that the body of the connector be designed for a force equal to one and one third times the required panel force and that the body be ductile. The code requires that all fasteners be designed for four times the required panel force. The anchorage to the concrete is required to engage the reinforcing steel in such a way as to distribute forces to the concrete and/or reinforcement and avert sudden or localized failure. The Code does recognize the advantage of this in calculating anchor strength. The engagement details are left to the designer. Since the force distribution philosophy is critical to seismic design and performance, it leads many designers to specify confining hoops (such as UCS5, Fig. 4.5.64, page 217), deformed bar anchors, or long reinforcing bars welded to plates, rather than headed studs or inserts. With appropriate orientation, the reinforcing anchors will act in tension with optimum efficiency. If studs are used and loaded near the edge of the concrete panel, it is recommended that they be enclosed in sufficient reinforcing steel to carry the loads back into the panel so a sudden tensile failure mode in the concrete is averted.

Fig. 4.5.4 Tall/narrow units

(a)

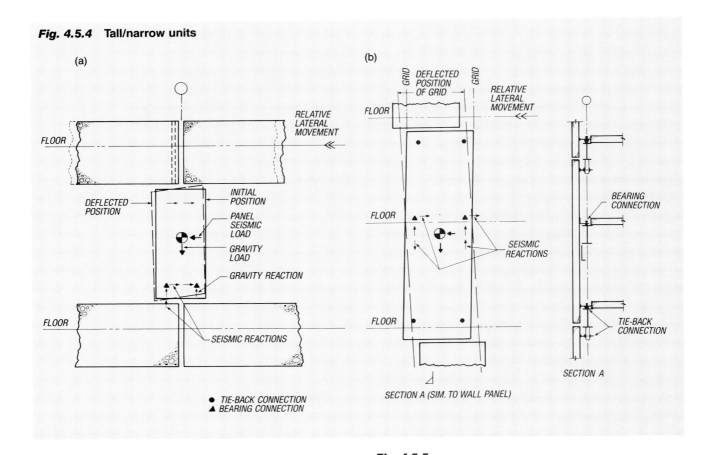

RELATIVE LATERAL MOVEMENT

FLOOR

DEFLECTED POSITION

INITIAL POSITION

PANEL SEISMIC LOAD

GRAVITY LOAD

GRAVITY REACTION

FLOOR

SEISMIC REACTIONS

● TIE-BACK CONNECTION
▲ BEARING CONNECTION

(b)

GRID
DEFLECTED POSITION OF GRID
GRID

RELATIVE LATERAL MOVEMENT

FLOOR

FLOOR

SEISMIC REACTIONS

FLOOR

BEARING CONNECTION

TIE-BACK CONNECTION

SECTION A

SECTION A (SIM. TO WALL PANEL)

When possible, it is advantageous to arrange concrete anchor studs so that the ones that carry tension due to gravity do not have to carry tension due to seismic forces. For example, if the horizontal leg of the angle bracket in EB6, Fig. 4.5.30, (page 208) and EB7, Fig. 4.5.31 were welded (via shear plate) to the structure plate to carry seismic load perpendicular to the panel, the bracket/stud arrangement of EB7, Fig. 4.5.31 would be preferred. In EB6, Fig. 4.5.30 the lower studs would be in tension due to gravity loads as well as seismic forces. In EB7, Fig. 4.5.31 the lower studs would carry the gravity tension, whereas the upper studs would carry the seismic force. Vertical shear would be carried by all studs in both cases.

In many cases, the wall panels are sufficiently outboard of the supporting frame, to require either outriggers off the beam or long panel brackets, as in EB 1, Fig. 4.5.25. For seismic forces in the plane of the panel, anchorage of the longer panel brackets to the panel can become quite cumbersome since the forces must be combined with gravity. If a shear transfer plate is added to the system, such as SP1, Fig. 4.5.56, the bracket anchorage problem becomes more manageable because the SP carries the seismic forces.

The panel shown in Fig. 4.5.5 illustrates load support connections for medium size units in earthquake Zone 3. This is an example of precast concrete units serving only as rain barriers, with the exterior cast-in-place shear wall serving as an airseal. The load supports were placed in recesses at the windows, making them readily accessible, Fig. 4.5.6. Following panel installation, these recesses were con-

Fig. 4.5.5

creted to complete the exterior airseal and fireproofing.

It is important to coordinate the design and detailing of connections with other functions, such as production, erection, tolerances and joints.

The architect should consider the location of horizontal joints between precast concrete panels as an integral part of the evaluation of economical fastening of the units. Different joint locations and corresponding connection concepts are shown in Fig. 4.5.7. (a) illustrates a common solution where joints occur just below floor level. This results in simple connections because this joint location normally allows sufficient concrete cover and anchorage. (b), (c),(d) and (e) show schematic solutions for connections with other joint locations. In (d) a brace, as shown dotted, will be required where dimension 'X' becomes too small to safeguard against possible rotation of the units. The condition in (e) will, in many cases, demand additional temporary bracing between floors, which may result in cumbersome erection procedures, since the connection at the top of the lower panel cannot be made directly to the building frame.

The proper accommodation of thermal movements in the wall is a major design consideration. In non-loadbearing units, such movements should be allowed to take place in individual units with no (or a minimum) effect on adjacent units. Movements caused by long term shrinkage (after a reasonable curing time) and creep are normally insignificant in cladding units, but these movements should be considered in the design of other precast concrete ele-

ments. The *PCI Design Handbook—Precast and Prestressed Concrete* supplies procedures for estimating such movements.

The flexibility required in the vertical and horizontal directions to prevent restraint at lateral connections may be accomplished with sliding through the use of connections with slotted or oversized holes. Low friction washers or sleeves slightly longer than the slotted receiver are sometimes used to ensure better performance. Long or medium length rods or bolts can bind instead of slide when load is applied at the far end. If nuts are used at sliding connections, they must be prevented from tightening or loosening with movement. This can be accomplished by using jamb nuts, patent nuts, or punched threads; by using a proprietary liquid thread locker; or by tack welding the nut to the tie-back rod (not at the stressed side) or a square plate washer large enough to have its rotation limited by an adjacent return; or by using a separate stub bar which would hit a stop if it were to turn too far. Alternatively, a connection that allows movement is a long ductile rod which has the potential to deform in vertical and horizontal directions. Long rods which flex with drift are the most common and the least prone to malfunction. However, care must be taken to ensure that the tie-back connection can satisfactorily resist panel forces perpendicular to the plane of the panel without excessive deflection or stress.

Connections and locations of connections should be designed to accommodate tolerances between the con-

Fig. 4.5.6

Fig. 4.5.7 Joint locations and corresponding connection concepts

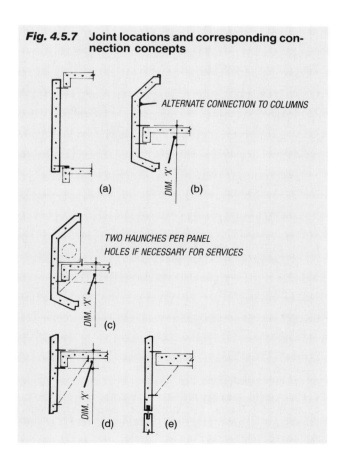

necting members consistent with industry manufacturing and erection tolerance standards and the specified tolerances for the supporting structure. These considerations may require clip angles and plates with slots or oversize holes to compensate for dimensional variation, as well as field welding or sufficient shim spaces to allow for elevation variations. Adequate clearance must be provided between precast concrete units and the supporting structure to allow for product and erection tolerances. Connection hardware should be designed to allow for stress occurring at maximum anticipated clearances and tolerances.

Ample tolerances must be provided in sizing connection materials to allow for both production and field tolerances. For example, to connect a 4 in. plate, it is recommended that the cast-in connection plates be oversized to 6 in. When detailing bolted connections, oversized holes in the connection plate or angle at least twice the size of the bolt but no less than 1 in. larger than the bolt diameter should be provided. A plate washer must also be used.

The three dimensional aspects of tolerances for structures—variations in length, height and plumbness—must be considered together with possible rotations and irregularities of the support surface. The clearance necessary for erection of panels will depend on the design, the dimensional accuracy of the support system and the limits of adjustment permitted by the connection details.

A conscious recognition of the difference between required adjustability for tolerance, and the need to prevent or allow subsequent movement is necessary. This sometimes appears to present conflicting requirements, but they can be provided for when treated individually. For example, adjustment can be accomplished with an oversized hole, and subsequent movement controlled with a plate washer welded over it. Washers used for this purpose can usually be welded on any two sides. If the washer is slotted, it can take load on one axis and allow movement on the other. The washer could be welded under the piece with an oversized hole to reduce bending in the bolt. A hole off center both ways in a plate washer allows maximum adjustment in a minimum space, since it can be rotated as necessary.

If connections must be located at the exterior face of spandrel beams or columns, more clearance will be needed to complete the connections than when the connections are located on the top surface of beams and the sides of the columns. Enough space must be provided to make the connection—sufficient room for welding or adequate space to place a wrench to tighten a bolt. Where a 1 in. clearance is needed but a 2 in. clearance creates no structural or architectural problems, the 2 in. clearance should be selected. When fireproofing is used, clearances should be planned from the face of the fireproofing material.

All connections should be provided with the maximum adjustability in all directions that is structurally and architecturally feasible. In larger more complex structures, vertical adjustments of 1 in. minimum and horizontal adjustments of 2 in. minimum should be provided to accommodate any misalignment of the support system and the precast concrete units. Slotted holes, welds, shim plates, leveling bolts and flexible steel sections are used to obtain connection adjustability at the time of installation of the cladding. The adjustability of the connections with regard to the support system should not change the location and width of the joints between the panels.

Connection details should consider the possibility of bearing surfaces being misaligned or warped from the desired plane. Adjustments can be provided by the use of drypack concrete, non-shrink grout, or elastomeric pads.

Where a unit is not erected within the tolerances assumed in the connection design, the structural adequacy of the installation should be checked, and the connection design should be modified, if required. No unit should be left in an unsafe support condition. Adjustments or changes in connections which could induce additional stresses in the unit or connection assembly (other than adjustments within the prescribed tolerances) should only be made after design analysis.

The importance of realistically assessing variations in the building lines and in the location of contractor's hardware prior to designing connections is illustrated in Fig. 4.5.8. In designing this connection for bending at the panel face, a theoretical cantilever arm of 3 in. is indicated. However, the clearance between the panel and the structure may become 2 in. and the erector may inadvertently place the shims at the extreme end of the bar (or angle) making the cantilever arm 5 in.. This increase in the arm would cause a significant increase in stresses in the bar.

The type of connections shown in Fig. 4.5.8 with bars of different shapes or angles (gusseted if required) are common load support connections. Such connections are anchored to the concrete using mild steel reinforcing dowels, studs, loops of flat bars, angles, or a combination of these. The connection assembly must be located accurately in the precast concrete unit to ensure proper functioning. Where angles are used for these connections and fastened to the concrete with bolts and inserts, the concrete surface behind the angle should be flat and well finished for proper bearing. The inserts should be placed perpendicular to this surface. Such angles may also be anchored with studs or welded to plates in the precast concrete units.

Embedded anchors, inserts, plates, angles, and other cast-in items should have sufficient anchorage and embedment for design requirements. Under no circumstances should main reinforcement be eliminated to accommodate hardware. Proper reinforcement (confinement) of concrete close to anchors, on which hardware units rely for anchorage, should also be provided and maintained.

When designing a connection, it is important to verify that the reinforcement (both prestressing strand and conventional bars) allows proper casting and vibrating of low-to medium-slump concrete in and around the connection region. When large quantities of reinforcement cross each other, honeycomb and voids must be prevented. Such production problems can be minimized by checking the connection region for dimensions and clearances prior to cast-

Fig. 4.5.8 Dimensions for design of connections

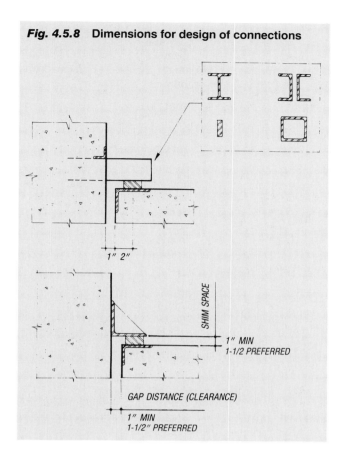

1" 2"

SHIM SPACE

1" MIN
1-1/2 PREFERRED

GAP DISTANCE (CLEARANCE)

1" MIN
1-1/2" PREFERRED

ing. Half-size or full size drawings are helpful in eliminating such potential problems. In making these drawings it is important that reinforcing bar bends be drawn to true scale.

Particular care must be taken in the design and detailing stage to properly size and locate reinforcing bar and stud anchorages. Large bars may be impractical due to their longer required development length, or the difficulty in obtaining proper bend geometry to conform to the connection hardware.

The design of the panel connection and the supporting frame are interdependent. For example, while structural steel frames tend to have better dimensional control, rotation and deflection of the support beams is much more common in steel frames than in concrete frames. Thus, connections for cladding should be designed to transmit the vertical loads to the longitudinal centerline of the steel beam, or provisions should be made to minimize torsional stresses induced into the supporting member by specifying heavier members, braces, or gussets. It is usually desirable to concentrate beam loads at or near the column to reduce deformations.

If the building frame members do not provide rotational resistance (this applies more to structural steel than to concrete), they must be braced to resist these forces or the connection at the precast concrete panel must develop moment capacity. Casting a structural steel section directly into the panel is an economical solution for connections

subjected to significant bending. As an alternate, a structural steel section can be welded to anchor plates which are bolted to or cast into the panel. Concrete haunches or corbels also provide a solution for heavy bending. Erection of panels with corbels is fast and simple, and tolerances can be easily accommodated. However, corbels take up space in the building.

If the deflection of the structural frame is sensitive to the location or eccentricity of the connection, limits should be given. Consideration should also be given to both initial deflection and long-term deflections caused by plastic flow (creep) of the supporting structural members.

Specific allowances cannot be assigned to erection deformations. Yet, by selecting ample clearances and fully considering realistic variations, the effects of erection deformations can be minimized. However, for multi-story or long-span structures, it is suggested that erection deformations be reviewed relative to the influence upon design of the members and connections.

Connections for loadbearing wall panels are an essential part of the structural support system, and the stability of the structure may depend upon them. Loadbearing panel connections should be designed and detailed in the same manner as connections for other precast structural members. Designers are referred to an extensive treatment of design methods in the PCI *Manual on Design and Typical Details of Connections for Precast and Prestressed Concrete* and *PCI Design Handbook—Precast and Prestressed Concrete.* Erected loadbearing walls may have horizontal and/or vertical joints across which forces must be transferred.

Horizontal joints in loadbearing wall construction usually occur at floor levels and at the transition to foundation or transfer beams. These joints may connect floors and walls or wall units only. Horizontal joint and connection details of exterior bearing walls are especially critical, since the floor elements are usually connected at this elevation and a waterproofing detail must be incorporated. Vertical joints may be designed so that the adjacent wall panels form one structural unit (coupled) or act independently. Typical loadbearing panel connections are shown in Section 4.5.10. The stability of the structure during construction must be considered when planning the erection procedures. Therefore, temporary guying and/or bracing must be provided until final structural stability is achieved in the completed structure. This bracing design is the responsibility of the precast concrete erector and should be shown on a bracing plan prepared by the erector. It sometimes requires review by the engineer of record and building official.

Load support connections are often made using steel base plates. Lateral connections can allow rotation (pin connections) or be rigid (moment connections), depending upon the structural system.

Fig. 4.5.9 shows one type of base plate connection. Fig. 4.5.10 illustrates a base connection with one arrangement of bars used in earthquake zones. This is an example of a connection where floors and roofs framing into precast concrete panels are cast-in-place concrete construction.

Fig. 4.5.9

The spandrel panels in Fig. 4.5.11 have simple concrete haunches cast as an integral part of the loadbearing panels; these haunches support the precast, prestressed concrete double tee floor and roof slabs. Blockouts in panels can also be used to support floor members, Fig. 4.5.12. Such pockets greatly decrease torsion stresses and minimize twist and eccentricity during erection, see Section 4.5.10 under Slab to Wall Connections.

The steel beam connection formed by two steel channels in Fig. 4.5.13 illustrates an example of exposed connections expressed as an architectural feature.

Fig. 4.5.11

Fig. 4.5.10

Fig. 4.5.12

Fig. 4.5.13

4.5.3 HANDLING AND ERECTION CONSIDERATIONS

In order to achieve the optimum overall economy of a precast concrete project, it is important to plan for a minimum of handling and erection time at the jobsite. This planning involves investigation of handling and erection procedures in order to minimize erection difficulties. Selection of suitable connections is also essential.

Even though the precaster may not be directly involved in the erection, the precaster is concerned with the overall economy of the project. The precaster should be allowed to play an active part in the design of connections for the purpose of achieving efficient erection, since the optimum solution may result in extra materials and/or special requirements in the manufacturing process. Efficient and economical plant manufacturing of precast concrete cannot be a goal in itself, but should be weighed equally with economic erection operations. These depend heavily on proper connections.

The architect should also understand how erection requirements may affect design. The type of connection details selected by the designer can drastically affect erection efficiency. Thus, it is important that the designer discuss the use of safe, efficient, and economical lifting hardware and connections to facilitate erection with the local precast concrete manufacturer and the erector as early in the project planning stage as possible. The precast concrete manufacturer or erector may prefer certain details or procedures not anticipated by the designer. Allowing alternate solutions will often result in more economical and better performing connections.

Hoisting and setting the precast concrete pieces is usually the most expensive and time-critical process of erection. It is desirable to have connections which are designed so that the erector can safely secure the member to the structure in a minimum amount of time without final aligning of members or totally completing the connections. If necessary, temporary bracing or connections should be used, with final adjustment and alignment in all directions relative to the structure and adjacent components completed after release of the crane. This is particularly important if the permanent connection requires field welding, grouting, drypacking, or placement of cast-in-place concrete. The temporary connection should not interfere with or delay placement of subsequent members. The temporary connections may have to be relieved or cut loose prior to completion of the permanent connections.

A certain amount of vertical adjustment for alignment at the connection is normal. Precast concrete erectors have a number of different solutions to this problem. Two common approaches for panels are illustrated in Fig. 4.5.14 as (a) and (b).

If the erector knows the exact panel dimension "D" for each panel and the vertical dimensions of panels are held to close tolerances, a survey crew may be used to pre-set shims to level "F" before lowering panels to floor level. Adjustments may still be required for some panels. But, with an instrument ready for checking the proper levels, the erector may have to hold the odd panels with the hoisting equipment long enough to make any required adjustment in the shims or if the panel isn't too heavy it can be lifted with a pry bar enough to adjust the shims. As part of the preliminary layout, the erector should also mark the theoretical centerlines of the joints so that the panels may be centered between these lines.

Another solution for vertical adjustment is the assembly shown in (b) of Fig. 4.5.14, where a leveling bolt is used to facilitate such adjustments. With leveling bolts the panel is usually set a little high and lowered to correct elevation with a wrench. The bolts may get quite large for heavy panels and require a rounded point to prevent "walking" when turned and also to ease horizontal adjustments. Sometimes a head down carriage bolt is used and adjusted with a wrench. The choice between these procedures and alternate ones should be at the option of the precaster and/or erector.

Connections should be planned so that they are accessible to the worker from a stable deck or platform. The type of equipment necessary to perform such operations as welding, bolting, post-tensioning, and pressure grouting should be considered. Operations which require working under a deck in an overhead position should be avoided, especially for welding. Any erection techniques that require temporary scaffolding should be avoided, if possible. Room to place wrenches on nuts and turn them through a large angle should be provided for bolted connections. Foundation piers should extend above grade so that the anchor bolts

Fig. 4.5.14 Vertical alignment methods

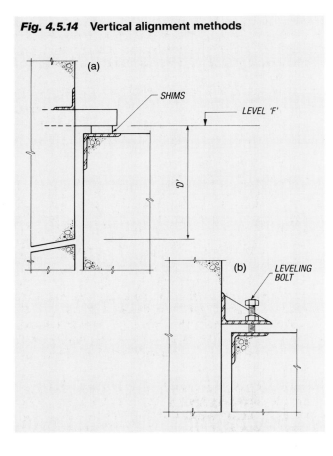

Fig. 4.5.15 Selection of hardware items

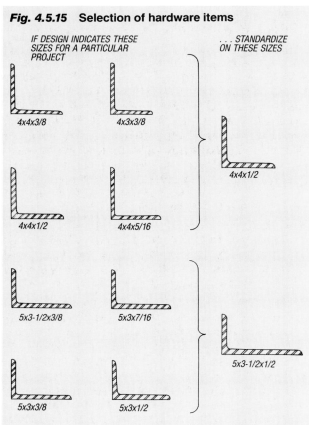

IF DESIGN INDICATES THESE SIZES FOR A PARTICULAR PROJECT

. . . STANDARDIZE ON THESE SIZES

4x4x3/8
4x3x3/8
4x4x1/2

4x4x1/2
4x4x5/16

5x3-1/2x3/8
5x3x7/16
5x3-1/2x1/2

5x3x3/8
5x3x1/2

can be readily adjusted during erection and the base plate need not be grouted in a hole, which can fill with water and debris. Properly drypacking column or wall panel bases in a narrow excavation is difficult.

All connections which serve similar functions within the building should be standardized as much as possible. As workers become familiar with the procedures required to make the connection, productivity is enhanced, and the potential for error is reduced. In general, heavier hardware should be used for a lightly loaded unit on the project if it will eliminate a special piece, and result in only slightly more material. This practice prevents the possibility of the lighter hardware being mistakenly used where the heavier one is required, Fig. 4.5.15. Furthermore, standardization of details facilitates selection and shipment of connection items to the project with fewer delays and added economies. Standardization can be used in still another context. With rare exception, all the items, materials, and procedures involved in making connections should be standard to the industry and readily available in the local area. Whenever possible, such items as field bolts and loose angles should be of common sizes for all connections. With bolted connections, 3/4 and 1 in. diameter bolts are most commonly used in the precast concrete industry. It is essential to use National Coarse standard threads or coil threads for bolted connections.

Standardization also applies to dimensioning of connection details. Little is gained by slight changes in dimensions, since the savings in materials may be more than offset by the extra labor involved in developing the modifications. If the changes in dimensions and materials in connections are not in increments large enough for visual recognition, there is greater chance that an improper component may be used at a given location.

Some types of connections require skilled craftsmen to accomplish, for example, welding and post-tensioning. The connection will be more economical, with fewer of these skilled trades required.

Efficient connections use a minimal number of pieces of hardware. For instance, a loadbearing connection should be utilized as the tie-back, when possible, rather than using a separate piece of hardware to perform that function.

Connection details should allow erection to proceed independently of ambient temperatures without temporary protective measures. Materials such as grout, drypack, cast-in-place concrete, and epoxies need protection or other special provisions when placed in cold weather. Also, welding is slower when the ambient temperature is low. If the connections are designed so that these processes must be completed before erection can continue, the cost of erection may be increased, and delays may result.

Connections that are not susceptible to damage in handling should be used, whenever possible. Reinforcing bars, steel plates, dowels, and bolts that project from the precast concrete unit will often be damaged during handling, requiring repairs to make them fit. It is often better to use items that are larger than required by design, so they will be rugged enough to withstand the rough handling often received. Connectors attached to the panel with threaded inserts may be removed during shipping to minimize damage. Threads on projecting bolts should be protected from damage and rust. The precast concrete manufacturer should clean out and cap or seal sleeves or inserts until used, to prevent dirt and water from entering and freezing.

When cast-in-place concrete, grout, or drypack is required to complete a connection, the detail should provide for self-forming, if possible. When not practical, the connection should allow for easy placement and removal of formwork. Field patching and finishing should be kept to a minimum.

Erection procedures should be evaluated in relation to the loading of the structure as erection progresses. The importance of establishing limitations for erection methods or sequence, due to design assumptions for the structure, has been covered in Section 4.2.9. It should also be considered in the design and detailing of connections. For example, it should be determined how far final connections in steel structures must be completed ahead of precast concrete erection, and what the minimum concrete strength should be for cast-in-place structures prior to loading them with precast concrete units.

Consideration should be given to the number of floors the erector wants to erect in a structure on temporary connections, before making final connections. This is especially important for grouted connections and cold weather conditions.

In the case of high-rise cast-in-place structures, delay in erecting precast concrete units will allow some shrinkage and creep to take place. Alternatively, connections and joints should be designed to accommodate this shrinkage and creep.

Stacking units with temporary erection shims is an erection procedure sometimes used to facilitate the installation of precast concrete panels. Unless such temporary loading of units has been specifically incorporated or allowed in the contract documents, it is recommended that the precaster/erector requesting temporary stacking be responsible for this temporary loading of the units.

Shims should be removed from joints of non-loadbearing units after connections are completed and before applying sealant unless the shim material itself is readily deformable. If left in place, shims should be non-corrosive or protected so that staining will not occur.

The indiscriminate use of shims, particularly steel shims, can sometimes lead to undesirable consequences, for example, when steel shims are used in conjunction with grouted joints in multi-story bearing wall construction. Even with well-compacted grout or drypack, the compressive modulus of elasticity of the steel shim is six times that of the drypack; consequently, the grout will deform (compress) more readily than the shim. The load transfer path will remain concentrated through the steel shims, rather than be distributed along the drypack. High load concentrations at shims can cause spalling at panel surfaces or vertical cracks in the panels.

In other circumstances, the unplanned use of shims may

interfere with the component interaction or movement intended by the designer. They may prevent the independent behavior of the connection mechanisms and/or cause transfer of load where none was intended.

4.5.4 HANDLING AND LIFTING DEVICES

The subject of handling and lifting devices is included in this section because their design and selection are determined by requirements similar to those governing connections.

The design of lifting devices, including checking of stresses in units during handling is normally the responsibility of the precaster, see Section 4.1.4. Erection inserts cast in precast concrete members vary depending on the member's use in the structure, the size and shape of the member, and the precast concrete manufacturer's preference.

The location of lifting devices can affect the ease of erection and connection of the precast concrete unit to the structure. Lifting points should be compatible with the method of shipping (flat or on edge) and be placed so that the structure does not interfere with the crane lines. The same lifting devices should be placed on the various precast concrete units so that frequent changes to the rigging can be avoided. Only one type and size of lifting device should be used in a precast concrete unit (i.e., don't use bolted devices at one end and lifting loops at the other end of the unit). The precast concrete manufacturer should clean out and cap or seal lifting inserts until used by the erector, to prevent dirt and water from entering and freezing.

The design should consider impact loads incurred during transit and handling of the units. The direction of pull which may vary in the case of separate handling operations and the strength of the concrete which may only be a fraction of the design strength, are also design considerations. Since lifting devices are subject to dynamic loads, ductility of the material should be part of the design requirements.

In view of the above design criteria and the possibility of low concrete strength at the time of stripping, connection hardware should not be used for lifting or handling unless carefully reviewed and approved by engineering personnel. The final connections should not be used for erection if this interferes with the process of attaching the unit to the structure.

Lifting devices can be prestressing strand or aircraft cable loops which project from the concrete, coil thread inserts, or proprietary lifting hardware. All require sensible engineering and adequate safety factors for safe, efficient usage.

When properly designed (for both insert and concrete strength), threaded inserts for lifting heavy units have many advantages. Correct usage is sometimes difficult to inspect during handling operations. Insufficient bolt thread engagement, improper sling angle, worn threads, and low concrete strengths could cause failure with any lifting device.

If possible, the placing of temporary lifting and handling

devices should be planned so that little or no patching will be required after use. When the temporary lifting and handling devices are located in finished edges or exposed surfaces, bolt or insert holes must be recessed, filled and patched. Often, these devices may be filled and finished at a later date. Specialized lifting equipment may also be used to eliminate the necessity of patching exposed lifting and handling devices, see Fig 4.2.17.

Details of lifting devices should include the consideration of corrosion protection where such hardware is left in the units. If the handling devices interfere with any other function, erection drawings should include removal instructions as well as possible field patching, see Section 4.2.9.

4.5.5 MANUFACTURING CONSIDERATIONS

Since working conditions and inspection in a plant are superior to field conditions and less dependent on climatic factors, operations demanding high quality standards are most efficiently and economically performed in the plant. Connections should be detailed so that the less complicated operations are completed in the field.

Optimum economy in the manufacturing and incorporation of hardware items into precast concrete units demands simplicity and repetition. These demands often result in the use of one connection detail for several units on one job. Although such connections must be designed for the most severe load conditions, the cost of the extra material required will often be less than the detailing, manufacturing, storing and scheduling costs for a new connection. As a safeguard against human error, changes in dimensions and materials in connection hardware should be in increments large enough for visual recognition. This is especially important for erection hardware, see Fig. 4.5.15.

It is not unusual to have so many reinforcing bars and connection anchors concentrated in one location that there is virtually no room for the concrete. Difficulties in placing and consolidating the concrete in these congested locations may lead to honeycombing within the connection, and congestion of reinforcing bars may result in their being improperly positioned. It should be remembered that precast concrete units are usually cast in forms with only the top accessible; if one item must be threaded in and around other items, labor costs can be significantly increased. Reinforcing bars, which appear as lines on drawings, have real cross sectional dimensions which are larger than the nominal dimension because of the deformation. Reinforcing bar bends require minimum radii, which can cause fit problems and leave some regions unreinforced. If congestion is suspected, it is helpful to draw large scale details of the area in question. In some cases, it may be economical to increase the unit size to make steel placement and concrete consolidation easier.

A major requirement for the incorporation of hardware in precast concrete units is the positioning of the hardware to required tolerances. It is often advisable for the precaster to provide jigs, positioning fixtures or mold brackets to ensure location of plant hardware in order to maintain tolerances and to avoid skewed or misaligned hardware.

Hardware placed in precast concrete panels should have provisions (holes, lugs, nuts, etc.) so that it may be secured to jigs, fixtures or brackets during placement and consolidation of the concrete, but these specific details should be left to the precaster. Hardware projections which require cutting through the forms are difficult and costly to place. Where possible, these projections should be limited to the top of the element as cast. Even this placement increases labor by inhibiting the finishing of the top surface. The second preference is for the projection to be on the side forms. The least desirable location for a projection is on the down face, unless it is welded or bolted on after stripping.

Avoid casting wood (nailers, lath) in precast concrete. If unavoidable, the most suitable wood is pressure treated lumber. Also, dissimilar metals should not be in contact with each other unless experience has shown that no detrimental galvanic action will occur; and casting aluminum into precast concrete should be avoided unless it is electrically insulated by a permanent coating such as bituminous paint, alkali-resistant lacquer, or zinc chromate paint.

The subject of safety for manufacturing and construction personnel, as well as the general public, should be the first consideration for all decisions concerning the design and detailing of precast concrete. This will demand close liaison between designer, precaster, erector and the general contractor in the critical area of connections, lifting devices and temporary bracing.

Design of connections should consider the ease of inspection during casting and after completion of installation. Connections should be protected from any hydrochloric (muriatic) acid used to finish or clean the units.

4.5.6 CONNECTION HARDWARE AND MATERIALS

A wide variety of hardware, including reinforcing bars, deformed bars headed studs, coil inserts, structural steel shapes, bolts, threaded rods, and other materials are used in connections. In order to achieve specified strength, the hardware must be properly anchored in concrete. Anchor plate connections are widely used in combination with flat metal straps, reinforcing bars, or metal studs welded to the plate. The exterior surface of the plate is normally flush with the concrete face and provides a weld area for connection to the support system. By replacing the flat weld plate with a structural steel shape or bent plate, the cast-in-portion can provide additional anchorage, see Fig. 4.5.8.

The most widely used hardware connection item is the threaded insert, and numerous variations are available. Inserts require careful engineering, starting with proper design and followed by correct application in the plant and safe usage in the field. This is particularly important where impact or earthquake conditions require ductile connections and for medium or heavy weight precast concrete units.

Inserts must be placed accurately because their capacity decreases sharply if they are not positioned perpendicular to the bearing surface or if they are not in a straight line with the applied force, Fig. 4.5.16.

Fig. 4.5.16 Placing of inserts

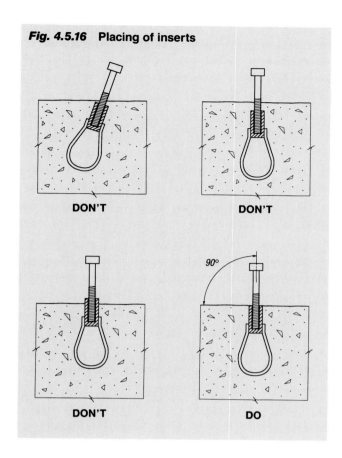

DON'T DON'T

DON'T DO

90°

Fig. 4.5.17 Wedge inserts

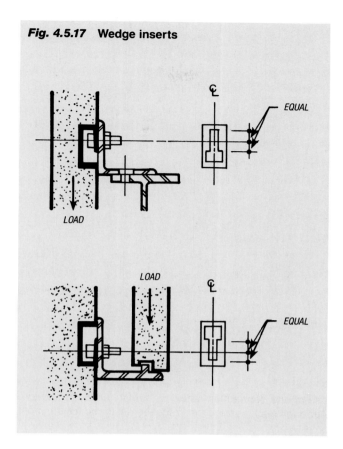

LOAD

EQUAL

LOAD

EQUAL

Inserts are easily located and held correctly by bolts and brackets during casting operations. Failure to achieve proper final positioning is either due to carelessness in set-up or failure to adjust the reinforcing cage to clear insert anchorages and still assure proper overlapping of anchorage and reinforcement. It is equally important to place inserts so that the depth of thread is constant for the same size insert throughout a particular job, Fig. 4.5.16. Otherwise, an erection crew may make mistakes in the field by not always engaging the full thread. Inserts should be kept clean of dirt or ice by protecting them temporarily with inexpensive plastic caps or other suitable devices installed in the plant after stripping.

Wedge inserts of the type shown in Fig. 4.5.17, made of malleable iron with an integral anchoring device, should be used with caution when connecting cladding elements to the structure. Although wedge inserts theoretically allow for 1 in. linear field adjustments, the position and fit of the bolt within the insert can determine its actual strength. The successful use of wedge inserts depends on a full engagement of the skew bolt head in the wedge, the full fit of the head with the wedge surface, and the snug tightening of the nut. If this cannot reliably be achieved, stress concentrations at the bolt head may lead to unpredictable behavior.

The problems connected with the use of wedge inserts often outweigh the advantages:

1. The insert must be oriented correctly.

2. The insert must be perfectly level.

3. No inspection of the fit of the skew head is possible after panel is erected.

4. Connection is very sensitive to over or under tightening of nut.

5. The connection is not able to resist force reversals or earthquake loadings.

Considering these limitations, it is recommended that wedge inserts only be used for light wind and gravity loads or secondary type connections.

Connections using bars or any standard steel section protruding from the precast concrete answer some of these considerations, but have requirements of their own for proper functioning. Careful placement of hardware to required tolerances including inclination of protruding bars or structural shapes is important, because the bearing surfaces of the panel hardware and the matching hardware on the structure must be parallel in order to obtain optimum bearing or load transfer.

Maintaining uniform connection locations requires careful tooling, which is a necessary initial cost for quality panel production. With any degree of repetition, such tooling will result in savings in daily manufacturing operations and will certainly reduce erection time. A connection positioned with the proper jig is fairly easy to inspect for proper location both prior to and during concrete placement.

Hardware to be placed at the project site by the general contractor or an agent should be detailed with provisions for simple and safe securing and be oversized to accommodate tolerances in placing. Angles with suitable nail or screw holes for fastening to side forms are superior to flat plates for accurate locations. Where large horizontal surfaces are required for hardware in cast-in-place structures, it is advisable to provide air release holds in the hardware to prevent air pockets and ensure proper concrete consolidation under such surfaces. Hardware for cast-in-place structures should be detailed to provide the proper anchorage. This may be done with bars, straps, studs, or reinforcing bars subject to appropriate selection of steel grades and welding. Anchorage details should allow placing with reasonable clearance from reinforcing steel in the structure.

To provide contractor's hardware continuously along floor edges or bases may be wasteful. Long pieces of hardware such as angles or plates with protruding anchors are likely to become bent. In addition, they are often cumbersome to place in the forms. Special circumstances may sometimes justify the expense.

The cost of contractor's hardware may be reduced if one item can be utilized for connecting two adjacent units. This requires that connections in the precast concrete units be located close to the edges which often facilitates production. This practice is recommended when it makes sense structurally.

The practical considerations for detailing of contractor's hardware, are equally applicable to plant hardware, but are normally part of the precaster's detailing.

4.5.7 PROTECTION OF CONNECTIONS

The degree of protection from corrosion required will depend on the actual conditions to which the connections will be exposed in service.

Industrial plants with severe exposures (high humidity and/or corrosive atmospheres) must have connection hardware protected from these elements. Such hardware must be completely encased in concrete or otherwise suitably protected where there is any danger of waste from plant operations, including water, reaching this connection hardware. The most common condition requiring protection is exposure to climatic conditions. Connection hardware generally needs protection against humidity or a corrosive environment.

Protection may be provided by:

1. Paint, with shop primer.

2. Coating with zinc-rich paint (95% pure zinc in dried film).

3. Zinc metallizing or plating.

4. Chromate or cadmium plating.

5. Hot dip galvanizing.

6. Epoxy coating.

7. Use of stainless steel for connections.

The cost of protection increases in the order of listing. Proper cleaning of hardware prior to protective treatment is important. It should be noted that the threaded parts of bolts, nuts or plates should not be galvanized or epoxy coated unless they are subsequently rethreaded prior to use.

Where connections requiring protection are not readily accessible for the application of zinc-rich paint or metallizing after erection, they should be metallized, galvanized or cadmium plated prior to erection and the connections bolted, where possible. If welding is required as part of the field assembly, the weld slag must be removed and the weld coated with zinc-rich paint or the epoxy repaired.

Special care should be taken when galvanized assembles are used. Many parts of connection components are fabricated using cold rolled steel or cold working techniques, such as bending of anchor bars. Any form of cold working reduces the ductility of steel. Operations such as punching holes, notching, producing fillets of small radii, shearing and sharp bending may lead to strain-age embrittlement of susceptible steels. The embrittlement may not be evident until after the work has been galvanized. This occurs because aging is relatively slow at ambient temperatures but is more rapid at the elevated temperature of the galvanizing bath.

The recommendations of the American Hot Dip Galvanizers Association and the practices given in ASTM A143, *Recommended Practice for Safeguarding Against Embrittlement of Hot Dip Galvanized Structural Steel Products and Procedure for Detecting Embrittlement,* and CSA Specification G164, *Galvanizing of Irregularly Shaped Articles,* should be followed.

Some designers have specified the use of stainless steel connections to prevent long-term corrosion. While this may appear to be the ultimate in corrosion protection, users are cautioned that the welding of stainless steel produces more heat than conventional welding. That, plus a higher coefficient of thermal expansion, can create adverse hardware expansion adjacent to the assembly being welded, thus causing cracking in the adjacent concrete and promoting accelerated long-term deterioration. When stainless steel connection plates are used, edges should be kept free from adjacent concrete to allow expansion during welding without spalling the concrete.

Fireproofing of connections is sometimes mandatory, depending on local codes and/or insurance regulations. Fireproofing will, in most cases, also provide corrosion protection. Many types of connections in precast concrete construction are not vulnerable to the effects of fire and, consequently, require no special treatment. For example, gravity-type connections, such as the bearing areas between precast concrete panels and concrete footings or beams which support them, do not generally require special fire protection nor do concrete haunches.

Concrete haunches increase the cost of the precast concrete unit itself, but the additional cost is likely to be less than the cost of any fireproofing of connections executed in the field. A drawback to this solution is the space requirement for the haunches, which often occur in areas needed for mechanical services. This may, however, be partly overcome by providing sleeves in the haunches.

If the panels rest on elastomeric pads or other combustible materials, protection of the pads is not generally required because deterioration of the pad will not cause collapse. Nevertheless, after a fire, the pads would probably have to be replaced so protecting the pads might prevent the need for replacement.

If the connections are to be fireproofed or concealed, this fact should be indicated in the contract documents.

Connections that can be weakened by fire, and thereby jeopardize the structure's integrity, should be protected to the same degree as that required for the structural frame. For example, an exposed steel bracket supporting a panel or spandrel beam may be weakened by fire and fail, causing the panel or beam to collapse. Such a bracket should be protected. The amount of protection depends on the stress-strength ratio in the steel at the time of the fire and the intensity and duration of the fire. The thickness of protection material required is greater as the stress level and fire severity increases.

Many connections simply provide stability and are under little or no stress in service. A fire could substantially reduce the strength of such a connection but it would still behave satisfactorily.

Generally, any parts of a panel connection which occur on the soffit of a slab and therefore above the fire, will require more fire protection than parts of a connection occurring on top of the slab.

Connections which require a fire resistance rating will usually have exposed steel elements encased in concrete by recessing the connection into the precast concrete element and/or floor slab, and drypacking or patching the recess with non-shrink grout after erection.

Anchoring the fire-protecting concrete or grout to steel surfaces is a problem that is often overlooked. Larger elements such as steel haunches can be wrapped with welded wire fabric or wire. For recessed plates and similar elements, connections such as those shown in Fig. 4.5.18 can be used.

There is evidence that the exposed steel hardware used in connections is less susceptible to fire-related strength reduction than other exposed steel members. This is because the concrete elements provide a "heat sink," which draws off the heat and reduces the temperature of the steel.

Some of the connections shown in Section 4.5.10 would be fairly easy to fireproof in the field except for the fact that concrete has to be cast-in-place at the connections. Where the placing of concrete floor fill is a separate operation, it will normally be done after erection of the wall units, and may be combined with concreting of the connections. Otherwise, connections may be sprayed with cementitious fireproofing, mineral fiber, or intumescent mastic compounds if the main structure is being spray-fireproofed, or enclosed with gypsum wall board.

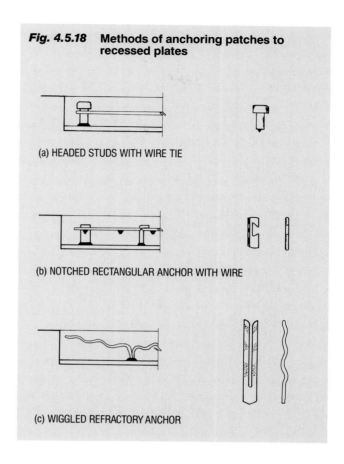

Fig. 4.5.18 Methods of anchoring patches to recessed plates

(a) HEADED STUDS WITH WIRE TIE

(b) NOTCHED RECTANGULAR ANCHOR WITH WIRE

(c) WIGGLED REFRACTORY ANCHOR

Fig. 4.5.19 shows the thickness of various commonly used fire protection materials required for fire endurances up to 4 hr when applied to connections consisting of structural steel shapes. The values shown are based on a critical steel temperature of 1000 deg. F, i.e., a stress-strength ratio (f_s/f_y) of about 65 percent. The values in Fig. 4.5.19b are applicable to concrete or drypack mortar encasement of structural steel shapes used as brackets or lintels.

When a rational analysis or design for fireproofing is not performed and concrete is used to fireproof the connections in the field, a conservative estimate would suggest that such concrete should have a thickness in inches corresponding to the specified hours of fire rating. Unless the nature of the detail supports this concrete, it should be reinforced with a light wire fabric.

4.5.8 FASTENING METHODS

Connections are usually bolted or welded. It is both difficult and expensive to weld overhead or in confined places; these situations should be avoided. Welding should also be avoided where hardware is galvanized. Thorough preremoval of the galvanizing is necessary in weld areas otherwise contaminations can occur creating poor weld quality. Cold galvanizing zinc-rich paint should be applied over the welded areas to replace the removed galvanizing.

Requirements for realistic tolerances often lead to sizable shim stacks which are not suitable for transmitting lateral forces of the magnitude developed in high seismic zones

Fig. 4.5.19 Thickness of protection materials applied to connections consisting of structural steel shapes

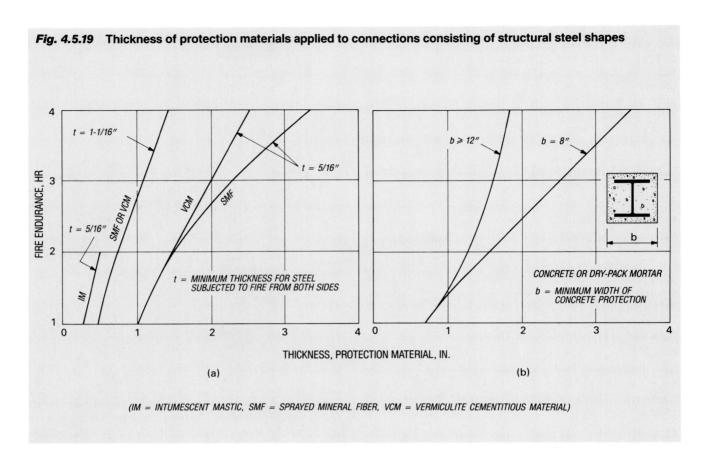

(a)

(b)

(IM = INTUMESCENT MASTIC, SMF = SPRAYED MINERAL FIBER, VCM = VERMICULITE CEMENTITIOUS MATERIAL)

or in locations with high winds. One solution is the addition of separate steel plates for horizontal loads as shown in Figs. 4.5.25 and 4.5.56 to 4.5.59. One-piece shim blocks may be used and are favored by some precast concrete erectors when they are combined with reliable panel dimensions and a survey of the structure prior to the placing of panels as described in Section 4.5.3.

Welded Connections for cladding panels are efficient and easily adjusted to suit varying field conditions. Their strength depends on reliable workmanship and the compatibility of welded materials with the metal to be joined. Where only a few field connections are to be welded, it is usually more economical to use an alternate method rather than require another trade on the project.

Hoisting and setting time is critical for economical erection. Structural welding which must be executed prior to the release of the unit from hoisting equipment should be minimized. If structural considerations indicate the necessity for welding, the precast concrete erector may transfer the precast concrete units to chain blocks for final positioning, or use monorails with several lifting trolleys. Provisions must be made to hold the unit safely in place until the final connections are completed. Welding should be performed by certified welders in accordance with the erection drawings. These should clearly specify type, size, length and location of welds, and, if critical, sequence, type of electrodes, minimum preheat and interpass temperatures. All welding, including tack welds, should be made in accordance with the applicable provisions of the American Weld-

ing Society (in Canada–The Canadian Welding Bureau).

When welding is performed on components that are embedded in concrete, thermal expansion and distortion of the steel may destroy the bond between the steel and concrete or induce cracking or spalling in the surrounding concrete.

The extent of cracking and distortion is dependent on the amount of heat generated during welding and the stiffness of the steel member. Heat may be reduced by:

1. Use of low-heat welding rods of small size.

2. Use of intermittent, rather than continuous welds.

3. Use of smaller welds and multiple passes.

Distortion can be minimized by using thicker steel sections; a minimum of 3/8 in. is recommended for plates. Some precasters line the metal with sealing foams, clay, or other materials to minimize the risk of the concrete cracking, especially if the metal is thinner than 3/8 in.

Cracks in concrete may be quite wide during welding, but will usually close significantly after the steel cools. If there is concern about potential structural damage due to the cracking, it may be advisable to provide supplementary deformed bar reinforcement in the concrete in the vicinity of the embedded components to be welded.

Bolted Connections often simplify and speed up the erection operation because an immediately positive connection is made. Final alignment and adjustment can be made later without tying up valuable crane time. In using bolted connections, it is desirable to standardize on the size of attachment hardware (clip angles and bolts). Standardization minimizes errors and inventory of hardware and improves productivity. With bolted connections, 3/4 in. or 1 in. diameter bolts with National Coarse threads are considered standard in the precast concrete industry and should be used. Regardless of the load requirements, a 1/2 in. diameter bolt should be the minimum size used for any precast concrete connection. Coil thread stock or coil bolts in 3/4 or 1 in. sizes are often used in lieu of National Coarse threaded stock to minimize the time required to make some connections and reduce the risk of thread damage.

Following erection of a precast concrete unit where slotted connections are used, a check of bolt tightness should be made. When movement is allowed, the bolt should be tight, but not so tight that it cannot move within the connection angle slot. Low friction washers (teflon or nylon) may be desirable to ensure movement. Roughness at sheared or flame cut edges may have to be removed. Washers, when used, should be large enough to overlap the sides of the slots, and allow full movement.

Bolted connections have to allow for reasonable erection tolerances. Where the connection cannot be made because the insert is out of place or missing altogether, alternative connections should be developed by the engineer of record.

Expansion Anchors are often used as corrective measures when cast-in inserts are misplaced or left out. They are inserted into predrilled or self drilled holes in hardened concrete. Proper performance of these anchors is largely dependent on field workmanship. Strength is obtained through expanding parts of the anchor which exert lateral pressure (wedge action) on the concrete. Lateral pressure is produced by proper tightening of the bolt or nut. Anchorage strength depends entirely on this lateral force. For connection reliability, the importance of correct installation and quality control cannot be overemphasized. The holes must be drilled straight, deep enough and with the proper diameter, and must be cleaned out. The minimum distance to the edge of the concrete and spacing between bolts should be based on the anchor manufacturer's recommendations. Care should be taken in placing the expansion anchors in the predrilled holes, so that expanding the anchor in the direction of the edge will be avoided.

The performance of expansion anchors when subjected to stress reversals, vibrations, or earthquake loading is such that their use for these load conditions should be carefully considered by the designer. Chemical anchors (resin capsule or epoxy anchors) could be considered for this application or for heavy loads. However, chemical anchors may degrade due to creep at temperatures in the 140 to 150 deg. F range. Such temperatures are readily experienced in warm climates, particularly in facade panels with dark aggregates. Also, chemical anchors may not be allowed in fire rated connection assemblies. Manufacturers' recommendations for installation must be followed.

Grouting or drypacking of connections is not widely used, apart from base plates for loadbearing units, because of the uncertainty of predicting load distribution. The difficulty in maintaining exact elevations and the inability to allow movements and still maintain weather-tightness must also be considered. Grouting should be used carefully when installed during temperatures below or near freezing. Grouting may be used where the above drawbacks do not exist or where the timing of the grouting and/or its exposure is not critical, as illustrated in Fig. 4.5.20.

Pressure grouting for loadbearing units was successfully employed in the horizontal joints in the office building pictured in Fig. 4.5.21. These panels were two-stories high, but were staggered to avoid continuous horizontal joints. In order to contain the pressure grout, special neoprene gaskets with suitable vents were designed specifically for the project.

In **Dowel Connections** the strength of the dowels depends on the diameter of the dowel, the embedded length, and the developed bond. Since placement of grout is required during erection, dowel connections usually slow down the erection and are often costly. Also, any precast concrete unit adjustment made soon after initial set of the grout may destroy dowel bond and reduce connection strength. It may be better to provide supplemental bolted connections to expedite erection.

Post-Tensioning may be used to make field connections between precast concrete members using either bonded or unbonded tendons. Bonded tendons are installed in preformed voids or ducts and are made monolithic with the member and protected from corrosion by grouting after the

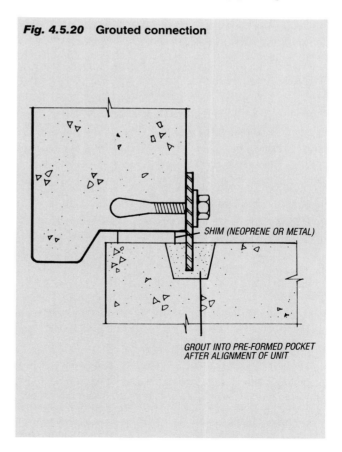

Fig. 4.5.20 Grouted connection

SHIM (NEOPRENE OR METAL)

GROUT INTO PRE-FORMED POCKET
AFTER ALIGNMENT OF UNIT

Fig. 4.5.21

stressing operation is completed. Unbonded tendons are protected against corrosion by a properly applied corrosion preventive organic coating. The unbonded tendons are connected to the member only through the anchorage hardware, which also must be protected from corrosion.

4.5.9 SUPPLY OF HARDWARE FOR CONNECTIONS

The responsibility for supplying **Contractor's Hardware** to be placed on or in the structure in order to receive the precast concrete units depends on the type of structure and varies with local practice. The furnishing and placing of hardware should be clearly defined in the contract documents. Hardware should be incorporated in the structure within the specified tolerances according to a predetermined and agreed upon schedule to avoid delays or interference with the precast concrete erection.

1. **Building frame of structural steel:** Hardware and bracing is normally supplied and installed as part of the structural steel contract. This requires sufficient time to provide proper hardware locations before detailing and fabrication.

2. **Building frame of cast-in-place concrete:** Hardware is placed by the general contractor, or the concrete subcontractor, to a hardware layout or anchor plan prepared by the precast concrete manufacturer. Drawings or specifications should state realistic tolerances for the placing of this hardware in relation to building lines

and levels as suggested in Section 4.6.3. Contractor's hardware may be supplied by the precaster, the general contractor, or be part of the miscellaneous steel subcontract. It is recommended that the precaster supply this hardware, since this usually reduces the problems of detailing and coordination. Efficient and economical details are facilitated by this approach, since modifications suggested by the precaster can be evaluated on their structural and functional merits, with no complicated contractual considerations involved. In any case, award of the precast concrete contract must be timely to allow design, detailing and fabrication of embedded items to be in sequence with casting schedules.

Field Installed Hardware consists of miscellaneous steel pre-welded or pre-bolted by the precast concrete erector prior to beginning the panel installation phase of the project. It is logical for the precast concrete manufacturer to supply this erection hardware even when he may not perform the actual erection. The reasons for this are similar to those mentioned for the supply of contractor's hardware. The alternative would be for the architect to detail and identify each piece, including the source of supply.

Plant Hardware cast into the unit is part of the precast concrete contract. Assurance that type and quantity of hardware items required to be cast into the precast concrete units for other trades are specified and not duplicated, is of greater importance than the supplier. Such items may be inserts for windows, fastenings for other trades, dovetails for flashing or masonry, etc. Specialty items, however, should be supplied from the trade requiring them.

Erection Hardware is the loose hardware needed in the field for final connection of the panel and should normally be the responsibility of the precaster to supply.

4.5.10 CONNECTION DETAILS

This section shows typical details for some of the more commonly used connections for cladding panels and load-bearing precast concrete walls, as well as other connections that may be useful in special applications. The details included are not exhaustive. They should not be considered as "standard," but, rather, as ideas on which to build. Detailed design information, such as component sizes, weld and anchorage lengths, joint sizes, bearing pad thicknesses, etc., is purposely omitted. The details may sometimes have to be combined to accomplish the intended purposes. For example, DB1 (Fig. 4.5.22) and DB2 (Fig. 4.5.23) are often combined. It is not the intent of the details to address the various ways of anchoring the connector to the concrete, nor the associated reinforcement. This is an engineering task required for each individual project. The connection details are not numbered in any order of preference.

There are many possible combinations of anchors, plates, angles, and bolts to form various connection assemblies. The details and final assemblies selected should be an optimum considering design criteria, production, erection, tolerances, and economy. Common practice by precast therefore additive to, the longitudinal applied forces.

concrete manufacturers in a given area may also influence the final selection of details on a particular project.

The examples shown for cladding panels cover the following broad categories:

1. Bearing (Direct and Eccentric)—DB, EB (Figs. 4.5.22 to 4.5.32).

2. Tie-Back (Bolted and Welded)—BT, WT (Figs. 4.5.33 to 4.5.43).

3. Alignment (Bolted and Welded)—BA, WA (Figs. 4.5.44 to 4.5.48).

4. Column and Beam Cover—CC, BC (Figs. 4.5.49 to 4.5.53).

5. Soffit Hanger—SH (Fig. 4.5.54).

6. Masonry Tie-Back—MT (Fig. 4.5.55).

7. Seismic Shear Plate—SP (Figs. 4.5.56 to 4.5.59).

8. Unique Conditions and Solutions—UCS (Figs. 4.5.60 to 4.5.66).

The loadbearing panel details cover the following categories:

1. Wall to Foundation—WF (Figs. 4.5.67 to 4.5.71).

2. Slab to Wall—SW (Figs. 4.5.72 to 4.5.76).

3. Wall to Wall—WW (Figs. 4.5.77 and 4.5.78).

Bearing (Direct and Eccentric) Connections are intended to transfer vertical loads to the supporting structure or foundation. Bearing should be provided at two points per panel, at one level of the structure. Bearing can be either directly in the plane of the panel along the bottom edge, or eccentrically located using a continuous or localized corbel on the panel. Localized corbels may consist of reinforced concrete corbels, cast-in steel shapes, or steel shapes welded to embedded plates after casting. Lateral load transfer capability can be provided by various tie-back arrangements. Tolerances in the support system generally necessitate the use of shims, leveling bolts, bearing pads, and oversized or slotted holes.

Direct bearing connections are used primarily for panels resting on foundations or rigid supports where movements are negligible. This includes cases where cladding panels are stacked and self-supporting for vertical loads with tie-back connections to the structural frame to resist forces perpendicular to the panel.

Eccentric bearing connections are usually used for panels above the first support level when movements of the support system are possible. Cladding panels are, by nature, fastened to and/or supported by a structure located in a different plane. Vertical self-supporting (stacked) panel systems excepted, practically all loads from the cladding react eccentrically on the supporting structure. Bending, combined tension, shear and torsion may have to be resisted by the connection, depending on the type of connection and load transfer details.

If leveling bolts are used, such as in EB1, Fig. 4.5.25, the final weld plates are proportioned for all lateral loads. The leveling bolt is usually left in place to carry the vertical load. If shims are used instead of leveling bolts, and lateral loads are to be carried, a similar weld plate is recommended, since the welding of shim edges is usually unreliable for transmitting significant forces. The erector's individual preference for shims or leveling bolts should be allowed.

Tie-Back (Lateral) Connections are intended to keep the precast concrete panel in a plumb or other desired position and resist wind and seismic loads perpendicular to the panel. Nearly every panel requires both tie-back connections and bearing (support) connections. The important characteristic of tie-back connections is their ability to carry tension and/or compression forces perpendicular to the panel. They may take forces in the plane of the panel, or allow movement vertically and/or horizontally.

Alignment Connections are used to adjust relative position with respect to adjacent elements; they do not usually transfer design (lateral) loads. Out-of-plane alignment of panels is often necessary, especially if they are very slender and flexible, or if prestressing or storage has caused warping or bowing.

Column and Beam Cover Connections are used when precast concrete panels serve as covers over steel or cast-in-place concrete columns and beams. The panels are generally supported by the structural column or beam and are themselves generally designed to transfer no vertical load other than their own weight. The vertical load of each length of column or beam cover section is normally supported at one elevation and tied back top and bottom for lateral load transfer and stability. Connections must have sufficient adjustability to compensate for tolerances of the structural system. Column cover connections are often difficult to reach, and once made, difficult to adjust. When access is available, consideration should be given to providing an intermediate connection for lateral support and restraint of bowing. "Blind" connections made by welding into joints between the precast concrete elements are sometimes employed to complete the final enclosure.

Soffit Hanger Connections can be made by modifying any of the tie-back connections previously discussed. If long, flexible hanger elements are used, a lateral brace must be provided in the form of a tie-back or cross brace to the support structure to achieve horizontal stability.

Masonry Tie-Back Connections are commonly used to attach concrete cladding panels to masonry walls. They are direct tie connections with an adjustable anchorage element cast into the panel and a bolt or strap anchor mortared or grouted into the masonry.

Seismic Shear Plates serve primarily to provide restraint for longitudinal forces in the plane of the panel. By their nature, they also carry loads perpendicular to the panel thus acting as a tie-back connection. Since seismic force is the most common in-plane force, these plates are often referred to as seismic shear plates. It is, in many cases, uneconomical to carry longitudinal forces on cantilevered, eccentric bearing connections. Hence, the shear plate connection can be used. Longitudinal force transfer on spandrels, for example, should be accomplished near midspan of the member to minimize volume change restraint forces which would be in the same direction as, and therefore additive to, the longitudinal applied forces.

Unique Connection Conditions and Solutions are presented in Figs. 4.5.50 (UCS1) through 4.5.66 (UCS7). Since each of these is a special situation, description of each is given along with the corresponding detail.

Wall to Foundation Connections are used to tie load-bearing or non-loadbearing walls to the foundation.

Slab to Wall Connections are made to join precast or cast-in-place concrete floor or roof members to precast concrete walls. Connections joining the slabs and walls may require load transfer for bearing, diaphragm action and moment resistance, or they may require movement accommodation such as needed when long roof members are joined to non-loadbearing walls. Load transfer may be through a continuous corbel (ledge) or individual (spot) corbels or connection hardware.

Blockouts in wall panels can also be used to support floor members. Such pockets in spandrels greatly decrease torsion stresses, and also minimize twist and eccentricity during erection. These pockets require substantial draft on their sides (1/2 in. every 6 in. depth) and should have at least 2-1/2 in. cover to the exposed face. More cover (3 in. minimum) is required if the exterior surface has an architectural finish. In the case of a fine textured finish, there will be a light area (the approximate size of the blockout) shown on the face of the panel due to differential drying. This will usually be obvious, despite the uniformity of the texture. The initial cure of the 2-1/2 to 3 in. of concrete 8 to 9 in. in the surrounding area will make the difference.

When the slab functions as a diaphragm, the connections must transmit diaphragm shear and chord forces. Flat, stemmed or hollow-core precast concrete slabs may be used. When the slab is used with composite topping, some connections may be necessary to achieve stability of the structure during erection with the final diaphragm connection achieved with dowels from the wall to the topping.

Most designs result in some degree of continuity for these connections. However, a fully fixed connection is generally not desirable. The degree of fixity can be controlled by a judicious use of bearing pads in combination with clamping forces, or by welding to anchor plates placed in floor members. Reinforcing steel can also be doweled, threaded or welded to the walls.

Connections to shear walls along the (non-bearing) sides of floor or roof slabs should be able to transmit lateral loads and should either allow some vertical movements to accommodate camber and deflection movements of the floor units, or be designed to develop forces induced by restraining the units, see Fig. 4.5.75.

Wall to Wall Connections are primarily intended to position and secure the walls, although with proper design and construction, they are capable of carrying lateral loads from shear wall or frame action as well. The two locations of wall to wall connections are horizontal joints, usually in combination with floor construction, and vertical joints.

Fig. 4.5.22 DIRECT BEARING (DB1)

Design
- simple
- lateral restraint not provided

Production
- simple

Erection
- simple
- does have large tolerance
- joint may be caulked or dry-packed

2 SHIM STACKS/PANEL

Fig. 4.5.23 DIRECT BEARING (DB2)

Design
- more detailing
- provides lateral restraint
- shims must be placed to hold vertical alignment until grouting or drypacking is done
- realignment is not possible once connection has been completed

Production
- more measuring
- reasonable tolerance each way

Erection
- wet placement requires care
- grout problem in cold weather
- may be best to field drill oversized hole into foundation

Variation
- grout could be injected through tubes allowing more time for alignment

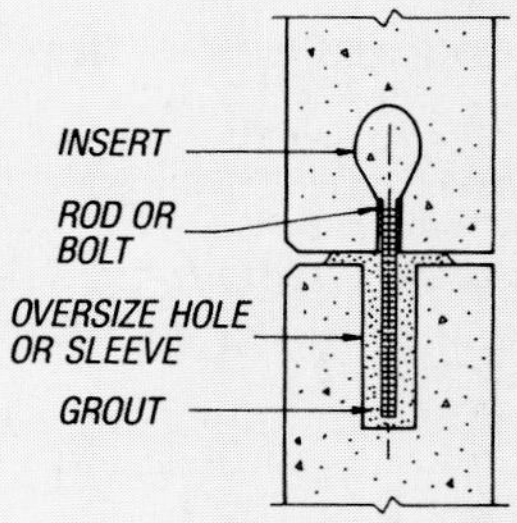

INSERT
ROD OR BOLT
OVERSIZE HOLE OR SLEEVE
GROUT

SHIM STACKS OCCUR AT 2 POINTS PER PANEL ADJACENT TO CONNECTION

Fig. 4.5.24   DIRECT BEARING (DB3)

Design
- preferable for bracket to be on contract drawing and shop installed
- may require restraint for shim stack

Production
- cost substantially more if column bracket field installed

Erection
- reasonable, if column bracket already there
- layout crew required if bracket not shop installed

Variations
- leveling bolt may be used in lieu of shims

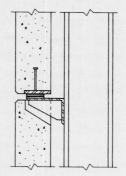

Fig. 4.5.26   ECCENTRIC BEARING (EB2)

Design
- hardware layout drawing required for General Contractor
- consider torque on projecting element if unsymmetrical section used
- panel must resist bending

Production
- simple
- requires early coordination with General Contractor
- requires additional space for storage and shipping

Erection
- simple

Variations
- W, I, channel, ST, flat bars, angle or TS may be used

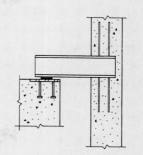

Fig. 4.5.25   ECCENTRIC BEARING (EB1)

Design
- weld all around may not be required
- keep bearing at centerline of beam to avoid torsion
- safety and sequence may dictate blockout to embed bracket in floor slab
- panel must resist bending

Production
- simple
- substantial shop fabrication
- leveling bolt is costly

Erection
- simple
- leveling bolt saves time

Variations
- different tie-back connection may be used in lieu of weld plate
- shims may be used in lieu of leveling bolt
- location and configuration of weld plate may vary

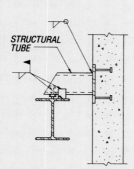

STRUCTURAL TUBE

Fig. 4.5.27   ECCENTRIC BEARING (EB3)

Design
- coordination drwg. for G.C.
- preferable for column bracket to be on contract drwg. and shop installed
- accommodates column size variation with same size panel bracket

Production
- simple
- requires early coordination with G.C.
- requires additional space for storage and shipping

Erection
- simple
- layout crew required if bracket not shop installed

Variations
- leveling bolt can be used by providing two projecting bars and welding a coupling nut between bars or at end of bracket
- panel bracket may be W, I, channel, TS or ST

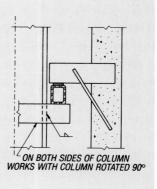

ON BOTH SIDES OF COLUMN WORKS WITH COLUMN ROTATED 90°

Fig. 4.5.28 ECCENTRIC BEARING (EB4)

Design
- hardware layout drwg. required for G.C.
- eliminates or reduces moment in panel
- simple reinforcement

Production
- simple
- requires early coordination with G.C.

Erection
- difficult
- requires layout crew before erection
- panel must be removed to change shims

Variations
- bracket may be another structural shape

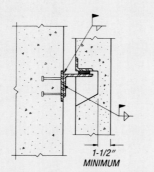

1-1/2" MINIMUM

Fig. 4.5.30 ECCENTRIC BEARING (EB6)

Design
- hardware layout drwg. required for G.C.

Production
- simple
- requires early coordination with G.C.

Erection
- simple

Variations
- leveling bolt may be used in lieu of shims
- weld plate may be used in lieu of separate tie-back connection

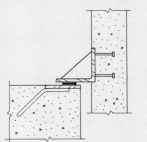

Fig. 4.5.29 ECCENTRIC BEARING (EB5)

Design
- hardware layout drwg. required for G.C.
- complex haunch reinforcement

Production
- involved
- extra forming, or haunch made separately and set in form
- proper location of reinforcing steel in haunch is critical
- requires early coordination with G.C.

Erection
- simple

Variations
- plate or angle may be used in haunch
- welded plate or insert is optional
- haunch may be continuous or intermittent
- plate washer may require welding for seismic load conditions

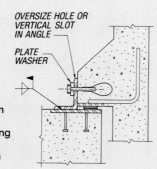

OVERSIZE HOLE OR VERTICAL SLOT IN ANGLE

PLATE WASHER

Fig. 4.5.31 ECCENTRIC BEARING (EB7)

Design
- hardware layout drwg. required for G.C.
- confinement steel around studs in panel may be required

Production
- simple
- requires early coordination with G.C.

Erection
- simple

Variations
- weld plate may be used in lieu of separate tie-back connection

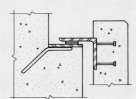

Fig. 4.5.32 ECCENTRIC BEARING (EB8)

Design
- important for shaped panels; can eliminate overturning moment from dead load when centerline of shim is at c.g. of panel

Production
- complex forming especially if location of haunch changes

Erection
- simple

Variations
- forming made easier by substituting a bolt-on steel bracket especially if haunch location changes

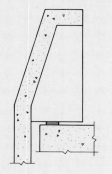

Fig. 4.5.34 BOLTED TIE-BACKS (BT2)

Design
- simple
- edge distance must be considered

Production
- simple

Erection
- simple
- must coordinate with steel in foundation
- accommodates large tolerance with exp. anchor

Variations
- if pre-set insert is used in place of exp. bolt, a slotted hole is necessary in the horizontal leg of the angle

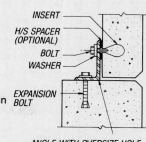

INSERT
H/S SPACER (OPTIONAL)
BOLT
WASHER
EXPANSION BOLT

ANGLE WITH OVERSIZE HOLE ON VERTICAL LEG

Fig. 4.5.33 BOLTED TIE-BACKS (BT1)

Design
- slenderness ratio of rod must be considered for compression load

Production
- simple
- adequate tolerance is provided when slotted insert set in opposite direction of slot in angle

Erection
- quick
- connection hardware is prewelded, thus panels are erected without welding
- panel alignment can be completed after release from crane

Variations
- if threaded insert is used, the in-plane movement may be achieved by flexibility of the rod, or by an oversized hole at the opposite end
- field weld angle to structure
- bolt angle to structure

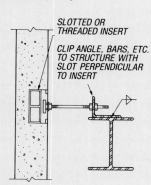

SLOTTED OR THREADED INSERT

CLIP ANGLE, BARS, ETC. TO STRUCTURE WITH SLOT PERPENDICULAR TO INSERT

Fig. 4.5.35 BOLTED TIE-BACKS (BT3)

Design
- slotted hole or oversized hole may be used to accommodate erection tolerance and any required movement
- consider clearances

Production
- simple

Erection
- to avoid volume change restraint, bolt should not be overtightened
- shims may be required

Variations
- field weld angles; requires bracing which may be achieved by another connection
- use threaded rod as in BT1

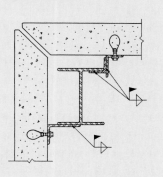

Fig. 4.5.36 WELDED TIE-BACKS (WT1)

Design
- volume change of panel and live load deflection of steel beam must be considered
- consider staggering studs to minimize magnification of the force on headed stud due to misalignment of plate
- rigid connection
- possible volume change restraint problems

Production
- simple

Erection
- requires bracing until welded; bracing may be achieved by another connection
- ample adjustment allowance

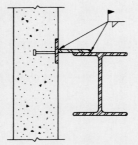

Fig. 4.5.38 WELDED TIE-BACKS (WT3)

Design
- if strap is used, volume change restraint in the plane of panel must be considered
- slenderness ratio of rod must be considered for compression load

Production
- simple

Erection
- requires bracing until welded; bracing may be achieved by another connection
- threaded rod should not be overtightened if future movement at slotted insert is expected

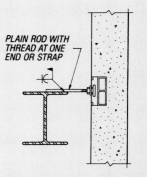

PLAIN ROD WITH THREAD AT ONE END OR STRAP

Fig. 4.5.37 WELDED TIE-BACKS (WT2)

Design
- rigid connection
- possible volume change restraint problems
- connection is difficult to inspect

Production
- simple

Erection
- requires bracing until welded; bracing may be achieved by another connection
- ample adjustment allowance
- alignment and welding must be completed before panel above is erected

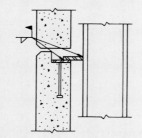

Fig. 4.5.39 WELDED TIE-BACKS (WT4)

Design
- live load deflection of superstructure must be considered
- if bracing angle is designed as an axially loaded member, the vertical component of force must be accounted for in the design of other connections on the same panel

Production
- simple

Erection
- slots and bolts are used for temporary erection connection
- weld after final alignment

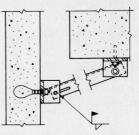

Fig. 4.5.40   WELDED TIE-BACKS (WT5)

Design
- good solution to avoid problems caused by structure deflection

Production
- simple

Erection
- if hardware is assembled prior to erection, oversized holes and plate washers are required

Variations
- use stiffer vertical members and eliminate the diagonal

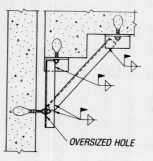

OVERSIZED HOLE

Fig. 4.5.42   WELDED TIE-BACKS (WT7)

Design
- good for seismic parallel and perpendicular forces

Production
- simple

Erection
- tolerances require various diameter rods

Variations
- change angle in panel to weld plate
- anchorage of plates may vary

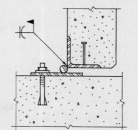

Fig. 4.5.41   WELDED TIE-BACKS (WT6)

Design
- a minimum bolt penetration into insert should be specified and ensured

Production
- simple

Erection
- quick
- adjustment allowance limited by ferrule and bolt lengths
- must have adequate clearance for welding

Variations
- weld may not be required if connection transfers only compression
- could be reversed

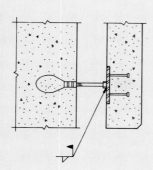

Fig. 4.5.43   WELDED TIE-BACKS (WT8)

Design
- good for seismic parallel forces
- hardware layout drwg. required for G.C.

Production
- simple
- requires early coordination with G.C.

Erection
- simple
- considerable adjustment

Variations
- change loose weld plate to loose angle

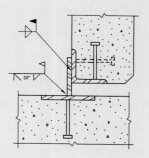

Fig. 4.5.44 BOLTED ALIGNMENT (BA1)

Design
- can also serve as a tie-back connection for light loads

Production
- simple
- requires close thickness tolerances

Erection
- quick
- good adjustment allowance
- to avoid volume change restraint, bolts should not be overtightened
- may require horseshoe shim spacers

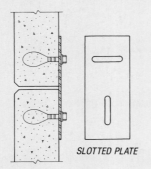

SLOTTED PLATE

Fig. 4.5.46 WELDED ALIGNMENT (WA1)

Design
- can also serve as tie-back connection for light normal load

Production
- simple
- face of panel to face of plate dimension is critical

Erection
- quick
- good solution when connection is not accessible after erection
- plate can be shop welded prior to fabrication or field welded prior to erection
- erection sequence should be considered and may be governed by this connection

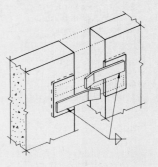

Fig. 4.5.45 BOLTED ALIGNMENT (BA2)

Design
- volume change relief is provided unless necessary to weld plate washers for specific loads

Production
- simple

Erection
- quick
- good adjustment allowance
- to avoid volume change restraint bolts should not be overtightened

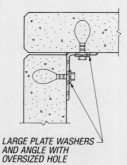

LARGE PLATE WASHERS AND ANGLE WITH OVERSIZED HOLE

Fig. 4.5.47 WELDED ALIGNMENT (WA2)

Design
- good shear transfer
- rigid connection
- possible volume change restraint problems

Production
- simple
- face of panel to face of plate dimension is critical

Erection
- quick, easy
- ample adjustment allowance

Variations
- various embedded plates or shapes may be welded together
- one side could be bolted with slotted or oversized hole

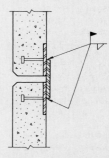

Fig. 4.5.48   WELDED ALIGNMENT (WA3)

Design
- rigid connection
- possible volume change restraint problems

Production
- simple and relatively inexpensive
- unusual mounting or attachment to side forms
- inexpensive hardware

Erection
- quick
- requires close joint tolerance
- different size bars required to accommodate different joint widths

Variations
- angles may be eliminated and field weld made directly to U shaped panel bar as shown

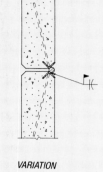

VARIATION

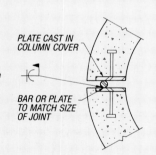

Fig. 4.5.50   COLUMN COVER CONNECTION (CC2)

Design
- can carry horizontal tie-back forces
- requires sufficient clearance between column and cover
- preferable for bracket to be on contract drwg. and shop installed

Production
- inserts must be attached to the back form for casting accuracy

Erection
- alignment must be done when erecting first column section
- oversized holes allow for adjustment and alignment

Variations
- can be used on any shape column covers
- can be used for connecting both half column covers if near top and accessible

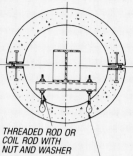

THREADED ROD OR COIL ROD WITH NUT AND WASHER

INSERTS CAST IN PANEL

Fig. 4.5.49   COLUMN COVER CONNECTION (CC1)

Design
- provides a rigid connection between column cover segments
- can be used where connection to column or beam would be difficult due to limited access
- 3/4″ minimum joint size is recommended

Production
- allows reasonable tolerances for alignment
- if the column section is thin, placement and coverage of plate is difficult

Erection
- panel joint must be sufficient to allow for welding
- care must be taken in preventing welding stain on exterior concrete
- care must be taken not to apply excess heat that would crack the concrete

PLATE CAST IN COLUMN COVER

BAR OR PLATE TO MATCH SIZE OF JOINT

Fig. 4.5.51   COLUMN COVER CONNECTION (CC3)

Design
- can be used only at top of column cover where access is available for welding
- used for lateral stability and alignment

Production
- the weld plates must be placed on the end form

Erection
- need access to top of column cover to make connections

Variations
- can be used on any shape column cover
- can be changed to bolted

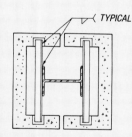

TYPICAL

Fig. 4.5.52 COLUMN COVER CONNECTION (CC4)

Design
- can be used for both vertical and tie-back loads with welded washer
- can be used where access is difficult

Production
- requires that the angle bolt assembly be cast in a manner so as to keep the bolt parallel to the face of the panel

Erection
- requires bracing until welded
- requires that the panel be properly aligned and set prior to welding and setting of the panel above

Variations
- use insert and bolt in lieu of bolt welded to angle
- use bolt cast close of c.g. of panel instead of bolt welded to angle

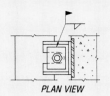

PLAN VIEW

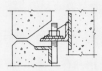

SECTION VIEW

Fig. 4.5.54 SOFFIT HANGER (SH1)

Design
- allows for adjustment and movement
- may require additional bracing for lateral loads

Production
- ease in casting inserts into panel

Erection
- allows for final alignment after panel is released
- may be difficult to get to areas requiring bolting

Variations
- angles or other shapes maybe used instead of threaded rods

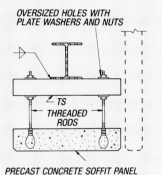

OVERSIZED HOLES WITH PLATE WASHERS AND NUTS

TS
THREADED RODS

PRECAST CONCRETE SOFFIT PANEL

Fig. 4.5.53 BEAM COVER CONNECTION (BC1)

Design
- beam must be designed to prevent excessive rotation during erection
- rigidity provided by welded connections must be considered

Production
- requires careful casting to match finishes on faces
- requires a close casting tolerance on the doweled connections for the cap piece

Erection
- requires that the erector place pieces in proper sequence
- may require a combination of bolting, welding and grouting
- care must be taken to prevent staining of exposed surface during welding

Variations
- alternate top conditions are shown but only one type should be used (watertightness of top condition should be considered)

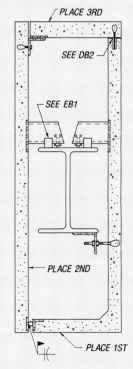

PLACE 3RD

SEE DB2

SEE EB1

PLACE 2ND

PLACE 1ST

Notes
- refer to EB1, BT1 and CC1

Fig. 4.5.55 MASONRY TIE-BACK CONNECTION (MT1)

Design
- the masonry may need to be reinforced

Production
- the slotted insert with masonry tie-back allows for ease of casting and considerable tolerance

Erection
- precast concrete member must be set, aligned and braced prior to layup of masonry
- temporary bracing is required

Variations
- threaded or slotted inserts or dovetail anchor slots in masonry and precast concrete may be used in lieu of strap anchors

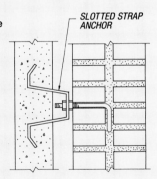

SLOTTED STRAP ANCHOR

Fig. 4.5.56   SEISMIC SHEAR PLATE (SP1)

Design
- normally one used at centerline of panel
- takes seismic force parallel to panel to minimize lateral load on bearing connections
- assume fixed at beam, pinned at panel
- particularly advantageous when panel to beam dimension is large
- also takes force perpendicular to panel
- thin plate allows some vertical movement

Production
- panel plate tolerance large

Erection
- welding required
- cannot be installed until panel fully aligned
- large tolerance

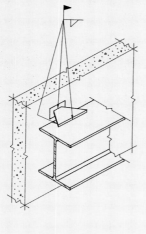

Variations
- connection to panel can be made with angle and slot perpendicular to panel to allow movement perpendicular to panel
- is sometimes accomplished with a pair of angles or flat bars
- could be changed to bolted fastenings
- simplest version is small rectangular plate to floor slab embedment when panel is close to slab edge

Fig. 4.5.58   SEISMIC SHEAR PLATE (SP3)

Design
- at mid-height of column covers to eliminate inertial overturn
- if not welded to column, must be used in pairs and column cover rotates in plane of wall with story drift so bearing connections must allow lift off
- if welded to column, the column cover translates in plane of wall which other connections must tolerate
- items above require careful integration of entire connection system and panel joint widths for inter-story displacements

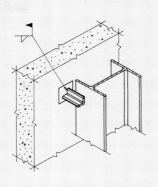

Production
- panel plate tolerance large

Erection
- welding required
- large tolerance
- can not be installed until panel fully aligned

Fig. 4.5.57   SEISMIC SHEAR PLATE (SP2)

Design
- similar to SP1 except combined with bearing connector rather than separate
- takes seismic force parallel and normal to panel to minimize requirements on bearing connector
- also takes perpendicular force so bearing connector need not be welded
- see also SP1

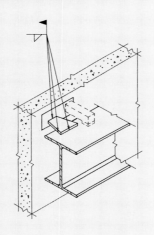

Production
- panel embedment serves dual purpose; so an additional one is not required

Erection
- since shear plate cannot be installed until panel is fully aligned, a temporary safety tie-back may be required during erection
- welding required
- large tolerance

Variations
- could be welded to outstanding arm of bearing connector

Fig. 4.5.59   SEISMIC SHEAR PLATE (SP4)

Design
- shims carry full panel weight
- shims should be immediately adjacent to welded angle
- can not be installed until unit fully aligned so temporary tie may be required during erection
- orientation of angle provides maximum capacity both parallel and perpendicular to wall

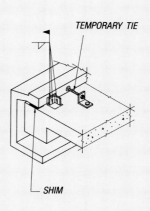

TEMPORARY TIE

SHIM

Production
- simple
- large tolerance
- separate embedment may be required for temporary tie

Erection
- can not be installed until panel fully aligned

Variations
- any type of plate, angle or T may be used for field plate
- leveling bolt could be recessed in sill for ease of alignment in lieu of shims

Fig. 4.5.60 UNIQUE CONDITIONS & SOLUTIONS (UCS1)

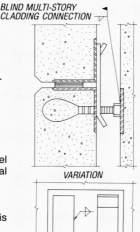

BLIND MULTI-STORY CLADDING CONNECTION

VARIATION

Design
- shims transfer weight to bottom of vertical run in stacked panel situations
- weld can only be achieved on upper part of bolt head - see variation
- creep (including of shims) will transfer some indeterminate load to bolt

Production
- dimension from face of panel to embedded angle is critical

Erection
- alignment can not be adjusted after upper panel is placed
- shims must be placed in joint
- requires care, not safe to install shims with fingers

Variations
- shop weld a plate to bolt head for greater field weld length
- provide through bolt from wall insert and grout face pocket to eliminate weld

Note
- this is an example of how connections can be combined (WT6 and WA1) and adapted to different conditions

Fig. 4.5.62 UNIQUE CONDITIONS & SOLUTIONS (UCS3)

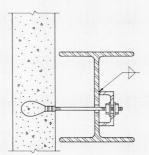

TIE-BACK WITH NO ACCESS BETWEEN PANEL & BEAM

Design
- use when tie-back well above beam bottom
- requires oversize hole in beam web and channel
- preferable for channel to be on contract drwg. and shop installed

Production
- insert location and beam bracket can be held at constant distance to floor (greater panel standardization)

Erection
- where no access between beam flange and panel

Variations
- use MC, L or split TS

Fig. 4.5.61 UNIQUE CONDITIONS & SOLUTIONS (UCS2)

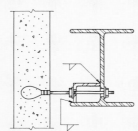

TIE-BACK @ LIMITED ACCESS AROUND BEAM

Design
- requires oversize hole in beam web and angle
- use to limit unsupported length of rod
- preferable for angle and holes to be on contract drwgs. and shop installed
- may have to allow for beam deflection

Erection
- use where panel to beam space does not allow reaching around beam flanges to install nut at web

Fig. 4.5.63 UNIQUE CONDITIONS & SOLUTIONS (UCS4)

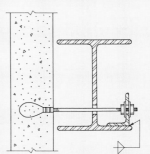

TIE-BACK WITH NO ACCESS BETWEEN PANEL & BEAM

Design
- requires oversize hole in beam web and angle
- preferable for angle to be on contract drwg. and shop installed

Production
- insert location must vary with beam depth

Erection
- use where no access between beam flange and panel

Variations
- use MC,C or TS

Fig. 4.5.64 UNIQUE CONDITIONS & SOLUTIONS (UCS5)

Design
- has very high load capacity when it engages and confines panel reinforcement
- good for dynamic loads, i.e. seismic
- size variability makes it adaptable to many panel configurations

Production
- expensive fabrication but alternates for equal capacity may be more costly

Variations
- bearing lug may be desirable to reduce shear on loop anchors

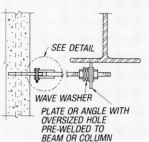

HIGH CAPACITY PANEL EMBEDMENT ANCHORAGE

PLAN SECTION

TYPE OPTIONAL SEE EB SERIES

VERTICAL SECTION

TYP.

(✳ BEARING LUG)

Fig. 4.5.65 UNIQUE CONDITIONS & SOLUTIONS (UCS6)

Design
- need for blind connection to precast concrete panel
- allow for tolerance
- requires layout drwg. to be provided to G.C.
- face of panel needs no patching

Production
- no special production problems

Erection
- requires temporary bracing if angle not welded until after alignment
- simple, welded slotted tie-back connection

Variations
- insert could be slotted

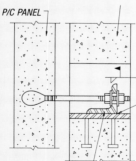

CAST-IN-PLACE OR MASONRY WALL

P/C PANEL

ANGLE WITH OVERSIZED HOLE AND PLATE WASHERS

Fig. 4.5.66 UNIQUE CONDITIONS & SOLUTIONS (UCS7)

Design
- tolerates high seismic drift without complications of sliding or flexing of rod
- intermediate length rods often bind rather than slide
- length/diameter ratio of rod may not take adequate compression or allow sufficient flexing
- wave washer flattens under nominal movement; prior to that, rod is pinned both ends, subsequently pinned left end only

Production
- simple
- economical flat bar embedment

Erection
- fast; carries load immediately yet allows subsequent alignment
- wave washer (spring, etc.) must be installed on side which is not loaded under dead load only but should not be over tightened
- wave washer is standard off the shelf hardware
- ample tolerance

Variations
- for full pivot at beam end, see variation
- coil spring or neoprene washer could be substituted for wave washer
- compression capacity can be increased with loose pipe over rod since it limits rod buckling

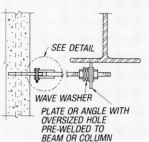

ARTICULATED TIE-BACK

SEE DETAIL

WAVE WASHER

PLATE OR ANGLE WITH OVERSIZED HOLE PRE-WELDED TO BEAM OR COLUMN

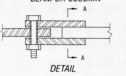

DETAIL

FLAT BARS

SECTION A-A

VARIATION

LOOSE WITH LOCK NUT OR TACK WELD TO ALLOW PIVOTING

Fig. 4.5.67   WALL TO FOUNDATION (WF1)

Design
- if connection is on exterior face of panel, it is susceptible to corrosion unless protected with mastic or grout
- hardware layout drwg. required for G.C.
- can be designed for horizontal shear and uplift; flexure in angle limits uplift capacity

2 SHIM STACKS/PANEL

Production
- simple
- embedded plates in wall may need to be jigged level if cast top-in-form to avoid tilting

VARIATIONS

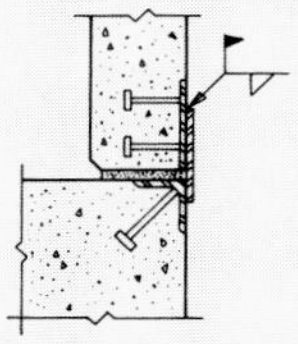

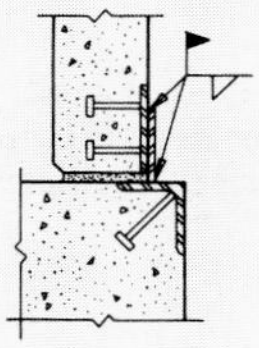

Erection
- quick and easy
- few tolerance problems if embedded plates are wider than angle
- welding may be difficult when connection is below grade
- space under wall usually filled with grout

Variations
- connections may be placed on both sides of wall to develop nominal moment resistance
- angles may be bolted to wall and/or foundation
- plates may be used in place of connection angles

Fig. 4.5.68   WALL TO FOUNDATION (WF2)

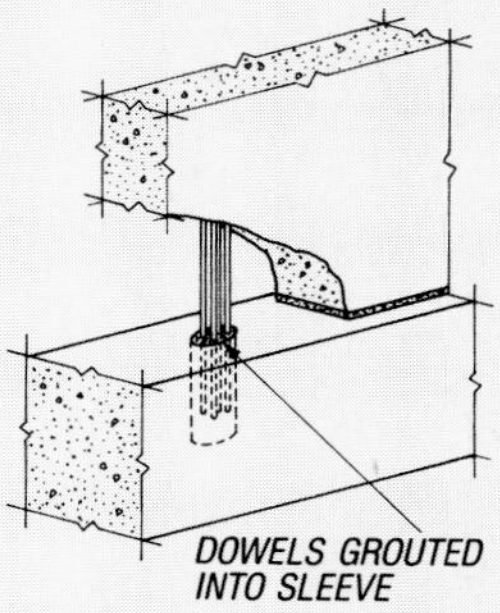

Design
- shear resistance is achieved
- capacity can be increased by use of confinement reinforcement around sleeve and bars

Production
- projecting dowels from panel can cause difficulties in storing and transporting panel
- location and alignment of dowels is critical

DOWELS GROUTED INTO SLEEVE

Erection
- grouting coordination required
- location and alignment of sleeve is critical
- no connection for panel during erection; necessary to brace
- use grout under panel
- alignment of panel must be made before initial set of grout
- must weather protect sleeve to prevent ice, water or debris from filling cavity

Variations
- sleeve may be placed in panel to receive dowels from foundation
- grout can be pumped into sleeve after alignment or before panel erection
- proprietary sleeve systems
- use insert or coupler and add threaded dowel in the field to reduce production and transportation problems
- single dowel is most commonly used

Fig. 4.5.69   WALL TO FOUNDATION (WF3)

Design
- hardware layout drwg. required for G. C.
- size joint for welding access

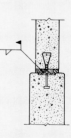

Production
- simple
- insert must be jigged so that bolt is plumb

Erection
- quick and easy
- allows vertical adjustment without crane

Variations
- bolt head may be welded for tensile and shear capacity
- plate may be eliminated but adjustment becomes more difficult
- use embedded bolt and projecting sleeve nut
- insert may be in foundation

Fig. 4.5.70 WALL TO FOUNDATION (WF4)

Design
- two directional stiffness
- headware layout drwg. required for G.C.

Production
- care required in jigging angles in form to ensure proper alignment

Erection
- may require temporary bracing

Variations
- if wall needs to be tied to floor slab, weld reinforcement to connection

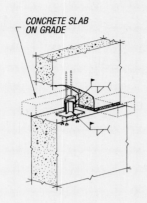

CONCRETE SLAB ON GRADE

Fig. 4.5.71 WALL TO FOUNDATION (WF5)

Design
- develops moment resistance at base
- can be used to resist uplift forces
- no positive connection until bar is tensioned
- hardware layout drwg. required for G. C.

Production
- duct placement tolerance in wall panel is critical
- grout vents may be required

Erection
- may require temporary bracing
- bar, duct and hardware placement tolerance in foundation and wall panel is critical
- requires drypack to reach design strength prior to tensioning
- post-tensioning equipment necessary

Variations
- shim under panel
- bars may be coupled at top of foundation
- post-tensioned bar may or may not be grouted

POST-TENSIONING BAR

2 SHIM STACKS/PANEL

VARIATIONS

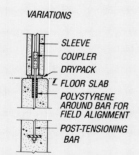

SLEEVE
COUPLER
DRYPACK
FLOOR SLAB
POLYSTYRENE AROUND BAR FOR FIELD ALIGNMENT
POST-TENSIONING BAR

Fig. 4.5.72 SLAB TO WALL (SW1)

Design
- welding at bottom of slab is not recommended as excess restraint results
- no moment capacity
- must consider eccentricity of loads
- top connection transfers horizontal shear forces or provides nominal torsion restraint for spandrel

Production
- special forming required for corbel
- corbel may be precast and set in form

Erection
- quick and easy
- allows adequate tolerances
- temporary bracing may be necessary

Variations
- steel corbel; may use inserts in panel to position angle while welding
- flag shaped plate (g) welded to embedded plate in wall can be used in hollow-core joints
- variation of (d) and (g), dowel may be in topping

COIL INSERT AND FIELD PLACED ROD

(a)

VARIATIONS

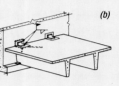

(b)

OR

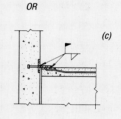

(c)

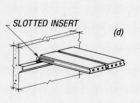

SLOTTED INSERT

(d)

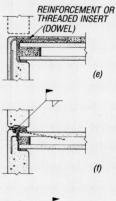

REINFORCEMENT OR THREADED INSERT (DOWEL)

(e)

(f)

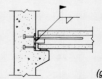

(g)

Fig. 4.5.73 SLAB TO WALL (SW2)

Design
- minimizes eccentricity of load on wall
- axial shortening of slab due to volume change should be considered when designing depth of recess
- pocket dimensions and tee end must be planned so that slab can "swing" into place; pocketed connection should not be used at both ends of slab
- top connection similar to connection SW1 (a) or (b) may be used

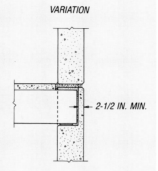

VARIATION

2-1/2 IN. MIN.

Production
- minimum of embedded hardware
- special forming required to allow stems to fit into pockets
- pockets in wall difficult to locate and form, usually do not follow tee taper
- pockets require adequate tolerance
- pocket may telegraph through gray conrete; exterior surface finish, e.g. retarded, sandblasted will help conceal

Erection
- do not drypack pocket around tee stem to allow stem freedom to rotate
- for ease of erection, pockets should not be used at both ends of slab

Variations
- pocket may be at top of panel

Fig. 4.5.74 SLAB TO WALL (SW3)

Design
- develop a rigid, moment connection
- avoid use of this detail at both ends of slab to prevent excessive restraint
- rotation of wall elements and effects on bracing wall connections and volume changes must be considered
- arrangement of weld plates must allow for welding access
- avoid overhead welding, if possible

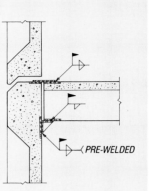

PRE-WELDED

Production
- plate jigging is necessary since embed is top-in-form as cast
- steel congestion must be well thought out

Erection
- welding must be completed before setting panel above

Variations
- wall corbel in lieu of angle seat

Fig. 4.5.75 SLAB TO WALL (SW4)

Design
- connection allows movements caused by temperature changes
- positive horizontal force transfer
- connection (c) allows vertical movement by flexing of plate and welds
- connection (d) allows vertical movement through flexibility of double tee flange

THREADED INSERT IN WALL PANEL

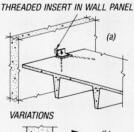

(a)

VARIATIONS

(b)

Production
- insert must be plumb and true
- washer must be oversize so it does not bind in the slot
- simple

Erection
- quick and easy
- tolerance problems minimized
- do not overtighten bolt in (a)

(c)

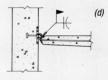

(d)

Fig. 4.5.76 SLAB TO WALL (SW5)

Design

- slabs and wall panels should be shaped to create self-forming elements and require a minimum of temporary support and bracing
- crack control is critical to avoid leakage and damage to outside surface of precast concrete panel
- if precast shell and cast-in-place concrete are to act compositely bond ties should be considered — but must be positioned so that reinforcement can be placed
- develops lateral load resisting wall and column sections
- reinforcement that passes across the interface should be adequate to support required shear forces using shear friction
- must be reinforced to transfer horizontal and vertical shear forces without undue deformations

Production

- placement of panel reinforcing steel must be held to close tolerances to minimize cracking
- good system but precast concrete difficult to cast

Erection

- temporary shoring is often required

Variations

- provide inserts in panel to facilitate forming for cast-in-place concrete

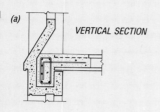

(a) VERTICAL SECTION

HORIZONTAL SECTION

VARIATIONS

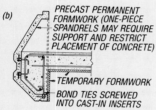

(b) PRECAST PERMANENT FORMWORK (ONE-PIECE SPANDRELS MAY REQUIRE SUPPORT AND RESTRICT PLACEMENT OF CONCRETE)

TEMPORARY FORMWORK

BOND TIES SCREWED INTO CAST-IN INSERTS

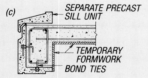

(c) SEPARATE PRECAST SILL UNIT

TEMPORARY FORMWORK

BOND TIES

Fig. 4.5.77 WALL TO WALL (WW1)

Design

- continuity through the connections
- connection is concealed and protected
- no connection between walls until splice sleeves or ducts are grouted
- sleeve and sleeve grout are proprietary

Production

- hardware placement is critical
- projecting dowels can cause difficulties in storing and transporting panels if dowels project from bottom of panel

Erection

- may be necessary to heat grout in cold weather
- temporary brace required
- requires a grout crew in addition to setting crew

Variations

- sleeve connector can be placed in either upper or lower panel — upper panel is preferred

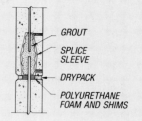

GROUT

SPLICE SLEEVE

DRYPACK

POLYURETHANE FOAM AND SHIMS

Fig. 4.5.78 WALL TO WALL (WW2)

Design

- can be used to withstand uplift forces
- connection is hidden and protected
- connection is not developed until tensioning is completed (bars are anchored)

Production

- duct and hardware placement in panels is critical
- tolerance on slab length critical
- thin panel outer lip projection subject to damage during handling

Erection

- temporary bracing is required
- drypack, tensioning, grouting sequence may limit erection to one story at a time
- grouting requires care to ensure complete filling

Variations

- bars may or may not be post-tensioned

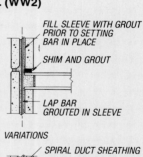

FILL SLEEVE WITH GROUT PRIOR TO SETTING BAR IN PLACE

SHIM AND GROUT

LAP BAR GROUTED IN SLEEVE

VARIATIONS

SPIRAL DUCT SHEATHING

THREADBAR COUPLER

SHIM & DRYPACK

GASKET AT SLEEVE

GROUT NOT SHOWN FOR CLARITY

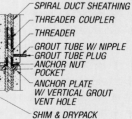

SPIRAL DUCT SHEATHING

THREADER COUPLER

THREADER

GROUT TUBE W/ NIPPLE

GROUT TUBE PLUG

ANCHOR NUT POCKET

ANCHOR PLATE W/ VERTICAL GROUT VENT HOLE

SHIM & DRYPACK

GROUT NOT SHOWN FOR CLARITY

4.6 TOLERANCES

4.6.1 GENERAL

The designer of architectural precast concrete must recognize that erection and manufacturing tolerances apply to this product as they do to other building materials. When tolerances are understood and allowances made for them in the design stage, the task of determining and specifying them becomes fairly simple.

By stating tolerances, the architect has strengthened and simplified his standards for acceptance, provided that tolerances have been realistically specified.

The architect must recognize that unrealistic and close tolerances are costly, particularly for custom produced elements. The cost of manufacturing to close tolerances decreases with increased repetition, see Section 2.2.2.

Tolerances set the limits of size and shape within which the actual precast concrete units must lie. The immediate reason for specifying tolerances is to establish construction criteria that will ensure that the parts will fit together without having to be modified. A more long term reason is to be certain the structure will perform as intended. The architect establishes the tolerances required to make the building concept work and must temper the desire for close tolerances with knowledge of what can be attained in the field at a reasonable cost.

Three groups of tolerances should be established as part of the precast concrete design: product tolerances, erection tolerances, and interfacing tolerances. Contrary to some beliefs, precast concrete tolerances in manufacture and erection do not usually cause site problems. The interfacing of precast concrete with other building materials is the prime area of contention.

Tolerances should be established for the following reasons:

1. *Structural* — To ensure that structural design properly accounts for factors sensitive to variations in dimensional control. Examples include eccentric loading condition, bearing areas, hardware and hardware anchorage locations, and locations of reinforcing or prestressing steel.

2. *Feasibility* — To ensure acceptable performance of joints and interfacing materials in the finished structure.

3. *Visual* — To ensure that the variations will be controllable and result in an acceptable looking structure.

4. *Economic* — To ensure ease and speed of production and erection by having a known degree of accuracy in the dimensions of precast concrete products.

5. *Legal* — To avoid encroaching on property lines and to establish a standard against which the work can be compared in the event of a dispute.

6. *Contractual* — To establish a known acceptability range · and also to establish responsibility for developing, achieving and maintaining mutually agreed tolerances.

The architect should be primarily responsible for coordinating the tolerances for precast concrete work with the requirements of other trades which adjoin the precast concrete construction.

While the responsibility for specifying and maintaining tolerances of the various elements may vary among projects, it is important that this responsibility be clearly assigned. In all cases the tolerances must be reasonable and realistic, within generally accepted limits. Manufacturing and erection costs are directly proportional to the tolerance refinement required. It is therefore economically desirable and more practical to design with maximum flexibility and to keep tolerance requirements as liberal as possible. The architect must evaluate the costs associated with the product and erection tolerances and connection details, paying special attention to unique and unusual situations, difficult erection requirements, connections which are very tolerance sensitive, and product or erection tolerances which are more stringent than normal practice.

When tighter-than-normal (industry standard) tolerances are to be specified, consideration should also be given to specified (or achievable) tolerances in the supporting structure. These must be compatible with those required for the precast concrete units. The specification of tighter-than-normal tolerances should be restricted to situations where a high degree of accuracy is important. Achievement of them will necessitate a very high standard of mold construction and close supervision and inspection at all stages of manufacture, with consequent increased cost.

It should be understood by those involved in the design and construction process that tolerances shown in this Manual must be considered as guidelines for an acceptability range and not limits for rejection. If these tolerances are met, the member should be accepted. If these tolerances are exceeded, the member may still be acceptable if it meets any of the following criteria:

1. Exceeding the tolerances does not affect the structural integrity or architectural performance of the member.

2. The member can be brought within tolerance by structurally and architecturally satisfactory means.

3. The total erected assembly can be modified economically to meet all structural and architectural requirements.

The enforcement of tolerances should be based on the technical judgment of the designer. This design professional is able to decide whether a deviation from the allowable tolerances affects safety, appearance, or other trades. In building construction very little out of tolerance work, whether it is concrete, masonry, cast-in-place concrete, steel, or precast concrete, has been rejected and removed solely because it was "out of tolerance."

4.6.2 PRODUCT TOLERANCES

Product tolerances define the limits of the variations of the size and shape of the individual precast concrete members which compose the building or structure. They are a measure of the dimensional accuracy required on individual members to ensure, prior to delivery, that the members will fit the structure without difficulty. Product tolerances are applied to physical dimensions of units such as thickness, length, width, squareness, and location and size of openings. They are normally established by economical and practical production considerations, and functional and appearance requirements. The actual tolerances should , however, be related to the amount of repetition, and the size and other characteristics of the precast concrete unit. The cost of manufacturing to close tolerances decreases with increased repetition.

The architect must specify product tolerances within generally accepted limits, as they relate to each individual project, or require performance within a generally accepted limit. The architect must account for the function of the member, its fit in the construction, and the compatibility of the member tolerances to those of the interfacing materials, Section 4.6.4. Tolerances for manufacturing are standardized throughout the industry and should not be made more exacting, and therefore most costly, unless absolutely necessary. Areas that might require more exacting tolerances could include special finish or appearance requirements, glazing details, and certain critical dimensions on open shaped panels, see Sections 3.5, 5.2. and 3.3.1.

For example, a special appearance requirement may be honed or polished flat concrete walls where bowing or warping tolerances might have to be decreased to 75 or 50% of tolerances proposed in Section 4.6.2 in order to avoid joint shadows. Another special case might be tolerances for dimensions controlling the matching of open shaped panels. These tolerances may have to be smaller than the standard dimensional tolerance (75% or 50%), unless the architect has recognized and solved the alignment problem as part of his design.

The architect should specify tolerances as they relate to each individual project. Architectural precast concrete can be manufactured to very close tolerances, but unless these tolerances reflect actual requirements, unnecessary cost factors may be introduced. Since manufacturing tolerances depend mainly on tooling, and these costs relate to repetition of identical castings, the cost of close tolerances is determined in part by repetitiveness in production. Envelope type molds (Section 2.2.4) are well suited to maintain close tolerances.

Where a project involves particular features sensitive to the cumulative effect of generally accepted tolerances on individual portions, the architect/engineer should anticipate and provide for this effect by setting a cumulative tolerance or by providing escape areas (clearances) where accumulated tolerances or production errors can be absorbed. The consequences of all tolerances permitted on a particular project should be investigated to determine whether a change is necessary in the design or in the tolerances applicable to the project or individual components. For example, there should be no possibility of minus tolerances accumulating so that the bearing length of members is reduced below the required design minimum. The designer should specify the nominal allowable bearing dimensions for each connection. These bearing dimen-

sions and their tolerance should be shown on the erection drawings.

Careful inspection of the listed tolerances will reveal that many times one tolerance will override another. The allowable variation for one element of the structure should not be applicable when it will permit another element of the structure to exceed its allowable variations. Restrictive tolerances should be reviewed to ascertain that they are compatible and that the restrictions can be met.

The product tolerances for architectural precast concrete panels have the following significance:

1. Length or width dimensions and straightness of the precast concrete will affect the joint dimension, opening dimensions between panels, and perhaps the overall length of the structure. Tolerances must relate to unit size and increase as unit dimensions increase.

2. Panels out-of-square can cause tapered joints and make adjustment of adjacent panels extremely difficult.

3. Thickness variation of the precast concrete unit becomes critical when interior surfaces are exposed to view. A non-uniform thickness of adjacent panels will cause offsets of the front or the rear faces of the panels.

Suggested Product Tolerances for architectural precast concrete panels are as follows:

Warping and bowing tolerances have an important effect on the edge match up during erection and on the visual appearance of the erected panels, both individually and when viewed together.

Warping is generally an overall variation from planeness in which the corners of the panel do not all fall within the same plane. Warping tolerances are stated in terms of the magnitude of the corner variation, as shown in Fig. 4.6.1.

Bowing is an overall out-of-planeness condition which differs from warping in that while the corners of the panel may fall in the same plane, the portion of the panel between two parallel edges is out of plane. Several possible bowing conditions are shown in Fig. 4.6.2. Differential temperature effects, differential moisture absorption between the inside and outside faces of a panel, and possible prestress eccentricity should be considered in design to minimize bowing and warping.

Fig. 4.6.1 Warping definitions for panels

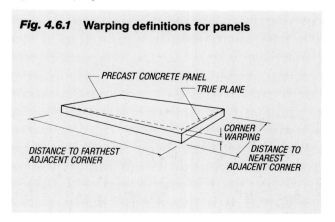

PRECAST CONCRETE PANEL
TRUE PLANE
CORNER WARPING
DISTANCE TO FARTHEST ADJACENT CORNER
DISTANCE TO NEAREST ADJACENT CORNER

Fig. 4.6.2   Bowing definitions for panels

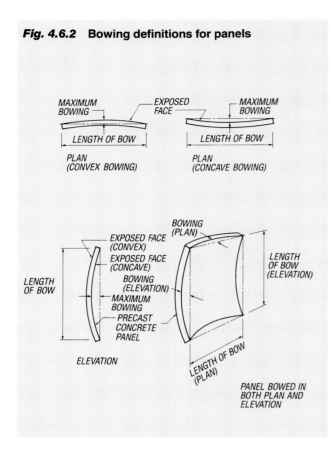

ELEVATION

PANEL BOWED IN
BOTH PLAN AND
ELEVATION

Table 4.6.1. Guidelines for panel thickness for overall panel stiffness consistent with suggested normal panel bowing and warping tolerances. *

Panel dimensions**	8'	10'	12'	16'	20'	24'	28'	32'
4'	3″	4″	4″	5″	5″	6″	6″	7″
6'	3″	4″	4″	5″	6″	6″	6″	7″
8'	4″	5″	5″	6″	6″	7″	7″	8″
10'	5″	5″	6″	6″	7″	7″	8″	8″

* This table should not be used for panel thickness selection.

** This table represents a relationship between overall flat panel dimensions and thickness below which suggested bowing and warpage tolerances should be reviewed and possibly increased. For ribbed panels, the equivalent thickness should be the overall thickness of such ribs if continuous from one end of the panel to the other.

Fig. 4.6.4   Measuring local smoothness variations

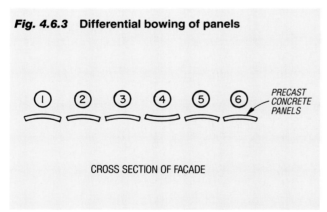

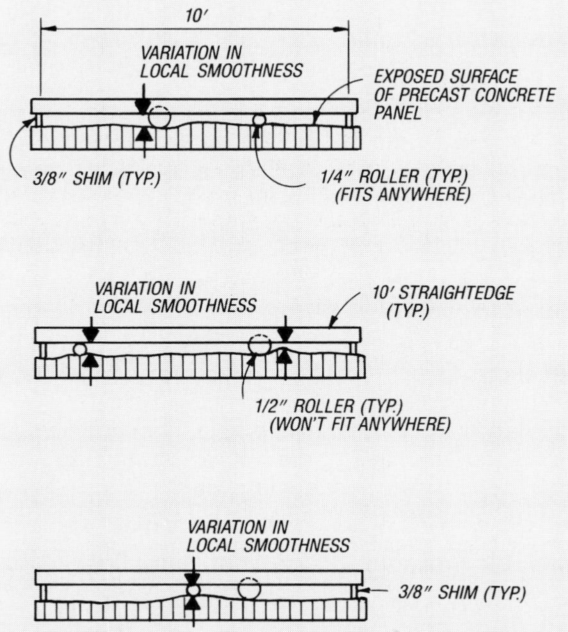

Fig. 4.6.3   Differential bowing of panels

CROSS SECTION OF FACADE

Note that bowing and warping tolerances are of primary interest at the time the panel is erected. Careful attention to pre-erection storage of panels is necessary, since storage conditions can be an important factor in achieving and maintaining panel bowing and warping tolerances.

Differential bowing is a consideration for panels which are viewed together on the completed structure. If convex bowing is positive (+) and concave bowing is negative (−), then the magnitude of differential bowing can be determined by adding the bowing values. For example, in Fig. 4.6.3, if the maximum bowing of Panel 3 is + 1/4 in. and the maximum bowing of Panel 4 is − 1/4 in. then the differential bowing between these adjacent panels is 1/2 in.

The likelihood that a panel will bow or warp depends on the design of the panel and its relative stiffness or ability to resist deflection as a plate member. Panels which are relatively thin in cross section, when compared to their overall plan dimensions, are more likely to warp or bow as a result of a number of design, manufacturing, and environmental conditions.

Slender panels should not be automatically subjected to the standard tolerances for bowing and warping. Table 4.6.1 represents a relationship between overall flat panel dimensions and cross sectional thicknesses below which the standard warping tolerances should be reviewed and possibly increased. Note that the thickness values in this table should not be considered as limiting values, but rather as

Fig. 4.6.5   Tolerances for columns

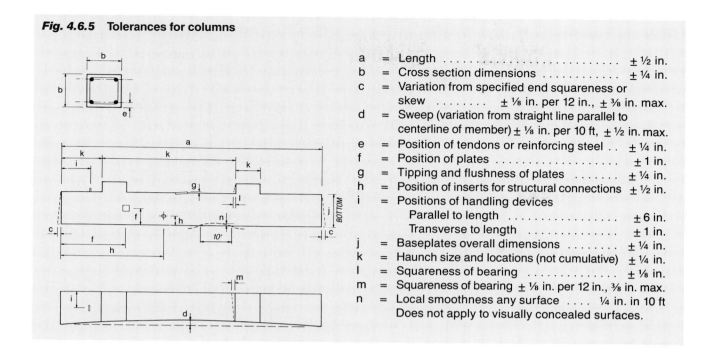

a	=	Length . . . . ± ½ in.
b	=	Cross section dimensions . . . . ± ¼ in.
c	=	Variation from specified end squareness or skew . . . . ± ⅛ in. per 12 in., ± ⅜ in. max.
d	=	Sweep (variation from straight line parallel to centerline of member) ± ⅛ in. per 10 ft, ± ½ in. max.
e	=	Position of tendons or reinforcing steel . . ± ¼ in.
f	=	Position of plates . . . . ± 1 in.
g	=	Tipping and flushness of plates . . . . ± ¼ in.
h	=	Position of inserts for structural connections ± ½ in.
i	=	Positions of handling devices
		Parallel to length . . . . ± 6 in.
		Transverse to length . . . . ± 1 in.
j	=	Baseplates overall dimensions . . . . ± ¼ in.
k	=	Haunch size and locations (not cumulative) ± ¼ in.
l	=	Squareness of bearing . . . . ± ⅛ in.
m	=	Squareness of bearing ± ⅛ in. per 12 in., ⅜ in. max.
n	=	Local smoothness any surface . . . . ¼ in. in 10 ft Does not apply to visually concealed surfaces.

an indication that more detailed consideration of the possible magnitude of warping and bowing is warranted. The major criteria for maintaining or relaxing bowing and warping tolerances will be the appearance requirements, the required type, as well as spacing, of connections, and the advice of the local precaster regarding overall economic and construction feasibility.

For ribbed panels, the equivalent thickness used in Table 4.6.1 should be the overall thickness of the ribs, if they run continuous from one end of the panel to the other. Similarly, panels which are manufactured using large aggregate concrete mixes (above 3/4 in. aggregate) or units which are fabricated from nonhomogeneous materials (such as two significantly different concrete mixes, natural stone or clay product veneers, insulating mediums, and the like) also require more careful consideration of warping and bowing tolerances.

Surface out-of-planeness, which is not a characteristic of the entire panel shape, is defined as a local smoothness variation rather than a bowing variation. Examples of local smoothness variations are shown in Fig. 4.6.4. The tolerance for this type of variation is usually expressed in fractions of an inch per 10 ft.

Fig. 4.6.4 also shows how to determine if a surface meets a tolerance of 1/4 in. in 10 ft. A 1/4 in. diameter by 2 in. long roller should fit anywhere between the 10 ft long straightedge and the element surface being measured when the straightedge is supported at its ends on 3/8 in. shims as shown. A 1/2 in. diameter by 2 in. long roller should not fit between the surface and the straightedge.

The dimensional tolerance requirements for products which may be used as architectural precast concrete elements are given in Figs. 4.6.5 and 4.6.6. It must be emphasized that these are guidelines only and that each project

must be considered individually to ensure that the stated tolerances are applicable.

Groups of inserts or cast-in items which must be located in close tolerance to each other should not be separated into two panels by a joint. Cast-in grooves, reglets, or lugs, that are to receive glazing gaskets, should be held relatively close to their correct location. Misalignment of these reglets at corners or casting these in a warped or "racked" position will restrict proper installation of the glazing gasket. In addition, gasket manufacturers place severe restrictive tolerances on the groove width and surface smoothness necessary to obtain a proper moisture seal of the gasket.

4.6.3 ERECTION TOLERANCES

Erection tolerances are those required for the efficient and acceptable matching of precast concrete members with the building structure. Erection tolerances control the position of the individual precast concrete members as they are located and placed in the assembled structure. They normally involve the general contractor and various subcontractors, such as the precast concrete erector. The basis for erection tolerances is determined by the characteristics of the building structure and site conditions. Tolerances are used to achieve uniform joints, level floor elevations and wall conditions having a plane surface. The specified erection tolerances affect the work of several trades and must be consistent with the tolerances specified for those trades.

Erection is both equipment and site dependent. There may be good reason to vary from the recommended tolerances to account for unique project conditions. The erection tolerances should be carefully reviewed by the designer and the involved contractors and adjusted, if necessary, to

Fig. 4.6.6 Tolerances for architectural panels, spandrels, and column covers

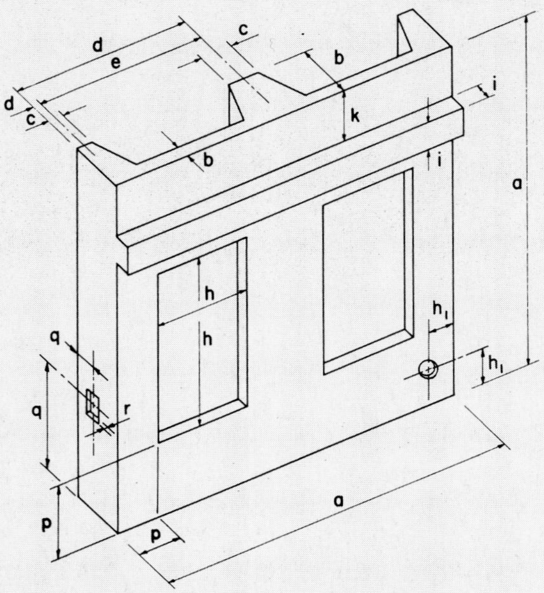

a = Overall length and width (measured at
neutral axis of ribbed members)
- 10 ft or under . . . . ± 1/8 in.
- 10 to 20 ft . . . . + 1/8 in., − 3/16 in.
- 20 to 40 ft . . . . ± 1/4 in.
- Each additional 10 ft . . . . . . ± 1/16 in. per 10 ft

b = Total thickness or
flange thickness . . . . − 1/8 in., + 1/4 in.

c = Rib thickness . . . . ± 1/8 in.

d = Rib to edge of flange . . . . ± 1/8 in.

e = Distance between ribs . . . . ± 1/8 in.

f = Angular variation of plane
of side mold . . . . ± 1/32 in. per 3 in.
of depth or ± 1/16 in.
whichever is greater

g = Variation from square or designated
skew (difference in length of the
two diagonal measurements) . . ± 1/8 in. per 6 ft
of diagonal or ± 1/2 in.
whichever is greater*

h = Length and width of blockouts
and openings within one unit . . . . ± 1/4 in.

h_1 = Location and dimensions of blockouts hidden
from view and used for HVAC and utility
penetrations . . . . ± 3/4 in.

h_2 = Some types of window and equipment frames
require openings more accurately placed. When
this is the case, the minimum practical tolerance
should be defined with input from the producer.

i = Dimensions of haunches . . . . ± 1/4 in.

j = Haunch bearing surface deviation
from specified plane . . . . ± 1/8 in.

k = Difference in relative positon of
adjacent haunch bearing surfaces
from specified relative position . . . . ± 1/4 in.

l = Bowing . . . . ± L/360
max. 1 in.

m = Differential bowing between adjacent
panels of the same design . . . . 1/2 in.

n = Local smoothness . . . . 1/4 in. in 10 ft
Does not apply to visually concealed surfaces.
(Refer to Fig. 4.6.4 for definition.)

o = Warping . . . . 1/16 in. per ft
of distance from nearest adjacent corner

p = Location of window opening within panel ± 1/4 in.

q = Position of plates . . . . ± 1 in.

r = Tipping and flushness of plates . . . . ± 1/4 in.

Positions tolerances. For cast-in items measured from
datum line location as shown on approved erection
drawings:

Weld plates . . . . ± 1 in.
Inserts . . . . ± 1/2 in.
Handling devices . . . . ± 3 in.
Reinforcing steel and welded wire fabric . . . . ± 1/4 in.
where position has structural implications or
affects concrete cover, otherwise ± 1/2 in.
Tendons . . . . ± 1/8 in.
Flashing reglets . . . . ± 1/4 in.
Flashing reglets at edge of panel . . . . ± 1/8 in.
Reglets for glazing gaskets . . . . ± 1/16 in.
Groove width for glazing gaskets . . . . ± 1/16 in.
Electrical outlets, hose bibs, etc. . . . . ± 1/2 in.
Haunches . . . . ± 1/4 in.

*Applies both to panel and to major openings in the panel.

meet the project requirements. The effects of adjusted tolerances on specific details at joints, at connections, and in other locations in the structure should be evaluated by the designer. Different details may have varying amounts of sensitivity to tolerances. If the final tolerances are different from those originally planned, the new tolerances should be stated in writing and noted on the erection drawings. The erector is responsible for erecting the members within the specified tolerances and completing the connections in the manner specified. Appropriate surveying and layout procedures should be followed to ensure accurate application of tolerances.

The primary control surfaces or features on the precast concrete members are erected to be in conformance with the established erection tolerances. The clearance is generally allowed to vary so that the primary control surface can be set within tolerance. Product tolerances are not additive to the primary surface erection tolerances. Secondary control surfaces which are toleranced from the primary control surfaces by the product tolerances are often not directly set during the erection process. An example would be the elevation of a second-story corbel on a multi-story column whose first-story corbel is selected as the primary elevation control surface.

The erection tolerances and product tolerances for some secondary control surfaces of a precast concrete member may be directly additive. Because of this, the erection drawings should make it clear to the erector which surfaces should control the erection. In instances where the tolerance of both primary and secondary control surfaces must be controlled during erection, the design should include provisions for adjustment which allow this to occur. This may occur with window openings between two spandrels when the critical elevation, top or bottom, as indicated on the erection drawings must be maintained. If more than one critical line is indicated, the erector should balance any deviations between the two edges. Requirements for controlling the various surfaces or features should be clearly outlined in the plans and specifications.

Erection tolerances are of necessity largely determined by the actual alignment and dimensional accuracy of the building foundation and frame. The general contractor is responsible for the plumbness, level, and alignment of the foundation and structural frame including the location of bearing surfaces and anchorage points for precast products.

The architect should recognize the critical importance of controlling foundation and building frame alignment tolerances and should include, on the contract drawings, clearance dimensions which make allowances for building frame tolerances so that the structure can be erected safely and economically. If the precast concrete units are to be installed reasonably plumb, level, square, and true, the actual location of all surfaces affecting their alignment, including the levels of floor slabs and beams, the vertical alignment of floor slab edges, and the plumbness of columns or walls must be known before erection begins. The general contractor is expected, and should be required to establish and maintain control points, benchmarks and lines in areas that remain undisturbed until the completion and acceptance of the project.

Tolerances in any building frame must be adequate to prevent obstructions that may cause difficulty with panel installation procedures. The structural frame must also provide for the use of standardized connections. Specifically, whenever possible, beam elevations and column locations should be uniform in relation to the precast concrete units with a constant clear distance between the precast concrete and support elements.

When determining tolerances, attention should also be given to possible deflection and/or rotation of structural members supporting precast concrete. This is particularly important for bearing on slender or cantilevered structural members. Consideration should be given to both initial deflection and to long-term deflection caused by plastic flow (creep) of the supporting structural members. If the deflection of the structural frame is sensitive to the location or eccentricity of the connection, limits should be given. Specific tolerances cannot be assigned to erection deformations. Yet, by selecting ample clearances and fully considering realistic tolerance variations, the influence of erection deformations can be minimized or eliminated for practical purposes.

Structural Steel Framing tolerances should be specified to conform with the American Institute of Steel Construction (AISC) *Code of Standard Practices*, Section 7.11. Particular attention is directed to the "Commentary" included in this Code, which provides a detailed explanation of the specified erection tolerances.

The allowable tolerances for steel frame structures make it impractical in tall structures to maintain precast concrete panels in a true vertical plane. Based on the allowable steel frame tolerances, it would be necessary to provide for a 3 in. adjustment in connections up to the 20th story and a 5 in. adjustment in connections above the 20th story if the architect/engineer insists on a true vertical plane. These adjustment in connections are not economically feasible. Therefore, precast concrete walls should follow the steel frame.

In other respects, a structural steel frame building presents different erection and connection problems than a concrete frame building. For example, structural steel beams, being relatively weak in torsion when compared to concrete, generally require that the load be applied directly over the web or that the connection be capable of supporting the induced moment. This, in turn, can place a greater structural requirement on the connection and create problems during erection if any rolling behavior occurs in the steel beam. When detailing precast concrete units for attachment to steel structures, allowance must be made for sway in tall, slender steel structures with uneven loading, and movements due to sun or wind on one side or seasonal thermal expansions and contractions.

Designs must provide for adjustment in the vertical dimension of precast concrete panels supported by the steel frame, because the accumulation of shortening of stressed steel columns will result in the unstressed panels supported at each floor level being higher than the steel frame connections to which they must attach. Observations in the field have shown that where precast concrete panels are erected to a greater height on one side of a multistory building than on the other, the steel framing will be pulled out of alignment. Precast concrete panels should be erected at a relatively uniform rate around the perimeter of the structure or the designer of the structural frame should determine the degree of imbalanced loading permitted.

Cast-in-Place Concrete Frame tolerances are given in ACI 117, *Standard Tolerances for Concrete Construction and Materials.* However, greater variations in height between floors are more prevalent in cast-in-place concrete structures than in other structural frames. This may affect the location or mating of the inserts in the precast concrete with the cast-in connection devices. Tolerances for cast-in-place concrete structures may have to be increased further to reflect local trade practices, the complexity of the structure, and climatic conditions. As a result, it is recommended that precast concrete walls should follow concrete frames in the same manner as for steel frames, if the details allow it and appearance is not affected.

The following tolerances, in addition to ACI 117 requirements, should be specified for cast-in-place concrete when precast concrete units are to be connected:

1. Footings, Caisson Caps, and Pile Caps

    a. Variation of bearing of surface from specified elevation: ± 1/2 in.

Table 4.6.2 Relevant Erection Tolerances for Cast-In-Place Concrete Structures and Structural Steel

VARIATION OR TOLERANCE	CAST-IN-PLACE CONCRETE	STEEL
Variations from the plumb, or column tolerances	¼ inch per 10 feet but not more than 1 inch Valid to 100 feet height. No tolerances suggested above 100 feet	1 to 1000, no more than 1″ towards building nor 2″ away from building line in the first 20 stories; plus 1/16″ for each additional story up to a maximum of 2″ towards building or 3″ away from building line
Tolerances in levels	In 10 feet . . . . ± ¼″ Up to 20 ft bay . . . . ± ⅜″ In 40 ft or more . . . . ± ¾″	Erection tolerances for levels normally not stated, as levels should be governed by close manufacturing tolerances
Variations from the linear building lines in relation to columns and walls	In any bay . . . . ± ½″ In bay 20 ft max. . . . . ± ½″ In bay 40 ft max. . . . . ± 1″	As set by column alignments Closer for elevator columns
Tolerances in beams and columns	Cross section ⎤ dimensions ⎦ − ¼″ + ½″	1 to 1000 in alignment Section tolerances are close
Tolerances for placing or fastening of other materials such as connection hardware in relation to building lines	From specified location . . . . ± ¼″	Not established

2. Piers, Columns, and Walls

  a. Variation in plan from straight lines parallel to specified linear building lines:

    1/40 in./ft adjacent members less than 20 ft apart or any wall or bay length less than 20 ft.

    1/2 in. adjacent members 20 ft or more apart or any wall or bay length of 20 ft or more.

  b. Variation in elevation from lines parallel to specified grade lines:

    1/40 in./ft adjacent members less than 20 ft apart or any wall or bay length less than 20 ft.

    1/2 in. adjacent members 20 ft or more apart or any wall or bay length of 20 ft or more.

3. Anchor Bolts

  a. Variation from specified location in plan: ± 1/4 in.

  b. Variation center to center of any two bolts within an anchor bolt group: ± 1/8 in.

  c. Variation from specified elevation: ± 1/2 in.

  d. Anchor bolt projection: − 1/4 in., + 1/2 in.

  e. Plumb of anchor bolts: ± 1/16 in./ft.

It should be recognized that ACI 117 applies only to reinforced concrete buildings, and the AISC Code only to steel building frames. Neither of these standards applies to buildings of hybrid construction (e.g., concrete floor slabs carried by steel columns or to concrete encased structural steel members, or fireproofed frames). Obviously, the location of the face of the fireproofing on the steel, as well as that of the steel member itself, are both critical. As the alignment of composite members, fireproofing and masonry work are not controlled by referencing these standards, the architect/engineer should require that the location of all such materials contiguous to the precast concrete units be controlled within tolerances which are, at most, no greater than those specified in ACI 117. Should there be some doubt as to what these tolerances should be, the precast concrete manufacutrer should be consulted for advice.

The recommended tolerances for steel and cast-in-place concrete structures are generous, Table 4.6.2. Consequently, it is generally poor practice to design gaps or joints between site work and precast concrete as an architectural feature.

A case in point would be the designing of individual panels to fit between cast-in-place columns and beams with either of these structural members exposed. This is illustrated in Fig. 4.6.7. Unless the cast-in-place structure is executed to well above normal, economic site tolerances, the width of joints must be allowed with a large tolerance (± 1/2 in. in the case of a 20 ft opening). The actual joint width may differ in each bay, and will certainly require sealants with corresponding flexibility. Joint widths may be adjusted to enable them to be equal at either end of a panel, but efforts toward equalizing the joints on either side of a column cannot be attempted unless panels can be adjusted horizontally after erection. The problems this could cause are avoided where the cladding passes in front of the columns and the jointing is between the panel edges.

Erection tolerances are less critical in buildings consisting entirely of precast concrete elements than for cast-in-place concrete or steel frame structures.

Fig. 4.6.7 **Design concepts to accommodate site tolerances**

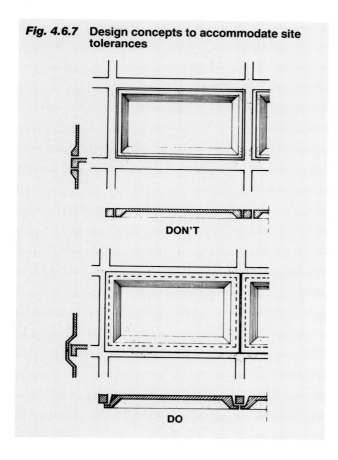

DON'T

DO

The erection tolerances are architectural precast concrete are given in Figs. 4.6.8 and 4.6.9. These are guidelines only and each project must be considered individually to ensure that the stated tolerances are applicable. After precast concrete erection and before other trades interface any material with the precast concrete members, it should be verified that the precast concrete elements are erected within recommended tolerances.

If reasonable tolerances and adjustments have been designed into the construction details and complied with, the erector should be able to:

1. Avoid joint irregularities such as tapered joints (panel edges not parallel), jogs at intersections, non-uniform joint widths.

2. Maintain proper opening dimensions.

3. Properly execute all fastening connections.

4. Align the vertical faces of the units to avoid offsets.

5. Prevent the accumulation of tolerances.

A more precise installation and general improvement in appearance is thus achieved.

Whenever possible, the precast concrete erector should perform a survey of the building as constructed and set out joint locations prior to actual product installations. This will keep the differential variation in joint width to a minimum, as well as identifying problems caused by building frame columns or beams being out of dimension or alignment.

The resulting joint width may vary considerably, within specified tolerance, from the theoretical width, but a smaller variation between adjacent joints should be possible.

Variations from true length or width dimensions of the overall structure are normally accomodated in the joints or, where this is not feasible or desirable, at the corner panels, in expansion joints, or in joints adjacent to other wall materials. A liberal joint width should be allowed if variations in overall building dimensions are to be absorbed in the joints. This may be coupled with a closer tolerance for variations from one joint to the next for appearance purposes. It is apparent that the individual joint width tolerance should relate to the number of joints over a given building dimension. For example, to accommodate reasonable variations in actual site dimensions a 3/4 in. joint may be specified with a tolerance of ± 1/4 in. but with only a 3/16 in. differential variation allowed between joint widths on any one floor, or between adjacent floors.

In a situation where a joint has to match an architectural feature (such as a false joint) a large variation from theoretical joint width may not be acceptable and tolerances for building lengths will have to be accommodated at the corner panels. A similar condition often occurs where precast concrete is interspersed with glass or curtain wall elements, as in precast concrete mullion projects. Close tolerances are often mandatory between the mullion and the glass or curtain wall. This condition demands additional flexibility that may be provided by the corner details. A solution which permits two or more corner elements to overlap and thus take up irregularities in the overall building dimensions is recommended, see Section 3.3.6.

Clearance is the space provided between adjacent members. It is one of the most important factors to consider in erection. The clearance space should provide a buffer area where frame, erection and product tolerance variations can be absorbed.

The designer should provide clearance space between the theoretical face of the structure and the back face of the units in detailing the wall and its relationship to the building structure. The face of the structure may be precast concrete, cast-in-place concrete, masonry, or structural steel frame. Adjacent materials may include products such as glass or subframes that are installed after the precast concrete panels are in place. If sufficient space is not provided, alignment of the wall as specified will likely necessitate delays and extra costs, and may be impossible. The failure to provide adequate clearances is an all too common deficiency of wall designs. When determining clearances, the following primary basic considerations should be addressed:

1. Product tolerance.
2. Type of member.
3. Size of member.
4. Location of member.
5. Member movement.
6. Function of member.
7. Erection tolerance.
8. Fireproofing of steel.
9. Thickness of plates, bolt heads, and other projecting elements.

Fig. 4.6.8 Erection tolerances—columns

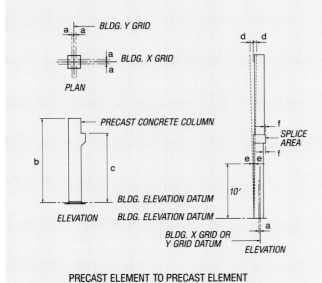

PLAN

PRECAST CONCRETE COLUMN

ELEVATION

BLDG. ELEVATION DATUM
BLDG. ELEVATION DATUM

BLDG. X GRID OR
Y GRID DATUM

ELEVATION

SPLICE AREA

PRECAST ELEMENT TO PRECAST ELEMENT

a = Plan location from building grid datum
    Structural applications . . . . $\pm$ ½ in.
    Architectural applications . . . . $\pm$ ⅜ in.

b = Top elevation from nominal top elevation
    Maximum low . . . . ½ in.
    Maximum high . . . . ¼ in.

c = Bearing haunch elevation from nominal elevation
    Maximum low . . . . ½ in.
    Maximum high . . . . ¼ in.

d = Maximum plumb variation over height of element
    (element in structure of maximum height of
    100 ft) . . . . 1 in.

e = Plumb in any 10 ft of element height . . . . ¼ in.

f = Maximum jog in alignment of matching edges
    Architectural exposed edges . . . . ¼ in.
    Visually non-critical edges . . . . ½ in.

Fig. 4.6.9 Erection tolerances—architectural wall panels

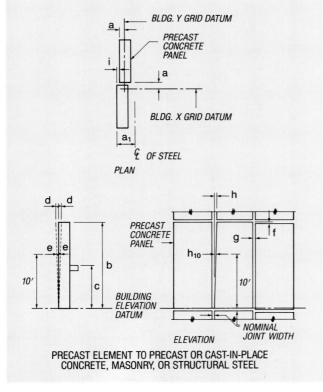

BLDG. Y GRID DATUM

PRECAST CONCRETE PANEL

BLDG. X GRID DATUM

C_L OF STEEL

PLAN

PRECAST CONCRETE PANEL

BUILDING ELEVATION DATUM

ELEVATION

NOMINAL JOINT WIDTH

PRECAST ELEMENT TO PRECAST OR CAST-IN-PLACE
CONCRETE, MASONRY, OR STRUCTURAL STEEL

a = Plan location from building grid datum* . $\pm$ ½ in.

a_1 = Plan location from centerline of steel** . $\pm$ ½ in.

b = Top elevation from nominal top elevation
    Exposed individual panel . . . . $\pm$ ¼ in.
    Nonexposed individual panel . . . . $\pm$ ½ in.
    Exposed relative to adjacent panel . . . ¼ in.
    Nonexposed relative to adjacent panel ½ in.

c = Support elevation from nominal elevation
    Maximum low . . . . ½ in.
    Maximum high . . . . ¼ in.

d = Maximum plumb variation over height of structure
    or 100 ft. whichever is less* . . . . 1 in.

e = Plumb in any 10 ft of element height . . . . ¼ in.

f = Maximum jog in alignment of
    matching edges . . . . ¼ in.

g = Joint width (governs over joint taper) . . . $\pm$ ¼ in.

h = Joint taper maximum . . . . ⅜ in.

h_{10} = Joint taper over 10 ft. length . . . . ¼ in.

i = Maximum jog in alignment of
    matching faces . . . . ¼ in.

j = Differential bowing or camber as erected between
    adjacent members of the same design . . ¼ in.

*For precast buildings in excess of 100 ft. tall, tolerances "a" and "d" can
increase at the rate of ⅛ in. per story to a maximum of 2 in.
**For precast concrete erected on a steel frame building, this tolerance
takes precedence over tolerance on dimension "a".

The design of this clearance should consider not only the possible variation in the precast concrete members, but also the possible variation in the support system (building frame). The clearances must enable the erector to complete the final assembly without field altering the physical dimensions of the precast concrete units.

The type of member is partially accounted for when product tolerances are considered. There are additional factors which should also be considered. An exposed-to-view member requiring small erection tolerances requires more clearance for adjustment than a non-exposed member with a more liberal erection tolerance. Similarly, a corner member should have a large enough clearance provided so it can be adjusted to line up with both of the adjacent panels.

The size and weight of the member is another consideration in determining erection clearances. Large members are more difficult to handle than smaller ones. Therefore, a large member being erected by a crane requires more clearance than the small member that can be hand erected or adjusted.

Clearance design should consider member deflection and rotation, and movements caused by temperature expansion and contraction, creep and shrinkage. The clearance between a vertical member and a horizontal member should allow for some movement in the horizontal member to prevent the vertical member from being pushed or pulled out of its original alignment. If not considered in the design, such movements can create waterproofing problems or roofing failure at the interface.

Consideration should be given to the limits of adjustment permitted by the connection details. If a connection is required to the faces of spandrel beams or columns or their fireproofing, more clearance will be needed to install fastenings than when the anchors are located on the top surface of beams and the sides of columns. Also, space should be provided to make the connection – sufficient room for welding or adequate space to place a wrench to tighten a bolt.

All connections should be provided with the maximum adjustability in all directions that is structurally or architecturally feasible. Where a 1 in. clearance is needed but a 2 in. clearance creates no structural or architectural problems, the 2 in. clearance should be selected. Less clearance is required for bolted connections than for grouted connections. In larger more complex structures, vertical adjustments of 1 in. minimum and horizontal adjustments of 2 in. minimum should be provided to accommodate any misalignment of the support system and the precast concrete units. Location of hardware items cast into, or fastened to the structure by the general contractor, steel fabricator, or other trades should be ± 1/4 in. in all directions (vertical and horizontal), plus a slope deviation of no more than ± 1/4 in. for the level of critical bearing surfaces.

Connection details should consider the possibility of bearing surfaces being misaligned or warped from the desired plane. Adjustments can be provided by the use of drypack concrete, non-shrink grout, or elastomeric pads.

Where possible, connections should be dimensioned to the nearest 1/2 in. The minimum clearance between various items within a connection should not be less than 1/4 in. with 1/2 in. preferred. The minimum clearance or shim space should be a minimum 3/4 in. for steel structures or 1 in. for cast-in-place concrete structures.

Where a unit is not erected within the tolerances assumed in the connection design, the structural adequacy of the installation should be checked and the connection design should be modified, if required. No unit should be left in an unsafe support condition. Any adjustments affecting structural performance, other than adjustment within the prescribed tolerances, should only be made after approval by the design engineer. Units should not be forced into place or installed by any method which would induce or impose any undue stress or overloads on the structure, on the units, or on the connections.

The following procedure is recommended for use in determining minimum clearances under normal conditions. It is a systematic approach for making a trial selection of a clearance value and then testing that selection to assure that it will allow practical erection to occur. It should be emphasized that engineering judgment must be included as part of the clearance determination process.

1. Determine the maximum size of the members involved (basic or nominal dimension plus additive tolerances). This should include not only the precast concrete members, but also other materials.

2. Add to the maximum member size the minimum space required for member movement.

3. Check to see if this clearance allows the member to be erected within the erection and interfacing tolerances, such as plumb, face alignment, etc. If the member interfaces with other structural systems, such as a steel frame or a cast-in-place concrete frame, check to see if the clearance provides for the erection and member tolerances of the interfacing system. Adjust the clearance as required to meet all of the needs. For example, if insulation is to be site applied on the interior face of the precast concrete, the insulation may be locally reduced to suit tight clearances.

4. Check to see if the member can physically be erected with the clearance determined above. Consider the size and location of members in the structure and how connections will be made.

5. Review the clearance to see if increasing its dimensions will allow easier, more economical erection without adversely affecting esthetics. Adjust the clearance as required.

6. Review structural considerations such as types of connections involved, sizes required, bearing area requirements, and other structural issues.

7. Check design to ensure adequacy in the event that minimum member size should occur. Adjust clearance as required for minimum bearing and other structural considerations.

8. Select the final clearance which will satisfy all of the conditions considered.

If the clearance provided is too small, erection may be slow and costly because of fit-up problems and the possibility of rework. A good rule of thumb is that at least 1/2 in. clearance be required between precast concrete members; 1 in. is the minimum clearance between precast concrete members and cast-in-place concrete with 1-1/2 in. preferred; and 1 in. is the minimum clearance required between precast concrete members and a steel frame. At least a 1-1/2 to 2 in. clearance should be specified in tall, irregular structures, regardless of the structural framing materials. The minimum clearance between column covers and columns must be 1-1/2 in.; 3 in. is preferred because of the possibility of columns being out of plumb or a column dimension causing interference with completion of the connection. If clearances are realistically assessed they will solve many tolerance problems. Where large tolerances have been allowed for a supporting structure, or where no tolerances are given, the clearance must be increased.

In determining clearances, it is suggested that product tolerances and member movement be the initial factors considered. This clearance should then be adjusted for other factors. It should be noted that it may not be practical to account for all possible factors in the clearance provided. Therefore, factors such as adjustment within the plumb and face alignment tolerances should be used as a way of correcting the remaining factors not accounted for in the initial clearance determination. The architect/engineer and the precast concrete manufacturer should consider the fact that the maximum variance from nominal dimensions given under product tolerances are not likely to occur as the standard throughout the project. Often, if this is done, the resulting clearance would be too great and components, such as haunches and connection angles, would be uneconomical.

The following example is given to show a thought process only. Situations are created to emphasize that judgment must be included as part of the clearance determination process. Therefore, the solution shown is not the only correct one for the situation described.

Given: A 36-story steel frame structure, precast concrete cladding, steel tolerances per AISC, member movement negligible. In this example precast concrete tolerance for variation in plan is $\pm$ 1/4 in. Refer to Fig. 4.6.10.

Find: Whether or not the panels can be erected plumb and determine the minimum acceptable clearance at the 36th story.

Procedure:

Step 1 — Product Tolerances
(Refer to Product Tolerances, Section 4.6.2)
Precast concrete cladding thickness . + 1/4 in., – 1/8 in.
Steel width . . . . + 1/4 in., – 3/16 in.
Steel sweep (varies) . . . . 1/4 in. assumption
Initial clearance chosen . . . . 3/4 in.

Step 2 — Member Movement
For simplification, assume this can be neglected in this example.

Step 3 — Other Erection Tolerances
Steel variation in plan, maximum . . . . 2 in.
Initial clearance . . . . 3/4 in.
Clearance chosen . . . . 2-3/4 in.

Step 4 — Erection Considerations
No adjustment required for erection considerations.

Step 5 — Economy
Clearance chosen . . . . 2-3/4 in.
Increasing clearance will not increase economy. No adjustment for economic considerations.

Step 6 — Structural Considerations
Clearance chosen . . . . 2-3/4 in.

Expensive connection but possible. No adjustment.

Step 7 — Check Minimum Member Sizes at 36th Story
(Refer to Product Tolerances)
Initial clearance . . . . 2-3/4 in.
Precast concrete thickness . . . . – 1/8 in.
Steel width . . . . – 3/16 in.
Steel sweep . . . . – 1/4 in.
Steel variation in plan minimum . . . . – 3 in.
Clearances calculated . . . . 6-5/16 in.

Step 8 — Final Solution
Minimum clearance used . . . . 2-3/4 in.

Note: When the minimum condition exists, the resulting

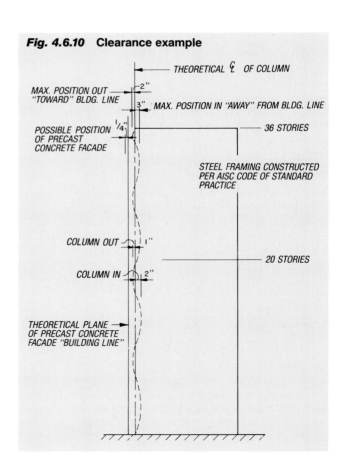

Fig. 4.6.10 Clearance example

THEORETICAL $\mathcal{C}$ OF COLUMN
MAX. POSITION OUT "TOWARD" BLDG. LINE — 2"
3" MAX. POSITION IN "AWAY" FROM BLDG. LINE
POSSIBLE POSITION OF PRECAST CONCRETE FACADE — 1/4"
36 STORIES
STEEL FRAMING CONSTRUCTED PER AISC CODE OF STANDARD PRACTICE
COLUMN OUT — 1"
20 STORIES
COLUMN IN — 2"
THEORETICAL PLANE OF PRECAST CONCRETE FACADE "BUILDING LINE"

clearance of over 6 in. produces an extremely expensive connection for the precast concrete. In addition, it produces a high torsional force that must be considered in the design of any steel horizontal supporting members. All together, this can substantially increase the cost of the steel perimeter members.

The 6 in. clearance is judged not practical, although the 2-3/4 in. minimal initial clearance is still needed. Therefore, the original erection tolerances need to be adjusted. Either the precast concrete member should be allowed to follow the steel frame or the frame tolerances need to be made more stringent. The most economical and recommended solution likely will be for the precast concrete member to follow the steel frame.

Another solution which has proven both practical and economical is to specify the more stringent AISC elevator column erection tolerances for steel columns in the building facade to receive the precast concrete panels. This type of solution should be settled as part of the design and specification process, or at least prior to finalization of the fabrication erection process.

4.6.4 INTERFACING TOLERANCES

Interfacing tolerances and clearances are those required for joining of different materials in contact with or in close proximity to precast concrete both before and after precast concrete erection, and for accommodating the relative movements expected between such materials during the life of the building. Typical examples include tolerances for window and door openings; joints, flashing and reglets; mechanical and electrical equipment; elevators and interior finishes; and walls and partitions.

Fabrication and erection tolerances of other materials must be considered in design as the precast concrete units must be coordinated with and must accommodate the other structural and functional elements comprising the total structure. Unusual requirements or allowances for interfacing should be stated in the contract documents.

Where matching of the different materials is dependent on work executed at the construction site, interface tolerances should be related to erection tolerances. Consideration should also be given to provision for adjustment of the materials after installation. Where tolerances are independent of site conditions, they will depend solely on normal manufacturing tolerances plus an appropriate allowance (clearance) for differential volume changes between the materials. For example, standard window elements have associated installation details which require a certain tolerance on window openings in a precast concrete panel. If the opening is completely contained within one panel, can the required tolerances on the window opening be economically met? If not, is it less expensive to the project to procure special windows or to incur the added cost associated with making the tolerances on the window opening more stringent? Also, openings for aluminum windows should allow room for some temperature expansion of the sash.

With interfacing tolerances it is important to note that the tolerances may be very system dependent. For example, windows of one type may have a quite different interface tolerance than windows of another type. If material or component substitutions are made after the initial design is complete, the responsibility for assuring that the interface tolerances are compatible logically passes to the party initiating the substitutions.

Adequate interface/erection tolerances are required for window openings, doors or louvers common to two or more panels and the cost of erecting the panels to achieve required window interface tolerances must also be considered. A similar condition often occurs where precast concrete is interspersed with glass or metal curtain wall elements, as in many precast concrete spandrel or mullion projects. Close tolerances are often mandatory between the mullion and the glass or curtain wall. This condition demands additional flexibility that may be provided in the corner details. A solution which permits two or more corner elements to overlap and thus take up irregularities in the overall building dimension is recommended. Also, the bow of spandrel panels may be critical, if windows are to be installed.

Product tolerances, erection tolerances, and interface tolerances together determine the dimensions of the completed structure. Which tolerance takes precedence on any given project is a question of economics, which should be addressed by considering fabrication, erection, and interfacing cost implications.

The precedence of product and erection tolerances raises similar questions of project economics. In this instance, the tolerance requirements and other costs associated with the connection details also must be considered. The architect/engineer must evaluate the costs associated with each of the three tolerances, paying special attention to unique and unusual situations, difficult erection requirements, connections which are very tolerance sensitive, and product or erection tolerances which are more stringent than normal practice.

Special tolerances or construction procedures require early decisions based on overall project economics. Once these decisions have been made, they should be reflected in the project plans and specifications. All special tolerance requirements, special details, and special procedures should be clearly spelled out in the specifications. The plans and specifications then define the established tolerance priority to be adhered to on the project.

4.7 JOINTS

4.7.1 GENERAL

The joints between precast concrete panels or between panels and other building materials must be considered the weakest link in the overall watertightness of the wall. The design and execution of these joints is therefore of the utmost importance and must be accomplished in a rational, economical manner. The joint treatment also has an effect on the general appearance of the project.

A joint will provide a degree of watertightness consistent with its design and exposure. In addition the purpose, size and function of the building will also determine design requirements for the joint. Design criteria for joints include:

1. Structural requirements (amount of movement to accommodate).

2. Architectural appearance.

3. Function of the building.

4. Exposure (orientation and climatic conditions).

5. Economics.

The following decisions must be made in response to the design criteria:

1. Width.

2. Type.

3. Location.

4. Number.

5. Architectural treatment.

6. Materials selection.

Ideally, joints should be located on elevations during the design development phase. Items affected by joint design are:

1. Panel size and dimensional accuracy.

2. Weathering.

3. Tolerances.

4. Transition between adjacent materials.

5. Location of openings.

Joints are required to accommodate changes in wall panel or structure dimensions caused by changes in temperature, moisture content, or load. Joint sealants must at the same time prevent water and air penetration through the building envelope. The joints between panels are normally designed to accommodate local wall movements rather than cumulative movements. Cumulative movements, as well as differential expansion movement of adjacent wall materials, are generally taken by specially designed expansion joints. Since an expansion joint may have to accommodate considerable movement, it should be designed as simply as possible. Although this might result in an appearance somewhat different from a normal joint, the architect is urged to either treat it as an architectural feature or simply leave it as a different, but honest, expansion joint. Fig. 4.7.1 shows a solution where the expansion joint does appear differently from the normal joints without disturbing the architectural integrity of the design.

Materials for expansion joints must be chosen for their ability to absorb appreciable movement while performing their primary function to control the movement of moisture and air. Fig. 4.7.2 shows a bellows-type expansion seal of neoprene that accommodates 1-1/2 in. of thermal movement. Joints must be designed first for weather protection, then for movement, and finally for appearance. In most cases, this requires that special gasket materials be used, rather than field molded sealants. Otherwise, the requirements for expansion joints are similar to those listed above for normal joints.

Some wall designs handle water properly in two-dimensional blueprints, but fail in three-dimensional reality. Isometric drawings should be used to show the proper intersection of horizontal and vertical seals and flashings. These intersections are a prime source of waterproofing problems. The second line of sealant placed must be run continuously across the intersection.

4.7.2 TYPES OF JOINTS

Joints between precast concrete wall units may be divided into two basic types: one-stage or two-stage joints.

ONE-STAGE JOINTS. As its name implies, the one-stage joint has a single line of defense for its weatherproofing capacity. This is normally in the form of a field molded sealant close to the exterior surface. See Figs. 4.7.3 and 4.7.11. Field molded one-stage joints are acceptable where joint width and movement are nominal.

The principal advantages of this type of joint are its simplicity and its almost universal suitability for normal joints between precast concrete panels. No grooves or special shapes are necessary. Thus, one-stage joints are normally the most economical with regard to initial cost. However, the economic picture may change when maintenance costs are included in the evaluation. The major disadvantages of one-stage field molded sealant joints are:

Fig. 4.7.1

Fig. 4.7.2 Expansion seal

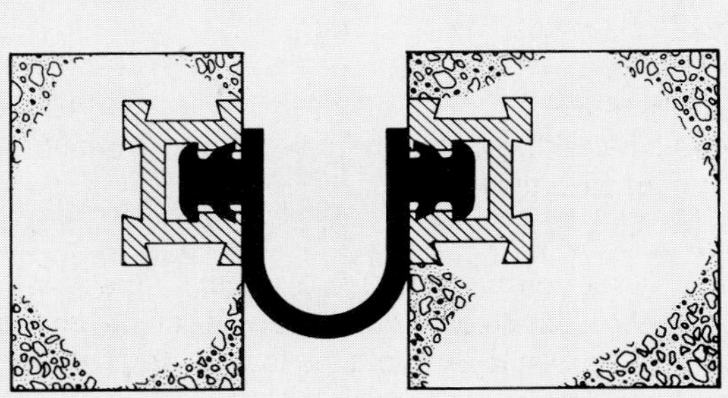

Fig. 4.7.3 One-stage joints

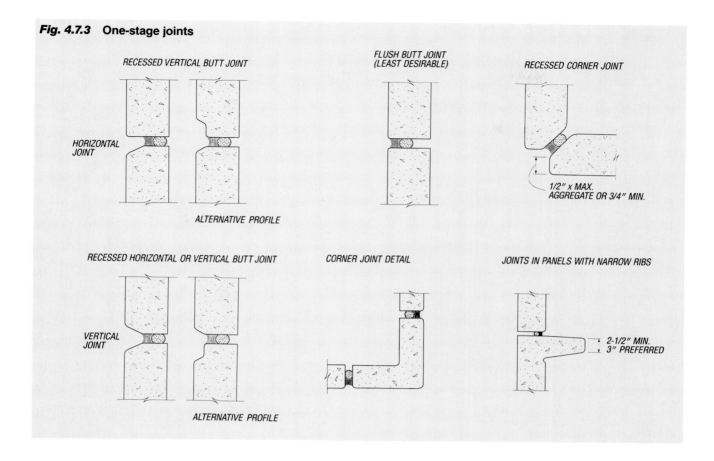

RECESSED VERTICAL BUTT JOINT

FLUSH BUTT JOINT
(LEAST DESIRABLE)

RECESSED CORNER JOINT

HORIZONTAL
JOINT

ALTERNATIVE PROFILE

1/2" x MAX.
AGGREGATE OR 3/4" MIN.

RECESSED HORIZONTAL OR VERTICAL BUTT JOINT

CORNER JOINT DETAIL

JOINTS IN PANELS WITH NARROW RIBS

VERTICAL
JOINT

2-1/2" MIN.
3" PREFERRED

ALTERNATIVE PROFILE

1. Workmen must have access to the face of the building using cradles or scaffolding.

2. The concrete surfaces must be smooth, free from laitance, clean, and dry to ensure good adhesion. Primers are also necessary with some sealant systems.

3. Even a small failure in adhesion of the compound may allow water penetration due to capillarity or pressure differentials.

4. The sealant is fully exposed to the major agents of aging and deterioration—ultraviolet light and the weather. Relatively expensive sealants must be used to ensure good long term performance.

The one-stage joint is the most common joint type used throughout North America. Although one-stage joints appear to perform satisfactorily in most climates, their performance depends heavily on the quality of materials, the condition of joint surfaces, proper field installation, and the overall wall design. One-stage joints should be regularly inspected and may demand fairly frequent maintenance to remain weathertight.

Since many sealants are subject to deterioration from the elements and ultraviolet (UV) exposure, double sealants have been installed around panels on a number of projects. The first layer is placed inside the joint, generally from the exterior, and recessed from the panel surface. This layer is fully protected from weather and UV exposure

by the second or outer layer of sealant, which is installed in the normal manner. Any leaks through the outer layer are contained by the inner layer and drained to the exterior, thus extending the life of the overall sealant installation. A primary advantage of this system is that performance does not rely on perfect workmanship. Even though two layers of field molded sealant are used, this is considered a one-stage joint.

TWO-STAGE JOINTS. These joints are based upon the open rainscreen principle. They are sometimes known as ventilated, or pressure equalization joints. A cavity wall design is considered a further application of the rainscreen principle. The rainscreen principle is based upon the control of the forces that can move water through small openings, rather than the elimination of the openings themselves.

These joints have two lines of defense for weatherproofing. The typical joint consists of a rain barrier near the exterior face and an airseal close to the interior face of the panel. The rain barrier is designed to shed most of the water from the joint, and the wind-barrier or airseal is the demarcation line between outside and inside air pressure. Between the two stages is an equalization or expansion chamber which must be vented and drained to the outside.

The simplest form of a horizontal two-stage joint is the well proven shiplap joint, Figs. 4.7.4(a) and (b). Typical details of two-stage vertical joints are shown in Figs. 4.7.4(d) and (e).

Fig. 4.7.4 Two-stage joints

(a) Two-stage joint—horizontal—using gasket

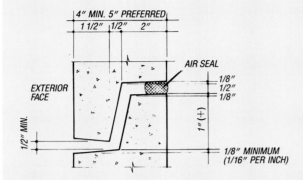

(b) Two-stage joint—horizontal—using field-molded sealants

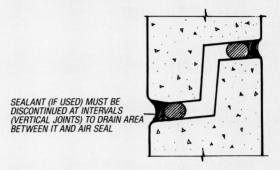

SEALANT (IF USED) MUST BE DISCONTINUED AT INTERVALS (VERTICAL JOINTS) TO DRAIN AREA BETWEEN IT AND AIR SEAL

(c) Minimum requirement for elementary two-stage joint

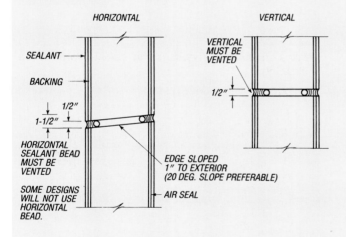

(d) Two-stage joint—vertical—using field-molded sealants

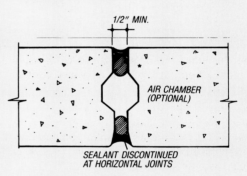

SEALANT DISCONTINUED AT HORIZONTAL JOINTS

* REVERSE SEALANT AND BACKER ROD WHEN INSTALLED FROM FRONT

(e) Modified two-stage joint with air chamber

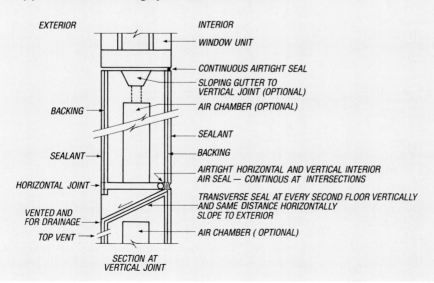

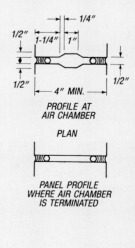

In the two-stage joint, the exterior sealant or rain barrier prevents direct entry of most airborne water. If airborne water (wind-driven rain) penetrates this barrier, it will drain off in the air chamber as the kinetic energy is dissipated by eddying and the air loses its ability to carry the water. The air chamber which is vented via the jointing to the outside environment forms a pressure equalizer space. Pressure equalization is achieved by using shiplapped horizontal panel joints, by venting the air space, or by leaving an open horizontal joint at the windows, which necessitates proper flashing details. With pressure equalization, water should not penetrate the wall system far enough to cause any problems.

The change in air pressure between the outside and inside atmospheres occurs at the wind barrier or airseal. The wind barrier must be located near to the interior side of the precast concrete panel. It prevents cold air from entering the building. Air moving from the building interior to the exterior usually carries moisture. Air must be prevented from flowing into cold spaces in the wall and contacting cold surfaces. In northern climates, thermal bridges can occur and allow condensation to form a buildup of frost in or on the walls, ultimately causing failure of the joint sealant. This frost later melts and runs back inside the building, giving the impression that the building is leaking. The wind barrier would normally be subjected to water penetration by capillary action. Since the outside air which reaches this seal has already been stripped of its water content, no moisture enters.

The airtight interior seal is the prime means in preventing warm, moist air movement and controlling the effectiveness of the air chamber and exterior rain-barrier. It is essential that the airseal be maintained. Failure of the airseal could result in a complete breakdown of the joint sealing mechanism. Therefore the interior airseal should be installed first.

Water either from penetration or condensation, should be drained from the joint by proper flashing installations. It is advisable to use these flashing details as dampers or vents to avoid vertical movement of the air in the expansion chamber caused by wind and outside air turbulence. Flashings should be installed at regularly spaced intervals along the height of vertical joints. Flashing using the sealant or with shiplapped joints using an elastomeric sheet is sometimes installed at each floor level, Fig. 4.7.4(e). A greater spacing of two or three stories may be sufficient for low-rise buildings and in areas of moderate wind velocities.

Vent or drain tubes should be 1/4 to 3/8 in. inside diameter polyvinyl chloride or other non-staining materials. The vent tube should slope down to the exterior face of the panel, penetrating the joint backing so it allows free movement of air between the outside environment and the cavity. Vents should project at least 1/8 in. beyond the sealant. They should be located at the floor levels with flashing and be spaced approximately 6 to 10 ft on centers. In most cases, the vents would be located at the junction of the horizontal and vertical joints.

The two-stage joint is the safest system for buildings sub-

ject to severe climatic exposure (greater than 5000 degree days). Some of the disadvantages of the two-stage drained joint system are:

1. Higher initial cost due to labor and materials required for their successful application.

2. The panel edge profile may involve relatively deep grooves, increasing the manufacturing cost. This edge profile renders the panel edge more vulnerable to damage during handling. Two-stage joints must be carefully detailed, fabricated, and installed if they are to succeed.

3. Installation of the components of this system are usually carried out during the erection of the cladding, with little opportunity for subsequent modifications or remedial work to correct omissions and poor workmanship. Sealants are not easily placed at the back of the two-stage joint. Resealing may be a difficult and expensive operation.

4. Conscientious workmanship or intensive supervision throughout the installation procedure is necessary, since inspection of the completed installation is difficult.

5. The geometry of the concept requires careful detailing, fabrication, installation and may place limitations on the architect with regard to the esthetic form of the cladding.

A minimum panel thickness of 4 in. with field-molded sealants or preferably 5 in. with gaskets and compression seals is required to accommodate both rain-barrier and airseal in a precast concrete wall panel. The joints must be accessible from the inside if the airseal is to be installed after the exterior rain barrier. If the sealant is applied from the inside, it is very difficult to achieve continuity around columns and floor slabs which block access. Four approaches for offsetting joint locations to provide accessibility are shown in Fig. 4.7.5. Alternatively, certain joint designs and horizontal joints inaccessible after erection may be sealed using preformed gaskets. The gasket material may be glued to the precast concrete unit to ensure that its location is maintained. However, it is recommended that pre-glued gaskets not be used in vertical joints, since proper compression of the gasket during erection may be difficult to achieve. It usually is necessary to move the precast concrete unit horizontally with come-a-longs after release from the crane and initial fastening of top connections. Many erectors like the buffer effect of the sponge neoprene, particularly for horizontal joints, which decreases spalling and/or other damage to panels inherent in such operations. Joints with closed-cell neoprene sponge airseals are suitable for erection in winter time.

Airseal material should be applied reasonably straight (especially if exposed on the interior of the panel), but should not be stretched during glueing or installation operations. Airseal gaskets should be spliced at 45 deg. in shiplap fashion. The gasket material should protrude beyond the precast concrete unit sufficiently to ensure a reasonably tight butt joint following installation of the precast concrete. Its dimensions should be chosen so that

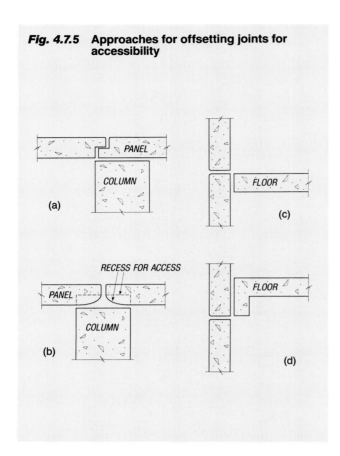

Fig. 4.7.5 Approaches for offsetting joints for accessibility

(a)

PANEL
COLUMN

(b)

PANEL
RECESS FOR ACCESS
COLUMN

(c)

FLOOR

(d)

FLOOR

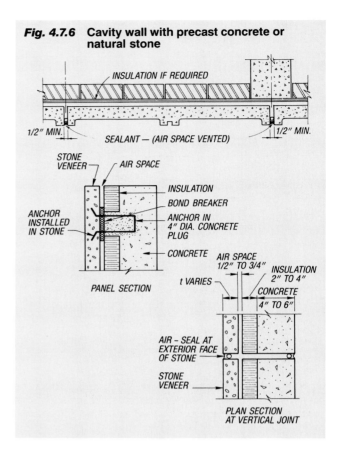

Fig. 4.7.6 Cavity wall with precast concrete or natural stone

INSULATION IF REQUIRED

1/2" MIN.   SEALANT — (AIR SPACE VENTED)   1/2" MIN.

STONE VENEER   AIR SPACE   INSULATION
BOND BREAKER
ANCHOR INSTALLED IN STONE
ANCHOR IN 4" DIA. CONCRETE PLUG
CONCRETE

PANEL SECTION

AIR SPACE 1/2" TO 3/4"   INSULATION 2" TO 4"
t VARIES   CONCRETE
4" TO 6"

AIR – SEAL AT EXTERIOR FACE OF STONE
STONE VENEER

PLAN SECTION AT VERTICAL JOINT

after completion of the wall, the gasket has 50 to 75 percent compression.

If panel configurations and joint sizes permit, a careful applicator may successfully install both the airseal and the rain screen from the exterior. The normal positions of the backing and sealant would be reversed for the interior airseal, Fig 4.7.4d. The special tools required may include an extension for the nozzle of the caulking gun and a longer tool for tooling the airseal. It should be noted that shiplap joints do not lend themselves to this type of installation.

The architect, precast concrete manufacturer, erector, and sealant applicator must all understand the function of the two-stage joints if optimum results are to be achieved. The most common mistakes in the installation of two-stage joints are leaving gaps in the airseal, making the rain-barrier airtight, and/or improperly venting or draining the expansion chamber.

A cavity wall can be very effective for separation and control of outside and inside air and humidity, Fig. 4.7.6. When precast concrete panels are used in cavity wall designs, they normally serve as the rain-barrier. An air space is maintained between the precast concrete and the interior wall, serving as the expansion chamber. Insulation, when required, is applied to the outside of the interior wall, eliminating condensation problems and making this wall subject only to constant interior temperatures. Although it is more expensive to build a cavity wall than a conventional

wall, the initial expense is justified for specific types of buildings and locations based on the lower maintenance costs and excellent performance records of cavity walls.

The construction of an insulated sandwich-stone veneer-faced precast concrete panel with 1/2 to 3/4 in. air space is shown in Fig. 4.7.6. In order to minimize bending of the stone wire anchors, the anchors are embedded in 4 in. diameter concrete plugs which penetrate the insulation. The plug is separated from the back side of the stone by a small section of a corrugated plastic formliner or voided plastic eggcrate to allow air circulation, or by a polyethylene foam pad. In most cases, it has been found that since the concrete plug is separated from the stone, it does not represent a serious thermal bridge and major condensation or discoloration of the exterior wall has not been reported. Fig. 4.7.7 illustrates a two-stage joint used for this type of panel on a project in Montreal.

4.7.3 NUMBER OF JOINTS

The number of joints in the architectural design should be minimized. This will result in a lower overall cost for the joints, potentially lower maintenance costs, and the economy of working with large panels.

Limitation of panel sizes to minimize movements in the joints is not recommended. It is generally more economical to select large panels and design the joints and sealants to allow for the anticipated movements. Optimum panel size should be determined by erection conditions, available

Fig. 4.7.7 Two-stage joint

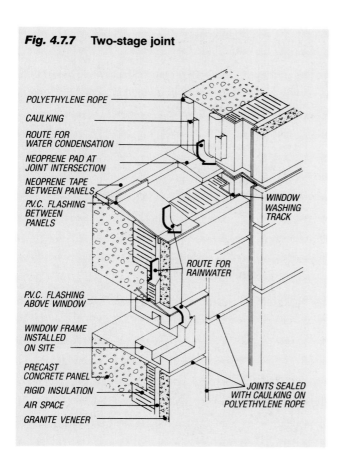

POLYETHYLENE ROPE

CAULKING

ROUTE FOR WATER CONDENSATION

NEOPRENE PAD AT JOINT INTERSECTION

NEOPRENE TAPE BETWEEN PANELS

P.V.C. FLASHING BETWEEN PANELS

WINDOW WASHING TRACK

ROUTE FOR RAINWATER

P.V.C. FLASHING ABOVE WINDOW

WINDOW FRAME INSTALLED ON SITE

PRECAST CONCRETE PANEL

RIGID INSULATION

AIR SPACE

GRANITE VENEER

JOINTS SEALED WITH CAULKING ON POLYETHYLENE ROPE

Fig. 4.7.8 Joints in panels with narrow ribs

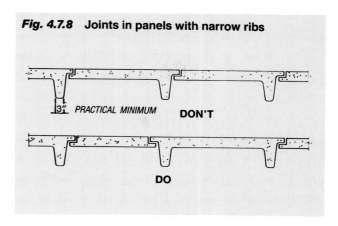

3" PRACTICAL MINIMUM   DON'T

DO

Fig. 4.7.9 Vertical joint

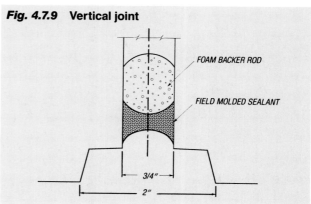

FOAM BACKER ROD

FIELD MOLDED SEALANT

3/4"

2"

Fig. 4.7.10 Unfortunate location of joints

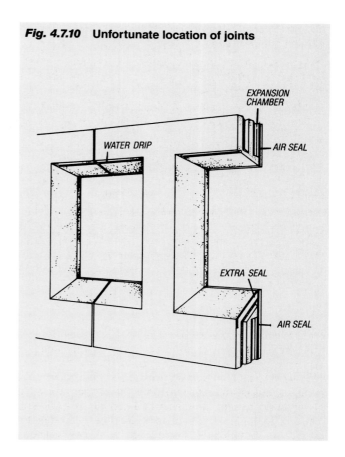

EXPANSION CHAMBER

AIR SEAL

WATER DRIP

EXTRA SEAL

AIR SEAL

handling equipment, and local transportation limitations as to panel weight and sizes. See Section 3.3.9 for details.

Only essential joints should be installed. Additional joints pose an unnecessary initial expense and maintenance burden. If the desired appearance demands additional joints, false joints may be used to achieve a more balanced architectural appearance. In order to match appearance of the two joints, the finish of the false joints should simulate the gaskets or sealants used in the real joints. Caulking the false joints adds unnecessary expense.

4.7.4 LOCATION OF JOINTS

Joints are simpler to design and execute if they are located at the maximum panel thickness. If there are any ribbed projections at the edges of the panels, joints should be placed at this location. Ribs at the edges improve the structural behavior of the individual unit. Also, panel variations—possible warping or bowing—are less noticeable when the joints are placed between ribs than when the joints are located in flat areas. However, complete peripheral ribs are not recommended since they are likely to cause localized water runoff resulting in unsightly weather staining. Instead, ribs should be placed at vertical panel edges.

If the ribs are too narrow to accommodate joints, Fig. 4.4.3, the full rib may be located in one panel only. Rigid erection schedules may be required if ribs overlap adjacent panels. A possible solution is to design every second panel with ribs at two parallel edges using the balance of the panels as infill units, Fig. 4.7.8. The best solution is no overlapping joints.

The predicted weathering pattern for a building is an important factor in locating and detailing joints. Joints should be made wide and recessed as shown in Fig. 4.7.9 to limit unexpected weathering effects. Recessed joints screen the joint from rain by providing a dead-air space which reduces air pressure at the face of the sealant. Also, the joint profile channels the rain runoff, helping to keep the building facade clean from unsightly runoff patterns. Set-backs should be provided at window perimeters and other vulnerable joints in the wall system to reduce the magnitude and frequency of water exposure.

Joints in forward sloping surfaces are difficult to weatherproof, especially if they collect snow or ice, Fig. 4.7.10. This type of joint should be avoided, whenever possible. When forward sloping joints are used, the architect should take special precautions against water penetration. This may be accomplished by sealing the front surface and using a two-stage joint. If a one-stage joint is used, the owner must regularly inspect the joints and perform any necessary maintenance. Finish requirements may also influence joint location, Section 4.7.7.

All joints should be aligned (rather than staggered throughout their length, see Fig. 4.7.13.) Non-aligned joints subject field-molded and pre-molded sealants to shear forces in addition to the expected compression or elongation forces. The additional stress may cause sealants to fail. In addition, non-aligned joints force panels to slide over each other, inducing high tensile forces. These forces may be great enough to crack the panels. In effect, the cracks relieve the induced stresses by forming alignment joints.

4.7.5 WIDTH OF JOINTS

Sealant life and performance are greatly influenced by joint width. Joints between precast concrete units must be wide enough to accommodate anticipated wall movements. Joint tolerances must be carefully evaluated and followed if the joint sealant system is to perform within its design capabilities. If units cannot be adjusted during erection to allow for proper joint size, saw cutting may be necessary. When joints are too narrow, failure of the joint will occur, and adjacent units may come in contact and be subjected to unanticipated loading, distortion, cracking and local crushing (spalling).

Joint widths should not be chosen for reason of appearance alone, but must relate to panel size, building tolerances, joint materials and adjacent surfaces. The required width of the joint is determined by the temperature extremes anticipated at the site location, the movement capability of the sealant to be used, the temperature at which the sealant is initially applied, panel size and fabrication tolerances of the precast concrete units. All of these factors take precedence over appearance requirements:

1. **Temperature Extremes and Gradients:** The temperature gradient used must reflect the differential between seasonal extremes of temperature and temperature at the time of sealant application. Concrete temperatures can and normally will vary considerably from ambient temperatures because of thermal lag. Although affected by ambient temperatures, anticipated joint movement must be determined from anticipated concrete panel temperature extremes rather than ambient temperature extremes. Consideration should be given to the orientation of the wall surface in relation to the sun. South facing walls will experience significantly higher temperatures than north facing walls.

2. **Sealant Movement Capability:** The minimum design of a panel joint must take into account the total anticipated movement of the joint (i.e., the concrete panels) and the movement capability of the sealant. All concrete is subject to volume changes from creep, shrinkage, and temperature variations.

3. **Application Temperature:** A practical range of installation temperatures considering moisture condensation at low temperatures and reduced working life at high temperatures, is from 40 to 90 deg F. This temperature range should be assumed in determining the anticipated amount of joint movement in the design of joints. A warning note should be included on the plans that, if sealing must take place for any reason at temperatures above or below the specified range, then a wider than specified joint may have to be formed. Alternately, changes in the type of sealant to one of greater movement capability or modifications to the depth-to-width ratio may be used to secure greater extensibility.

The *PCI Design Handbook—Precast and Prestressed Concrete* supplies figures for estimating volume changes directly related to the size of the panel. Most drying shrinkage occurs in the first weeks following casting, and creep normally levels out after a period of months. For these reasons, movements caused by ambient temperatures are more important than those caused by shrinkage. For load-bearing panels, the effect of creep may be cumulative and hence assume more importance.

Many factors may be involved in actual building movement. These include, but are not limited to: mass of material, color, insulation, building load, building settlement, method of fastening and location of fasteners, differential heating due to variable shading, thermal conductivity, differential thermal stress (bowing), building sway, and seismic effects. Material and construction tolerances that can produce smaller joints than anticipated are of particular importance. Such tolerances should be considered in the design calculations and considerations.

The larger the panel, the wider the theoretical joint should be in order to accommodate realistic tolerances in straightness of panel edge, edge taper, and panel width. For example, with a typical one story panel 5 feet wide, a joint width of 1/2 in. will accommodate tolerances suggested in Section 4.6. With an 8 foot wide panel, a 5/8 in. theoretical joint width is indicated.

Tolerances in overall building width and length are normally accommodated in panel joints, making the overall building size tolerance an important joint consideration. Two exceptions to this general rule are:

1. Projects where the designer has specified tolerances for the structure which match panel and joint tolerances.

2. Projects where variations may be accommodated at the corner panels, in expansion joints, or in joints adjacent to other wall materials.

If the joints must accommodate variations in building dimensions, a liberal tolerance should be allowed for the joint widths, refer to Section 4.6.3. For example a 3/4 in. joint may be specified with a ± 1/4 in. tolerance. This tolerance should accommodate reasonable variations in the actual site dimensions. For reasons of appearance, tolerances covering differential joint widths for a floor or an elevation may be specified. In the previous example, joint width may range from 1/2 in. to 1 in., but should be held to a differential variation of 3/16 in. The joint width should then be specified as follows: 3/4 in. ± 1/4 in. with a maximum 3/16 in. variation in width of adjacent joints. Alternatively, the jog in alignment of edge may be specified, see Fig. 4.6.9.

A practical calculation of panel joint size can be made as follows:

$$J = \frac{(100A)}{X} + B + C$$

where:

J = minimum joint width, in.
X = stated movement capability of the sealant in percent.
A = calculated movement of panel from thermal changes
  = (coefficient of thermal expansion) (change in temperature) (panel length)
B = material construction tolerances.
C = seismic or other considerations as appropriate.

The following example will illustrate the use of this equation. Concrete panels of 15 ft length, expecting a temperature change in the concrete of 120 deg. F, with a material or construction tolerance of 0.25 in., are to be sealed with a sealant having ±25% movement capability (as determined by ASTM C719). The coefficient of thermal expansion of the concrete is 6×10^{-6} in./in./deg. F. The calculated movement of the panel from thermal change is as follows:

$(6 \times 10^{-6}$ in./in./deg. F) (120 deg. F) (180 in.) = 0.130 in.
X = 25%
Construction tolerance = 0.25 in.
No seismic considerations
The calculated minimum joint width is as follows:

$$J = \frac{(100 \times 0.130 \text{ in.})}{25} + 0.25 \text{ in.} = 0.77 \text{ in.}$$

If joint width determined is too wide, another sealant having a greater movement capability should be selected. For example, if movement capability is ±50%, the joint width in the example becomes 0.51 in.

To provide optimum quality for the installation and performance of field-molded sealants, the architect should specify joint widths between 1/2 and 3/4 in. Corner joints may be 1 in. wide to accommodate the extra movement and

bowing often experienced at this location. A minimum joint width of 3/4 in. is recommended for two stage joints to allow sufficient space for insertion of the interior seal. Joints under 3/8 in. wide are undesirable. Narrow joints are considered to be a very high risk for any joint sealant installation.

When using neoprene, or other gasket or compression seals, the allowable joint width is increased. It should be remembered, however, that these materials are most commonly available for joint widths of 1/2 in. to 1 in. They are not practical for joints below 1/2 in. and may require special orders for joints above 1 in.

The required sealant depth is dependent on the sealant width at the time of application. The optimum sealant width/depth relationships are best determined by the sealant manufacturer. Since all manufacturers do not agree on the ideal proportion, however, generally accepted guidelines are:

1. For joints up to 1/2 in. wide: The sealant should have a minimum depth of 1/4 in. and a maximum depth equal to its width.

2. For joints from 1/2 in. to 1 in. wide: The sealant depth should be equal to one half the width, but not less than 1/2 in. The sealant should have a concave shape giving greater thickness at the panel faces.

3. For joints in excess of 1 in. wide: Sealant depth should be limited to 5/8 in. maximum.

The depth of the sealant can be controlled by using a suitable backup material. To obtain the full benefit of a well designed shape factor, the backup material must also function as a bond breaker, Fig. 4.7.11.

4.7.6 MATERIALS FOR JOINTS

The most common joint materials are field-molded sealants. These sealants are used in both one-stage and two-stage joints. If used as an airseal, they may be applied from the front provided joint width or depth permit, or from the interior if access to the joint is not blocked by edge beams or columns. If used as rain-barriers, they should be broken at suitable locations or the expansion chamber safely vented by other means.

Fig. 4.7.11 Ideal joint with field-molded sealant

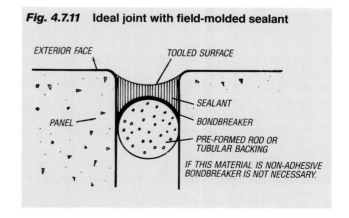

EXTERIOR FACE — TOOLED SURFACE
PANEL — SEALANT — BONDBREAKER — PRE-FORMED ROD OR TUBULAR BACKING — IF THIS MATERIAL IS NON-ADHESIVE BONDBREAKER IS NOT NECESSARY.

Table 4.7.1 Comparative Characteristics and Properties of Field-Molded Sealants

	Polysulfides		Polyurethanes		Silicones		
	One-Component	Two-Component	One-Component	Two-Component	One-Component	One-Component	Two-Component
Chief Ingredients	Polysulfide polymers, activators, pigments, inert fillers, curing agents, and nonvolatilizing plasticizers	Base: polysulfide polymers, activators, pigments, plasticizers, fillers, Activator: accelerators, extenders, activators	Polyurethane prepolymer, filler pigments & plasticizers	Base: polyurethane prepolymer, filler, pigments, plasticizers, Activator: accelerators, extenders, activators	Siloxane polymer pigments & fillers acetoxy system	Siloxane polymer pigments: alcohol or other non-acid cure	Siloxane polymer pigments: alcohol or other non-acid cure
Primer Required	usually	usually	usually	usually	usually	occassionally	occassionally
Curing Process	chemical reaction with moisture in air & oxidation	chemical reaction with curing agent	chemical reaction with moisture in air	chemical reaction with curing agent	chemical reaction with moisture in air	chemical reaction with moisture in air	chemical reaction with curing agent
Tack-Free Time (hrs.) (ASTM C679)	24	36-48	24-36	24-72	1	1-2	½-5
[1]Cure Time (days)	7-14	7	7-14	3-5	7-14	7-14	4-7
Max. Cured Elongation	300%	600%	300%	500%	300%	400-1600%	400-2000%
Recommended Max. Joint Movement	± 25%	± 25%	± 15%	± 25%	± 25%	± 25% to + 100, − 50%	± 12½% to ± 50%
Max. Joint Width	¾″	1″	1¼″	2″	¾″	1″	1″
Resiliency	high	high	high	high	high	moderate	high
Resistance to Compression	moderate	moderate	high	high	high	low	low
[2]Resistance to Extension	moderate	moderate	medium	medium	high	low	low
Service Temp. Range °F	− 40 to + 200°	− 60 to + 200°	− 40 to + 180°	− 40 to + 180°	− 60 to + 350°	− 60 to + 300°	− 60 to + 300°
Normal Application Temp. Range	+ 40 to + 120°	+ 40 to + 120°	+ 40 to + 120°	+ 40 to + 120°	+ 20 to + 160°	+ 20 to + 160°	+ 20 to + 160°
Weather Resistance	good	good	very good	very good	excellent	excellent	excellent
Ultra-Violet Resistance, Direct	good	good	poor to good	poor to good	excellent	excellent	excellent
Cut, Tear, Abrasion Resistance	good	good	excellent	excellent	poor	poor-excellent	excellent knotty tear
[3]Life Expectancy	20 years +	20 years +	20 years +	20 years +	20 years +	20 years +	20 years +
Hardness Shore A (ASTM C661)	25 - 35	25 - 45	25 - 45	25 - 45	30 - 45	15 - 35	15 - 40
Applicable Specifications	FS:TT-S-00230C ASTM C920 19-GP-13A (Canadian)	FS:TT-S-00227E ASTM C920 19-GP-24 19-GP-3B (Canadian)	FS:TT-S-00230C ASTM C920 19-GP-13 (Canadian)	FS:TT-S-00227E ASTM C920 19-GP-24 (Canadian)	FS:TT-S-00230C FS-TT-S-001543A ASTM C920 19-GP-9B (Canadian)	FS:TT-S-00230C FS-TT-S-001543A ASTM C920 19-GP-18 (Canadian)	FS:TT-S-00227E USASI A-116.1 ASTM C920 19-GP-19 (Canadian)

[1]Cure time as well as pot life are greatly affected by temperature and humidity. Low temperatures and low humidity create longer pot life and longer cure time; conversely, high temperatures and high humidity create shorter pot life and shorter cure time. Typical examples of variations are:

Two-Part Polysulfide

Air Temp.,	Pot Life	Initial Cure	Final Cure
50°	7-14 hrs.	72 hrs.	14 days
77°	3-6 hrs.	36 hrs.	7 days
100°	1-3 hrs.	24 hrs.	5 days

[2]Resistance to extension is better known in technical terms as modulus. Modulus is defined as the unit stress required to produce a given strain. It is not constant but, rather, changes in values as the amount of elongation changes.

[3]Life expectancy is directly related to joint design, workmanship and conditions imposed on any sealant. The length of time illustrated is based on joint design within the limitations outlined by the manufacturer, and good workmanship based on accepted field practices and average job conditions. A violation of any one of the above would shorten the life expectancy to a degree. A total disregard for all would render any sealant useless within a very short period of time.

For a comprehensive discussion of joint sealants used between wall panels, refer to ASTM C962, *Standard Guide for Use of Elastomeric Joint Sealants*. Table 4.7.1 provides a list of common field-molded sealants and their qualities.

The sealants used for specific purposes are often installed by different subcontractors. For example, the window subcontractor normally installs sealants around windows, whereas a second subcontractor typically installs sealants around panels. The designer must select and coordinate all of the sealants used on a project for chemical compatibility and adhesion to each other. In general, contact between different sealant types should be avoided, if possible.

The recommendations of the sealant manufacturer should always be followed regarding mixing, surface preparation, priming, application life, and application procedure. Good workmanship by qualified sealant applicators is the most important factor required for satisfactory performance.

The edges of the precast concrete units and the adjacent materials must be sound, smooth, clean, and dry. They must also be free of frost, dust, loose mortar or other contaminants that may affect adhesion such as form release agents, retarders, or sealers. It may be more economical and effective to prepare joint surfaces prior to erection if a large number of units require surface preparation. It may also be desirable to conduct adhesion or peel tests to determine the compatibility of the sealant with the contact surfaces.

Chemically curing type sealants should not be applied to wet or icy surfaces as they may cure or set before they can bond to the concrete surface. Some methyl methacrylate resin sealers inadvertently sprayed in the joints may peel away from the concrete surface leaving a void between sealant and concrete, while silicone water repellents in the joints will prevent adhesion of joint sealant to concrete surface. Use of a stiff wire bursh, light grinding or sandblasting followed by air blowing may be necessary to remove surface contaminants.

Also, before caulking, the joint should be wiped with a cloth dampened with an oil-free solvent such as xylol. Sometimes, smooth concretes which are very shiny exhibit a "skin" on the surface. The skin may peel off leaving a gap between it and the concrete after the joint sealant has been applied to the concrete. It may be necessary to remove the skin by using a stiff wire brush followed by a high pressure water rinse.

The caulking gun should have a nozzle of proper size and should provide sufficient pressure to completely fill the joints. Joint filling should be done carefully and completely, thoroughly working the compound into the joint. After all joints have been completely filled, they should be neatly tooled, slightly concave, to eliminate air pockets or voids, and to provide a smooth, neat appearing finish. Surface of sealant should be a full, smooth bead, free of ridges, wrinkles, sags, air pockets and embedded impurities.

A tooling agent may be used to help reduce drag in the process of tooling. Water is probably the best agent, however some applicators prefer soapy water. If soapy water is used, it should be used with caution as it can contaminate joints not yet sealed and adversely affect adhesion.

It is recommended that tools be used dry, or wet only with clean water, when tooling light colored sealants. Other solvents may be suitable as tooling agents, provided they are specifically approved by the sealant manufacturer.

Surfaces soiled with sealant materials should be cleaned as work progresses, as removal is likely to be difficult once the sealant has cured. Solvent or cleaning agent used should be recommended by the sealant manufacturer.

Backup materials help to shape sealants. When selecting a backup material and/or bond breaker, the recommendations of the sealant manufacturer should be followed to insure compatibility. The backup should not stain the sealant, as this may bleed through and cause discoloration of the joint. Backup materials, should be of suitable size and shape so that, after installation, they are compressed 30 to 50 percent. Proper selection and use of backup material is essential for the satisfactory performance of watertight joints. Length-wise stretching, twisting or braiding of the tube or rod stock should be avoided. When inserting a polyethylene foam backup rod, a blunt tool should be used to avoid skin puncture of the rod and possible out-gassing which may cause blistering of the sealant.

The principal functions of backup materials are:

1. Controlling the depth of the sealant in the joint (provide sealant dimensions).

2. Serving as a bond breaker. This prevents the sealant from bonding to the back of the joint and exposing itself to three dimensional stress.

3. Assisting in tooling of the joint.

4. Protecting the back side of the sealant from attack by moisture vapors trying to escape from the building. Use of double backer rods has been recommended where high vapor pressure occurs at the immediate back surface of the sealant and should be placed at about 25 percent of the panel depth behind the first.

Primers may be recommended by the sealant manufacturer for the following reasons:

1. To promote adhesion of sealants to porous surfaces or to reinforce the surface.

2. To promote adhesion of sealants to surfaces such as porcelain enamel, unusual types of glass, certain metals and finishes, and wood.

3. To promote adhesion of sealants to an existing surface treatment which is difficult to remove.

Special care should be exercised to avoid staining the outside face of the precast concrete unit since some primers will leave an amber colored stain if brushed along the surface. This stain will have to be mechanically removed which will be expensive. The primer should be allowed to cure before application of the sealant. The sealant and primer should always be supplied by the same manufacturer.

The following characteristics should be considered when making the final selection of sealants from those with suitable physical (durability) and mechanical (movement capability) properties:

1. Adhesion to different surfaces—concrete, glass, aluminum, etc.

2. Surface preparation necessary to ensure satisfactory performance — priming, cleaning, drying, etc.

3. Serviceable temperature range.

4. Drying characteristics — dirt pickup, susceptibility to damage due to movement of joint while sealant is curing.

5. Puncture, tear and abrasion resistance.

6. Color desired and color retention.

7. Effect of weathering — water and sunlight — on properties such as adhesion, cohesion, elasticity, etc.

8. Staining of surfaces caused by sealant or primer.

9. Ease of application.

10. Environment in which the sealant is applied.

11. Compatibility with other sealants to be used on the job.

Among gasket or compression sealants, the most commonly used for airseals is closed-cell neoprene sponge conforming to ASTM C509. The sponge is used in rectangular sections when pre-applied to individual panels and as a rope when installed after erection. Foam plastics have also been used. However, they are not recommended, because they have not yet proven sufficiently elastic or permanent. Neoprene compression seals must remain compressed approximately 15 percent of the uncompressed seal width at the maximum joint opening to maintain sufficient contact pressure to seal and resist displacement. Generally, the amount of compression is limited to 55 percent of the uncompressed seal width at maximum closing to prevent overcompression. For larger sizes of compression seals, the 55 percent value may be exceeded. The allowable movement is thus approximately 40 percent of the uncompressed seal width. The allowable movement for impregnated foams is much less, about 5 to 20 percent. The recommendations of the manufacturer of the compression seals relative to lubricants, adhesives, and conditioning of materials in cold weather should be followed in order to facilitate installation.

The most critical condition exists when the joint is fully open at low temperature, since compression set (or lack of low temperature recovery) may adversely affect sealant performance. The principle of size selection is similar to that used for field-molded sealants, in that the seal must be maintained within the specified compression range considering the installation temperature, width of formed opening, and expected movement.

When it is necessary to apply sealant below 40 deg. F, steps must be taken to ensure clean, dry, frost-free surfaces. The area to be sealed should be wiped with a quick-drying solvent such as xylol, just before sealing. Also a wider than specified joint may have to be formed, or changes in the type of sealant or depth-to-width ratio may be required to secure greater extensibility. The applicator must know the joint size limitation of the sealant selected. The area may be heated, if possible, or at least the sealant should be slightly warm when applied.

4.7.7 ARCHITECTURAL TREATMENT

Joints should be designed as a strong architectural feature of the wall. Recessing the joints will help to de-emphasize the variation in adjacent surfaces sometimes inherent in large wall panels. Complicated edge and fenestration profiles should be avoided for economy in manufacturing and erection. Complicated profiles are more vulnerable to damage in handling and more difficult to make watertight.

Detailing suggestions for typical architectural precast concrete panel joints, Fig. 4.7.12 include:

1. Allow either a chamfered or reveal joint, since these types of joints can accommodate tolerances required of panel thickness and the shadows formed within these joints will minimize any adverse effects on the esthetic appearance of the joinery system. By making the joints appear wider than they actually are, the visual differences in their width are proportionally reduced. This tends to make the difference more difficult to detect and masks slight misalignments of the joints which might otherwise be especially noticeable at intersections. Simplifying the profile of the joints by chamfering the panel edges also has the obvious advantage of making the edges less vulnerable to damage.

2. Avoid use of butt joint as the tolerances required of panel thickness will form shadow lines directly over the panels rather than within the joints. This will detract from the esthetic appearance of the panels.

Detailing suggestions for staggering architectural precast concrete wall panels, Fig. 4.7.13 include:

1. Check for excessive thermal bowing of panels and set

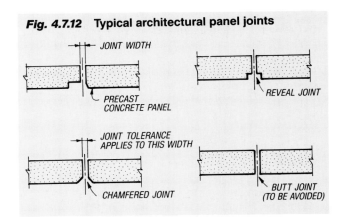

Fig. 4.7.12 Typical architectural panel joints

JOINT WIDTH

PRECAST CONCRETE PANEL

REVEAL JOINT

JOINT TOLERANCE APPLIES TO THIS WIDTH

CHAMFERED JOINT

BUTT JOINT (TO BE AVOIDED)

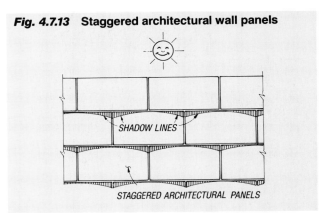

Fig. 4.7.13 Staggered architectural wall panels

SHADOW LINES

STAGGERED ARCHITECTURAL PANELS

panel tolerances to avoid unpleasant shadow marks as shown.

2. Consider joint configuration and joint tolerances to minimize shadow effects.

3. For loadbearing walls, there is a serious drawback to using horizontally staggered panels. If staggered panels are used, on every other floor, the floor slab must bear on two different panels. The floor slab connection problem created should be avoided, if at all possible.

Fig. 4.7.14 shows a panel, several stories high, incorporating a flat parapet section at the top. If the architect requires the parapet surface to appear as a flat continuous band surrounding the building, joints in the parapet may be de-emphasized by close matching and flush sealing. The erector has a better chance of accomplishing this, if he has larger tolerances for matching panel faces over the balance of the panel lengths. The additional tolerance is provided by recessing the joints below the parapet section.

An excellent architectural treatment of joints is the molding of panel edges. A small draft is required for the first 3/4 in. (or 1 in.) of the returns. The draft is normally in the order of 1/16 in. to 1/8 in., see Fig. 2.9.8. The joint openings will thus be slightly tapered, a desirable feature which prevents water and snow accumulations in the joint from exerting pressure during freezing conditions. Repetitive mold usage is essential to maintain a reasonable economy when providing molded edges.

Fig. 4.7.14 Good architectural features of joints

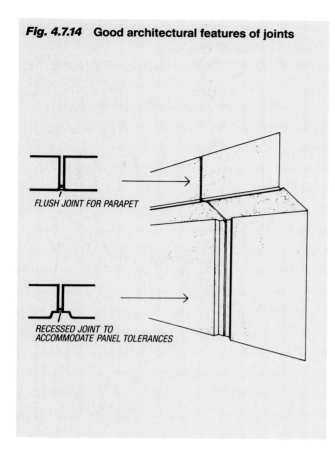

FLUSH JOINT FOR PARAPET

RECESSED JOINT TO
ACCOMMODATE PANEL TOLERANCES

Fig. 4.7.15 Sealing against uneven finishes

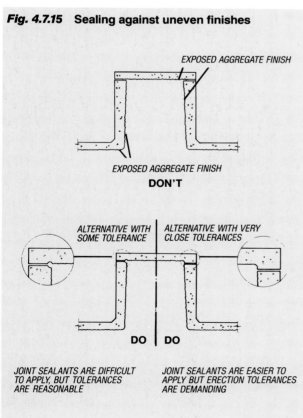

EXPOSED AGGREGATE FINISH

EXPOSED AGGREGATE FINISH

DON'T

ALTERNATIVE WITH
SOME TOLERANCE

ALTERNATIVE WITH VERY
CLOSE TOLERANCES

DO | **DO**

JOINT SEALANTS ARE DIFFICULT
TO APPLY, BUT TOLERANCES
ARE REASONABLE

JOINT SEALANTS ARE EASIER TO
APPLY BUT ERECTION TOLERANCES
ARE DEMANDING

Finish requirements may also influence joint details. The sealant must be applied to a relatively smooth surface. thus, the sealant must be held back from the face of the exposed aggregate panels. This requirement is simple to comply with when the design includes recessed external joints, Fig. 4.7.9. For optimum performance, sealants should be applied below the beveled area. When exposed aggregate surfaces come together at an inside corner, the situation is more difficult. Special attention must be paid to finish and joint details, Fig. 4.7.15.

4.7.8 FIRE PROTECTIVE TREATMENT

Joints between wall panels are similar to openings. Most building codes do not require openings to be protected against fire if the openings consititute only a small percentage of the wall area and if the spatial separation is greater than some minimum distance. In such cases, the joints would not require protection. In other cases, openings may have to be protected.

However, most codes permit openings to have a lesser degree of protection than the main walls. For example, the Uniform Building Code requires that when openings are permitted and must be protected, the "openings shall be protected by a fire assembly having a 3/4 hr fire-protection rating." In such cases, the joints should be provided with the same protection level as other openings. When no openings are permitted, the fire resistance required for the wall should also be provided at the joints.

Joints between wall panels should be detailed to prevent the passage of flames and hot gases. The details should ensure that the transmission of heat through the joints does not exceed the limits specified in ASTM E119 *Standard Methods of Fire Tests of Building Construction and Materials*. Concrete wall panels expand when heated, so the joints tend to close during fire exposure. Flexible, noncombustible materials such as ceramic fiber blankets provide thermal, flame, and smoke barriers. These fire resistive blankets and ropes must be installed under a minimum of 10 to 15 percent compression. When used in conjunction with caulking materials, they can provide the necessary fire protection and weathertightness while permitting normal volume change movements. Joints that do not require movement can be filled with mortar.

Fire tests of wall panel joints show that the fire endurance, as determined by a temperature rise of 325 deg. F over the unexposed joint, is influenced by joint type, joint treatment (materials), joint width, and panel thickness. By providing the proper thickness of insulating materials within the joint, it is possible to attain fire endurances essentially equal to those of the panels.

Fig. 4.7.16 shows the fire endurance of one-stage butt-joints and two-stage shiplap joints in which the joint treatment consisted of sealants and polyethylene foam backup rods.

Table 4.7.2 is based on results of fire tests of panels with butt joints and ceramic fiber felt in the joints. The tabulated values apply to one-stage butt joints and are conservative for two-stage and shiplap joints.

Table 4.7.2 Protection of joints between wall panels utilizing ceramic fiber felt

Panel thickness* (in.)	Thickness of ceramic fiber felt (in.) required for fire resistance ratings and joint widths shown							
	Joint width = ⅜ in.				Joint width = 1 in.			
	1 hr	2 hr	3 hr	4 hr	1 hr	2 hr	3 hr	4 hr
4	¼	N.A.	N.A.	N.A.	¾	N.A.	N.A.	N.A.
5	0	¾	N.A.	N.A.	½	2⅛	N.A.	N.A.
6	0	0	1⅛	N.A.	¼	1¼	3½	N.A.
7	0	0	0	1	¼	⅞	2	3¾

N.A. = Not application

Interpolation may be used for joint widths between ⅜ in. and 1 in.

*Panel equivalent thicknesses are for carbonate concrete. For siliceous aggregate concrete change "4, 5, 6, and 7" to "4.3, 5.3, 6.5, and 7.5". For sand-lightweight concrete change "4, 5, 6, and 7" to "3.3, 4.1, 4.9, and 5.7".

The tabulated values apply to one-stage butt joints and are conservative for two stage and ship-lap joints as shown below.

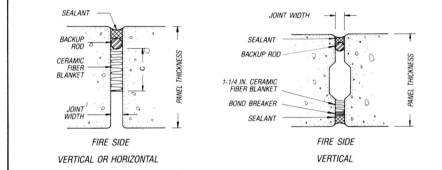

VERTICAL OR HORIZONTAL

VERTICAL

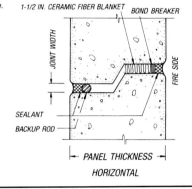

HORIZONTAL

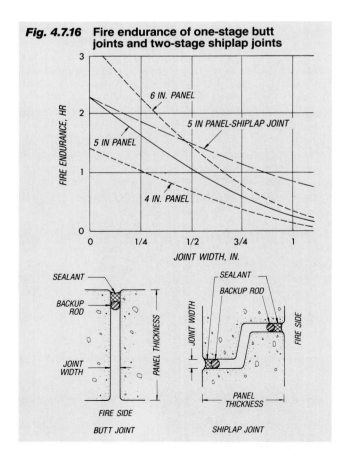

Fig. 4.7.16 Fire endurance of one-stage butt joints and two-stage shiplap joints

BUTT JOINT

SHIPLAP JOINT

4.7.9 JOINTS IN SPECIAL LOCATIONS

Below Grade joints between panels require special attention. If appreciable water pressures are expected, special gasket assemblies will be required. Other joints may be grouted unless temperature changes are expected to cause joint expansion or contraction, Fig. 4.7.17.

Stone Veneer-Faced Precast Concrete Panel jointing requires that each piece of stone be positioned and held within established tolerances during fabrication of the panel. Joints between veneer pieces on a precast concrete panel are typically a minimum of 1/4 in. although they have often been specified equal to the joint width between precast concrete elements, usually 1/2, 3/4 or 1 in., depending on the panel size. As actual joint width between precast concrete panels, as erected, depends largely on the accuracy of the main supporting structure, it is not realistic to require matching joint widths between stone pieces and panels.

Often, an invisible joint is specified, e.g., less than 3/16 in., especially on polished veneer. This is simply not possible because the joint must have the width necessary to allow for movements, tolerances, etc. Also, due to tolerances and natural warping, adjacent panels may not be completely flush at the joint and shadow lines will appear. Rather than attempting to hide the joint, the joint should be emphasized by finding an esthetically pleasing joint pattern with a complementary joint size.

In the form, the veneer pieces are temporarily spaced with

Fig. 4.7.18

Fig. 4.7.17 Joints between units below or near grade

PRECAST CONCRETE PANEL

PRECAST CONCRETE PANEL

FLOOR

INSULATION

FLOOR   PACK WITH SUITABLE INSULATION

FOUNDATION WALL   FOUNDATION WALL

a non-staining, compressible spacing material, such as rubber, neoprene, soft plastic wedges, or chemically neutral, resilient, non-removable gaskets (such as sealant backer rods) which will not stain the veneer or adversely affect the sealant to be applied later. Shore A hardness of the gasket should be less than 20. The gaskets should be of a size and configuration that will provide a pocket to receive the sealant and also prevent any of the concrete backup from entering the joints between the veneer units. Non-acidic based masking or duct tape has also been used to keep concrete out of the stone joints so as to avoid limiting stone movement. The spacer material should be removed after the panel has been removed from the form, unless it is a resilient sealant backup.

Caulking between stones or panels should be an elastomeric, usually polysulfide, polyurethane, or silicone, that will not stain the stone veneer material. Some grades of silicone sealants are not recommended by their manufacturers for application on high-calcite content materials (marbles) as they may stain light colored stones. In some projects, caulking between stone pieces on a panel may be installed more economically and satisfactorily at the same time as the caulking between precast concrete elements, while on other projects consideration may be given to caulking the veneer material at the plant. Plant caulking of stone to stone joints is recommended in areas subject to freezing and thawing, if panels will be left in prolonged storage during winter months.

In many cases, stone veneer is used as an accent or feature strip on precast concrete panels. The major difference in the stone attachment is that a 1/2 in. minimum space is left between the edge of the stone and the precast concrete to allow for differential movements of the materials. This space is then caulked as if it were a conventional joint. The 28-story building in Fig. 4.7.18 has 1-1/4 in. thick granite inset and anchored to the precast concrete with 3/16 in. diameter stainless steel anchors. There is a 5/8 in. space between the edges of the stone and the precast concrete on both the U-shaped column covers and the spandrel panels.

The bondbreaker between the stone veneer and concrete backup may function as a vapor barrier on the exterior face of the concrete and keep moisture in the veneer or at the interface, unless drainage provisions are provided. Also, after some period of time, gaps may develop between the

stone veneer and concrete backup at the bondbreaker which could become a location for moisture penetration due to capillary and gravity action, particularly where the window or roof design allows water to puddle on top of the panel. One solution that has been used for this problem is a modified rain-screen joint (two-stage joint) as shown in Figs. 4.7.4(e) and 4.7.19. The approach in Fig. 4.7.19 provides an airtight 1 in. wide urethane seal, bonded to the stone veneer and concrete backup, and continuous along both sides and top of the panel. Other designers have used a sealant applied to the top and side edges of the stone/concrete interface after the panels are cast. Care must be taken to ensure that the sealant used is compatible with the sealant to be applied to panel joints after erection of the panels.

Fig. 4.7.19 Seal at veneer/concrete interface

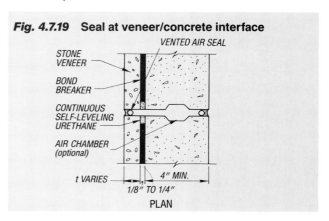

STONE VENEER

VENTED AIR SEAL

BOND BREAKER

CONTINUOUS SELF-LEVELING URETHANE

AIR CHAMBER (optional)

t VARIES   4" MIN.

1/8" TO 1/4"

PLAN

Fig. 4.7.20 Open drainage system for plazas

Fig. 4.7.21 Closed drainage system for plazas

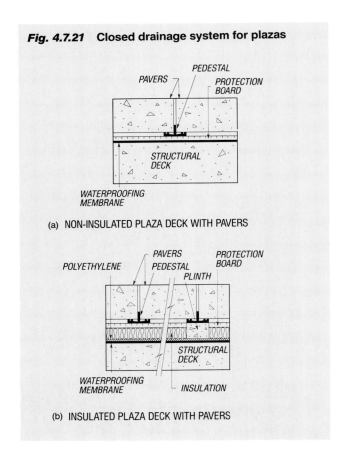

(a) NON-INSULATED PLAZA DECK WITH PAVERS

(b) INSULATED PLAZA DECK WITH PAVERS

The bondbreaker is not sealed at the bottom of the panel, so that any moisture which may get behind the stone veneer can freely drain. In the case of long panels, a sloping gutter is sometimes used, not only under the window, but at every horizontal joint.

4.7.10 PLAZA JOINTS

There are two basic types of plaza drainage systems:

1. **Open Drainage Systems.** This type of system employs open joints and pads supported by pedestals, Fig. 4.7.20.

2. **Closed Drainage Systems.** This type of system uses closed joints at the wearing surface, Fig. 4.7.21. The closed joints may consist of sealants, sliding plates, mortar or compressing seals. Sliding plate joints are not recommended for use on plaza drainage systems.

The advantages of an open drainage system are as follows:

1. The plaza wearing surface can be placed level, which simplifies surface drainage patterns. It is important, however, that each individual paver have a raised crown in the center, since warpage or creep can cause pavers to sag in the middle if there is no support at this location.

2. Pedestals and wearing pads can be placed directly over the waterproofing membrane, so that if a leak occurs, the surface to be repaired is more readily accessible.

3. Mortar and sealant joints are eliminated, reducing maintenance.

The disadvantages of an open drainage system are as follows:

1. Open joints less than 1/2 to 5/8 in. wide can become clogged with all forms of debris, and open joints over 1/4 in. wide are a potential tripping hazard for wearers of spiked heels.

2. Rectangular pavers require at least four pedestals, and it always seems that one of them is in a different plane from the other three. However, leveling plates or devices can be placed over pedestals to compensate for this unevenness.

3. The concentrated loads imposed on the membrane by the pedestals can create problems, depending on the type of membrane and how it is protected. Insulation is not usually effective in distributing the load, unless it is a suitable seamless, high density, nonabsorptive insulation placed under the membrane. Protection board should be used to decrease point load pressure on the membrane.

4. Keeping the open joints uniform and preventing the pavers from shifting under vehicular traffic requires some form of intermittent, resilient filler in the joints. This filler material can fall out, leaving the pavers vulnerable. The filler can also contribute to the clogging problem.

5. Expansion joints in an open drainage system are awkward to handle at the wearing surface.

The advantages of a closed drainage system are as follows:

1. The wearing surface can be sealed.

2. The spacing of joints is not as limited as in the pedestal system and joints will be more uniform.

3. More variety is possible in the choice of wearing surface materials, used both singly or in combination.

4. Large-heavy vehicular traffic that should be expected in any large, open plaza can be accommodated. (Fire trucks, parade floats, and garden and maintenance vehicles must be taken into consideration here.)

5. Insulation between the structural slab and the wearing surface can be accommodated more readily.

The disadvantages of the closed drainage system are as follows:

1. Mortar and sealant joints require maintenance.

2. Adequate drainage slope is required at the wearing surface, which may at times create a difficult design problem.

3. Water that leaks behind pavers can freeze and cause cracking of the mortar joints.

4. The sliding plate joints, often specified in plaza drainage systems, frequently fail to perform properly. The concept calls for a non-corrosive sliding plate or saddle anchored on one side of the joint and free to slide over the other side. Over a period of time, dirt, dust, and debris can restrict the movement of the plates. Initial alignment difficulties and differences in thermal movement between the metal plate and the paver to which it is attached also pose problems. Sliding plate joints are not recommended.

OTHER ARCHITECTURAL DESIGN CONSIDERATIONS

5.1 GENERAL

Architectural precast concrete may be in contact with many other materials under service conditions. Because the application or interfacing of such materials can be as important as the design of the individual components, this section discusses various types and properties of interfaces, including glazing, energy conservation, acoustical properties, fire resistance, and roofing. Joints between precast concrete panels and other systems or materials can be designed using the same principles that govern joints between precast concrete units, see Section 4.7. The combination of building materials selected must provide an esthetically pleasing facade while effectively separating the external and internal environments. If an efficient barrier or building envelope has been provided, the architect can isolate and control the interior environment with heating and air-conditioning systems. The building envelope is composed of the architectural precast concrete panel, joints, and other building materials discussed in this chapter.

5.2 WINDOWS AND GLAZING

5.2.1 DESIGN CONSIDERATIONS

Glass must be designed to take both wind and thermal loads and to transfer wind loads into the surrounding framing. The glass industry has established guidelines for bow, squareness, corner offset in framing, and variation of mullions from plumb or horizontals from level:

1. Bow: 1/16 in. in any 4 ft length of framing.

2. Squareness: 1/8 in. difference in the lengths of the diagonals of the frame.

3. Corner offset: 1/32 in. at each corner.

4. Plumb or level: 1/8 in. in 12 ft or 1/4 in. in any single run.

The glass industry has also established guidelines for the allowable deflection of glass framing members under design loads. The intention of the guidelines is to minimize the potential for glass breakage. They state that the glass surround should not impose any bending or highly concentrated compressive loads on the glass and that the framing should not deflect more than 3/4 in. or 1/175th of its span under loading, whichever is less. The guidelines also

Fig. 5.2.1

require that the glazing system chosen isolate the glass from the other parts of the wall. The glazing solution selected should relate to realistic and attainable tolerances for the precast concrete, see Section 4.6. For example, erection of spandrel panels requires close attention to tolerances because the window system must fit between upper and lower concrete spandrel panels.

Several methods are commonly used to incorporate window sash in precast concrete panels. Glazing may also be accomplished with the help of special gaskets, or by incorporating the glass directly into the panels. Each of these methods warrants discussion and illustration. Windows require careful detailing at their interface with the wall. Consideration needs to be given to: (1) mechanical connection of the window frame to panel; (2) watertightness; (3) airtightness; (4) flashings, rain deflectors, and drainage of infiltrated water; and (5) condensation.

The precast concrete panel may be pre-glazed in the precaster's plant, Fig. 5.2.1. This enables the building to be "closed-in" immediately upon installation, allowing interior work to proceed at an accelerated pace. Pre–glazing can allow mechanical, electrical, and finish trades to begin work months earlier than they would be able to with conventional construction due to the rapid completion of a weathertight exterior wall system.

It is primarily an economic decision whether the installation of windows should be done on the ground or following panel erection. The quality of weatherproofing should improve if installation is performed on the ground or inside a plant protected from inclement weather. When this method is used, proper protection of the windows during the subsequent construction stages must be considered.

It is recommended that the architect specify, as part of the glazing contract, that the window manufacturer install window(s) for approval in the initial production panel or mockup units. This practice has often proven advantageous. It affords the architect and the owner an opportunity to assess the appearance of the finished wall assembly and to approve all details relating to the assembly such as color and type of sealant and interior trim. The mockup sample may also pinpoint problems for the subcontractors, allowing them to be solved before large numbers of elements are affected. This provision is a worthwhile safeguard against later delays in the enclosure of the building.

5.2.2 SASH CAST INTO CONCRETE

The problems of replacing, at some future time, sash cast into concrete should be considered.

Steel Sashes can be cast directly into precast concrete panels. A sash with suitable built-in anchors is held in the mold while the concrete is placed and finished. The sash can then be cleaned and, ultimately, glazed.

Wood Sashes can be cast directly into precast concrete panels using a similar procedure, provided the wood is properly impregnated so that it will not absorb moisture from the wet concrete. Wood which is not impregnated will

shrink as it dries, reducing the effectiveness of the seal.

Aluminum Sashes have also been cast directly into concrete panels. However, this procedure is not recommended for large windows since the expansion coefficient of aluminum is much greater than that of concrete. Care must be exercised to prevent reinforcing steel and aluminum from touching. Contact between the two metals will lead to electrolytic action that can result in corrosion. The aluminum frame should be protectively coated throughout casting operations, since wet concrete will mar most aluminum finishes.

Plastic Window Frames have had only minor usage in precast concrete panels. It may be possible to incorporate this type of window sash during casting, subject to protection of some finishes, if assurance is provided that no differential expansion will occur between the two materials.

5.2.3 SASH ATTACHED AFTER CURING OF CONCRETE PANELS

Fig. 5.2.2 shows several examples where an aluminum sash is attached after the concrete has cured. Fig. 5.2.3 shows a wood casement installation. Fig. 5.2.4 shows fixed window installations using metal or wood framing.

The architect must specify the supplier of window hardware. This hardware is frequently furnished directly to the precast concrete plant by the window supplier. Hardware installed in the precast concrete units to provide fastening for the windows may consist of plastic blocks that are ta-

Fig. 5.2.2 Aluminum sash in precast concrete units

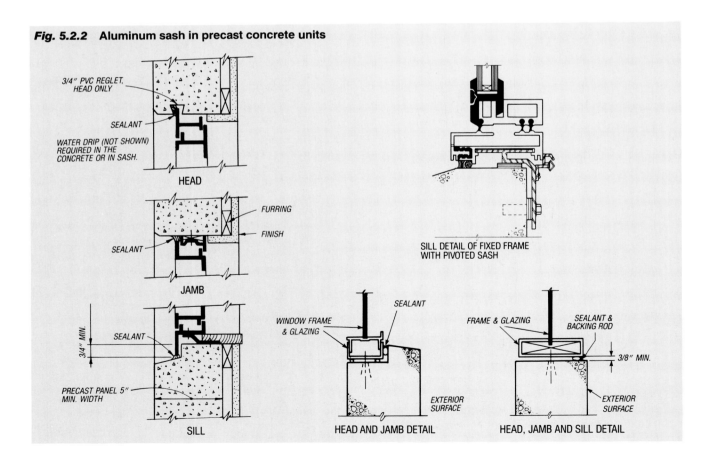

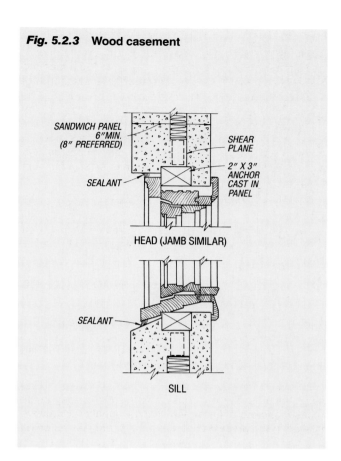

Fig. 5.2.3 Wood casement

SANDWICH PANEL
6" MIN.
(8" PREFERRED)

SHEAR PLANE

2" X 3" ANCHOR CAST IN PANEL

SEALANT

HEAD (JAMB SIMILAR)

SEALANT

SILL

pered to resist pullout, metal inserts, or impregnated tapered wood. If ferrous metal inserts are used for aluminum windows, they must be galvanized or cadmium plated, or plastic washers must be used to prevent contact between the steel and the aluminum.

Where windows are installed using a two-stage joint, it may be necessary to vent the air space between the rain-barrier and the airseal through the concrete. Refer to Section 4.7.2. This is normally accomplished with a small plastic tube. The amount of water to be drained should normally be minimal so that no freezing problems will occur provided the outlet is protected. Consequently, such venting may emerge in the panel joints behind the rain–barrier.

Often, a joint sealing system shows a lack of continuity; it fails to include the window perimeter. Fig. 5.2.5 indicates the application of the modified rainscreen in conjunction with window detailing.

5.2.4 GASKETED GLAZING

The watertightness of glazing with a gasket, which is one of the most common methods of glazing, relies solely on achieving a proper seal between the panel and the gasket, Fig. 5.2.6. The necessity of a smooth surface at this point cannot be overemphasized. Thus, this glazing method should only be used with a metal surface or the plastic reglet shown. The quality of such reglets on the market varies. It is therefore necessary that the architect specify the best quality reglet available by description or by name.

Fig. 5.2.4 Fixed window

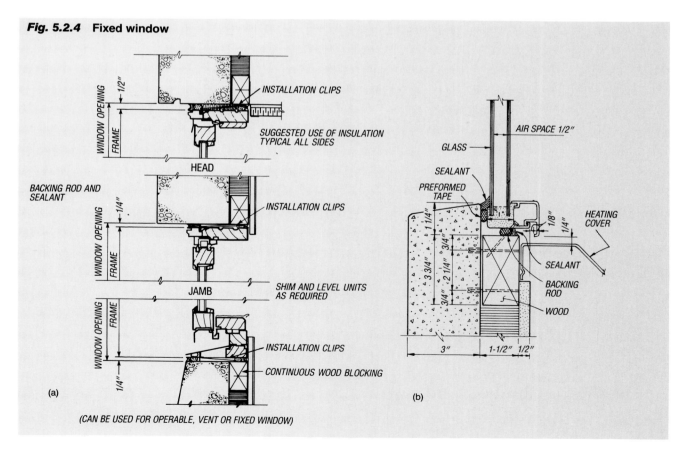

WINDOW OPENING
FRAME
1/2"
INSTALLATION CLIPS
SUGGESTED USE OF INSULATION TYPICAL ALL SIDES

HEAD

BACKING ROD AND SEALANT

WINDOW OPENING
FRAME
1/4"
INSTALLATION CLIPS

JAMB

SHIM AND LEVEL UNITS AS REQUIRED

WINDOW OPENING
FRAME
1/4"
INSTALLATION CLIPS
CONTINUOUS WOOD BLOCKING

(a)

(CAN BE USED FOR OPERABLE, VENT OR FIXED WINDOW)

AIR SPACE 1/2"

GLASS

SEALANT

PREFORMED TAPE

HEATING COVER

SEALANT

BACKING ROD

WOOD

3 3/4" 1 1/4" 3/4" 3/4" 2 1/4"

1/8" 1/4"

3" 1-1/2" 1/2"

(b)

ASTM C964 should be used as a guide to determine proper use of lockstrip gaskets.

Tight tolerances are also required for the width and alignment of the groove. The alignment of the groove is important, because a nearly perfect plane must be provided for the glass. The tolerance for practical construction should always be ± 1/8 in. of all frame dimensions at the sealing surface of the reglet. Splice and corner joints must be aligned to within a tolerance of ± 1/16 in. The corner angles in the plane of the glass should be held to ± 2 degree tolerance to properly receive the gasket lips. The minimum height or width dimension should be 18 in. Otherwise, it is virtually impossible to install the glass in the panel and zip the lockstrip into place.

Gasket glazing should not be used unless the window opening is completely contained within the panel, Fig. 5.2.7. In general, window blockouts should not be located across a panel joint since the tolerances required for panel plumbness, panel width, and the joint may cause problems with the proper fit of the window units.

For laminated or sealed insulating glass installations, glass manufacturers often require that weepholes be used in the gaskets as a precaution against entrapping any water that enters the gasket channel.

All repetitive questionable concrete frame conditions should be mocked up and water tested in accordance with ASTM E331 during the initial precast concrete production stages. This allows any necessary corrective action to be taken at the earliest possible time.

Fig. 5.2.6 Typical gasket glazing

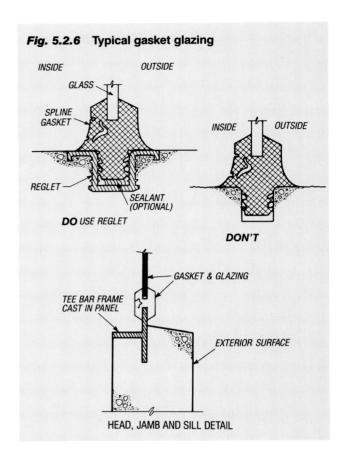

Fig. 5.2.5 Rain screen — window detail

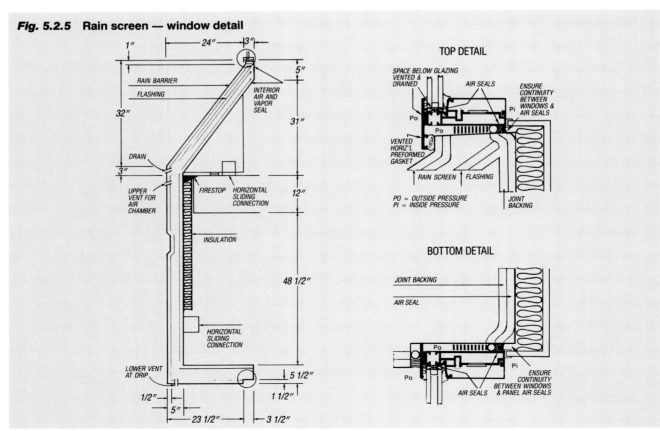

Fig. 5.2.7 Glazing in open and closed shape panels

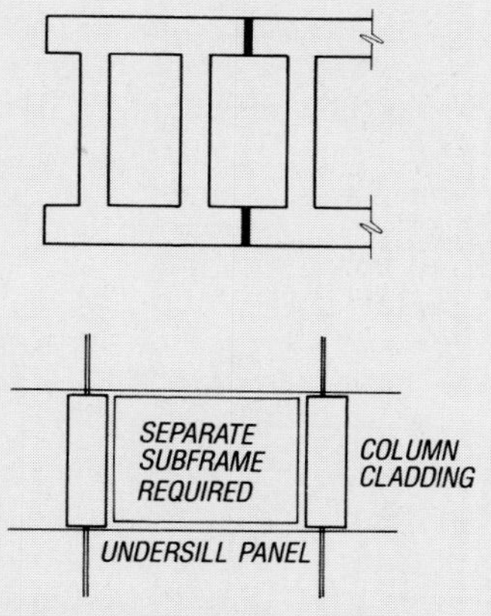

Don't *GLAZE WITHOUT PROPER SASH*

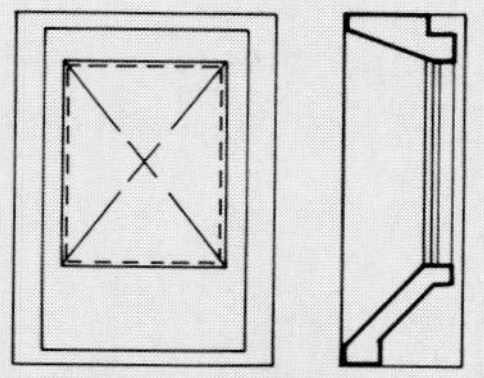

SEPARATE
SUBFRAME
REQUIRED

COLUMN
CLADDING

UNDERSILL PANEL

Do *GLAZE DIRECTLY OR WITH GASKETS*

5.2.5 GLAZING DIRECTLY TO CONCRETE

Glass to concrete glazing, without the need for a separate window frame, was pioneered in Canada and the USA. Significant cost savings and simplification of detailing and installation work can be achieved using this method on fixed glazed openings. It can be an economical solution, provided the proper details are carefully executed. Service records in excess of 25 years have been reported in severe climates. Fig. 5.2.8 show the most common solutions for this glazing. In method (b), fixing blocks should be cast into the rabbet about 3 in. from each corner and at 9 to 10 in. intermediate spacing to take fixing screws for the aluminum stop.

The aluminum angle or wood blocking asserts a slight pressure on the airseal and the glass. This pressure may induce stresses in the glass if the concrete recess is not a true plane. Close tolerances for the planeness of this recess, rigid molds, and accurate bearing and fastening of the mold piece forming the recess are all essential.

Glass has a coefficient of thermal expansion much nearer to that of concrete than those of the typical window frame materials. This minimizes the strain on the adhesive qualities of bedding materials. The usual prerequisite for direct glazing is that the opening be completely formed in a single wall unit. Nevertheless, openings for direct glazing can be formed by two or more separate precast concrete units, provided the method of fixing prevents undue differential movement. The alignment of rabbets, grooves, etc.,

Fig. 5.2.8 Glazing directly to concrete

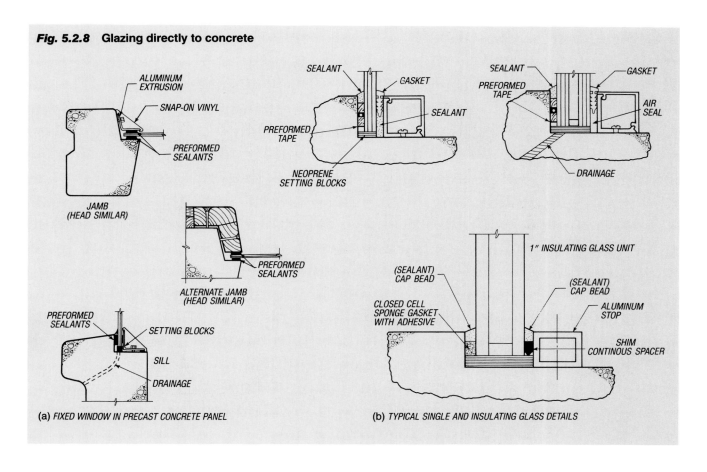

ALUMINUM EXTRUSION

SNAP-ON VINYL

PREFORMED SEALANTS

JAMB (HEAD SIMILAR)

PREFORMED SEALANTS

ALTERNATE JAMB (HEAD SIMILAR)

PREFORMED SEALANTS

SETTING BLOCKS

SILL

DRAINAGE

(a) *FIXED WINDOW IN PRECAST CONCRETE PANEL*

SEALANT

GASKET

PREFORMED TAPE

SEALANT

NEOPRENE SETTING BLOCKS

SEALANT

PREFORMED TAPE

GASKET

AIR SEAL

DRAINAGE

1" INSULATING GLASS UNIT

(SEALANT) CAP BEAD

(SEALANT) CAP BEAD

CLOSED CELL SPONGE GASKET WITH ADHESIVE

ALUMINUM STOP

SHIM CONTINOUS SPACER

(b) *TYPICAL SINGLE AND INSULATING GLASS DETAILS*

in seperate precast concrete units creates severe tolerance problems which must be offset by savings in other areas.

Glass lites should be set on two 80 to 90 Shore A durometer setting blocks positioned at quarter points. When this is impractical, the setting blocks can be installed to within 6 in. of the vertical glass edge.

Water should not be permitted to remain in the glazing rabbet. A drainage system should incorporate enough weep holes to ensure adequate drainage; usually this consists of three, 3/8 in. diameter holes equally spaced at the sill.

This glazing method is recommended only where the panel shape already provides a rigid frame and the repetition is sufficient to justify the high mold cost involved. Under these circumstances, the economies can be very real.

The 5-story building in Fig. 5.2.9 has 7 ft. 3 in. high load-bearing spandrel panels with butt-glazed insulated glass and spandrel glass partially covering the spandrel panels.

5.2.6 OTHER ATTACHED OR INCORPORATED MATERIALS

Other materials incorporated into the precast concrete units might include inserts or hardware used to attach other materials, reglets used to accommodate flashings, nailing strips, or similar continuous fastening strips.

Mechanical and electrical outlets and ornamental items can also be built into the units. Materials which react with concrete should not be used. If their use is unavoidable, they must be suitably protected. Galvanized and stainless steel and plastics are acceptable but, as indicated in Section 3.6.7, the weathering effects of some metals should be considered. Also, flashings must be galvanically compatible with the reglets or counterflashing receivers. Wood is also acceptable if preservative treated and impregnated to prevent water gain.

There has been great progress in window washing systems. More sophisticated safety equipment has been designed into the systems and more flexibility has been given to the designer of the wall. Historically, an I-beam profile has been used for mullions which would provide a track in which the guides from the washing system platform can roll. In recent years, these exposed tracks have been replaced by reveals in the panels that allow the track to be recessed. The track may be in the face of the panel, Fig. 5.2.10 or in the joints of the panel with a track in the edge of the abutting panels, Fig. 5.2.11. In many areas of the country, tiebacks or buttons are being used. These lend themselves to designs that are flush, with no projections in the facade.

The button system is, in effect, a reverse of the old track design. Originally, the track configuration was designed into the facade and a roller or trolley assembly engaged into the track to hold the platform to the side of the building. The button system employs small round discs attached to

Fig. 5.2.9

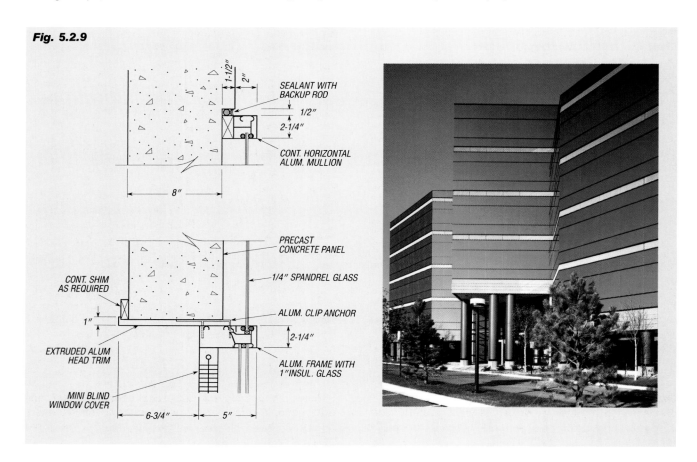

SEALANT WITH BACKUP ROD
CONT. HORIZONTAL ALUM. MULLION
PRECAST CONCRETE PANEL
1/4" SPANDREL GLASS
ALUM. CLIP ANCHOR
ALUM. FRAME WITH 1"INSUL. GLASS
CONT. SHIM AS REQUIRED
EXTRUDED ALUM HEAD TRIM
MINI BLIND WINDOW COVER

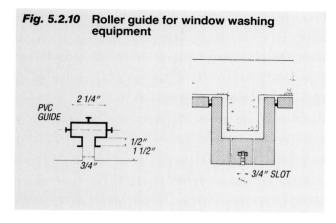

Fig. 5.2.10   Roller guide for window washing equipment

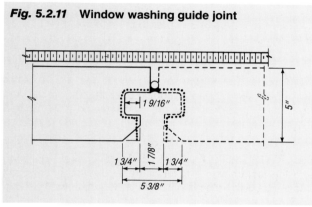

Fig. 5.2.11   Window washing guide joint

the facade and a track installed on the washing platform. The track length and button placement are selected so that there are always a minimum of two buttons engaged at any one time on each track. Each platform has two tracks. This system allows greater design flexibility and provides several economic advantages.

It is extremely important for the architect to locate attached or incorporated materials on the contract drawings and clearly indicate the supplier and installer of these materials in the specifications.

5.3 ENERGY CONSERVATION

5.3.1 GLOSSARY

British thermal units (Btu) — Approximately the amount of heat to raise one pound of water from 59 to 60 deg. F.

Degree day (D) — A unit, based on temperature difference and time, used in estimating fuel consumption and specifying nominal heating load of a building in winter. For any one day, when the mean temperature is less than 65 deg. F, there are as many degree days as there are degrees F difference in temperature between the mean temperature for the day and 65 deg. F.

Dew-point temperature (t_s) — The temperature at which condensation of water vapor begins for a given humidity and pressure as the vapor temperature is reduced. The temperature corresponding to saturation (100 percent relative humidity) for a given absolute humidity at constant pressure.

Film or surface conductance (f) — The time rate of heat exchange by radiation, conduction, and convection of a unit area of a surface with its surroundings. Its value is usually expressed in Btu per (hr) (sq ft of surface area) (deg.F temperature difference). Subscripts "i" and "o" are usually used to denote inside and outside surface conductances, respectively.

Heat transmittance (U) — Overall coefficient of heat transmission, air-to-air, the time rate of heat flow usually expressed in Btu per (hr) (sq ft of surface area) (deg.F temperature difference between air on the inside and air on the outside of a wall, floor, roof, or ceiling). The term is applied to the usual combinations of materials and also single materials such as window glass, and includes the surface conductances on both sides. This term is frequently called the U-value.

Perm — A unit of permeance. A perm is 1 grain per (sq ft of area)(hr)(in. of mercury vapor pressure difference).

Permeability, water vapor (μ) — The property of a substance which permits the passage of water vapor. It is equal to the permeance of 1 in. of a substance. Permeability is measured in perm inches. The permeability of a material varies with barometric pressure, temperature and relative humidity conditions.

Permeance (M) — The water vapor permeance of any sheet or assembly is the ratio of the water vapor flow per unit area per hour to the vapor pressure difference between the two surfaces. Permeance is measured in perms.

Two commonly used test methods are the Wet Cup and Dry Cup tests. Specimens are sealed over the tops of cups containing either water or desiccant, placed in a controlled atmosphere usually at 50 percent relative humidity, and weight changes measured.

Relative humidity (RH) — The ratio of water vapor present in air to the water vapor present in saturated air at the same temperature and pressure.

Thermal conductance (C) — The time rate of heat flow expressed in Btu per (hr)(sq ft of area)(deg.F average temperature difference between two surfaces). The term is applied to specific materials as used, either homogenous or heterogeneous for the thickness of construction stated, not per in. of thickness.

Thermal conductance of an air space (a) — The time rate of heat flow through a unit area of an air space per unit temperature difference between the boundary surfaces. Its value is usually expressed in Btu per (hr) (sq ft of area) (deg.F).

Thermal conductivity (k) — The time rate of heat flow, by conduction only through a unit thickness of a homogeneous material under steady-state conditions, per unit area, per unit temperature gradient in the direction perpendicular to the isothermal surface. Its unit is (Btu-in.) per (hr)(sq ft of area)(deg.F).

Thermal resistance (R) — The reciprocal of a heat transmission coefficient, as expressed by U, C, f, or a. Its unit is (deg.F)(hr)(sq ft of area) per Btu. For example, a wall with a U-value of 0.25 would have a resistance value of:

$$R = 1/U = 1/0.25 = 4.0$$

5.3.2 GENERAL

The planned design of an energy-conserving building requires the architect's understanding of the effects of design decisions on energy performance. Some effects (for example, of the building's orientation) are more or less obvious; others are more subtle and require rather complex analysis of many factors.

Precast concrete panels have many built-in advantages when it comes to saving energy. Their versatility leads to unique solutions for many energy conservation problems. The relative importance of particular types of systems in any given building depends to a large extent on its intended use. For instance, in office buildings, cooling and lighting loads far exceed heating loads in all climates. However, in residential buildings cooling is relatively less important than heating for most climates.

The designer should be aware that several factors, other than U-values, determine the actual performance of the envelope in conserving energy. Some of these factors are:

1. Building orientation and aspect ratio.

2. Color of exterior surfaces of envelope materials.

3. Thermal mass of exterior envelope materials.

4. Wind velocities.

5. Infiltration through the envelope.

6. Orientation, proportioning and external shading of glass areas.

Building orientation plays an important role in building energy consumption. If possible, the long axis of the building should be oriented in the east-west direction to help in controlling the effect of the sun on heating and cooling loads.

To maximize solar heating, glass should be located on the south wall. South-facing glass should be shaded to minimize solar exposure in the summer while allowing maximum solar exposure in the winter.

In the southern regions of the country, the primary emphasis is on cooling. Glass should be located on the north side of buildings in these regions to minimize heat gains from the sun.

It is more difficult to control the effects of solar exposure on east and west walls. The area of east and west walls and the amount of glazing in these walls should therefore be minimized, although east glazing will help warm an office building in early morning hours after night-temperature setbacks.

Building shape is another important consideration. Shape influences energy performance in two ways. First, it determines the area of the building skin. The skin area affects the gain or loss of heat by conduction. The larger the skin area, the greater the gain or loss. Second, shape influences how much of the floor area can be illuminated using natural light from the sun, called daylighting. The potential energy savings from daylighting is particularly significant in commercial buildings because of the large lighting requirements in these buildings.

Precast concrete panels are available in a variety of colors and textures. These two factors can be used to improve the energy conserving features of the walls.

On exterior surfaces, light colors decrease solar heat gain; dark colors increase solar heat gain. In most cases, a dark-colored north wall and light-colored east and west walls form the most energy-conserving arrangement. Light colors and high reflectivity are especially important where cooling dominates the energy requirements. It should be noted, however, that the color of the exterior walls has relatively little effect on energy consumption when the walls have low U-values and high thermal mass.

Increasing the surface roughness of the wall exterior causes an increase in the amount of sunlight absorbed and reduces the effect of wind on heat loss and gain. Ribbed panels act as baffles to wind, thereby reducing conductive heat loss and infiltration. Although this has a somewhat smaller effect than proper color selection, it can help to reduce total energy consumption.

Wind can decrease the exterior still-air film that usually surrounds a building and contributes to the insulating U-values of wall elements, thus increasing heating and cooling loads. This effect becomes less marked as the U-value decreases and thermal mass increases. Wind also carries solar heat away from a building and evaporates moisture on wet surfaces, thus cooling the skin to temperatures lower than the ambient air. High winds create pressure differences across walls which will tend to cause air leakage through the walls. Cold air leakage to the inside must be heated and probably humidified. This also requires an expenditure of energy.

Architectural precast concrete walls can be very energy efficient. Recessed window walls, vertical fins, and various other sculptured shapes facilitate the design of many types of shading devices for window areas, including vertical and horizontal sunshades. This passive solar design can be augmented by using insulation to design specific thermal characteristics into each face of the structure to suit its sun orientation and natural environment. To obtain a range of U-values, the precast concrete walls may have insulation applied to the back or the insulation may be fully incorporated into a sandwich wall panel.

The use of glass in buildings requires special consideration during the design stage. Window design can have profound effect on indoor thermal conditions. Heat gain through a sunlit glass area is many times greater than through an equal area of precast concrete and its effect is usually felt almost immediately. Properly designed shading devices can modify the thermal effects of windows to a very great extent.

Shading is a fundamental design strategy in the summer

for preventing solar heat gain and diffusing bright sunlight. In the cooler months, when the sun's angle of incidence is low, the shading devices may be angled to let the sunshine in and help reduce heating loads, Fig. 5.3.1. The shading approach selected can reinforce and enhance the design content and form of the building, in some cases becoming the prime form-giving element. Its primary function, shading, may even have to be modified or compromised in order to meet other important requirements. Fig. 5.3.2 shows preferred cross sections for economical use of precast concrete as shading elements. Note that in each case, the spandrel and sunscreening elements are integral and may be lifted into place in one operation.

Shading using horizontal or vertical plane(s) projecting out in front of or above a window can be designed to intercept the summer sun, admit much if not all the winter sun, and allow a view out. If the plane projects far enough from the building, a single projection may be sufficient, as in the case of generous roof overhangs or windows recessed deeply between vertical fins. Alternatively, more modest projections can be equally effective in shading but they must be more closely spaced. Closely spaced horizontal or vertical planes may begin to dominate the view out of a window and in any case change the scale of the window. The proportion of the space divided by the shading planes becomes as important as the overall window proportion in determining the esthetic effect of the fenestration.

In summer, vertical fins will shade the early morning and late afternoon sun while horizontal members keep out the high-altitude mid-day sun. In winter these shades will not interfere with the sun because of its low altitude and southerly azimuth at sunrise and sunset.

Horizontal shading is most effective on southern exposures, but if not extended far enough beyond the windows, it will permit solar impingement at certain times of the day. Designs may be flat or sloping; sloping versions may be of shorter length, but obstruct more of the sky view, Figs. 5.3.3 and 5.3.4. The detached screen panel parallel to the wall in Fig. 5.3.5 was used to block the rays of the sun, while still allowing light to enter the windows. Sun-shading may also be provided through the use of a free-standing perimeter structure set in front of the actual building enclosure, Fig. 5.3.6. By this method it is possible to combine control of glare and solar gain, with enhanced long-term weathering performance through the removal of rainwater at each floor level.

LeCorbusier was insistent on the controlled daylighting of his buildings and used the term 'brise-soleil' for external balcony screens at the Marseilles Unite. In windy areas, the solar screens can be made to serve the double purpose of windbreaks.

Sunscreen panels, which have pockets to receive precast double tees, form the south, east and west faces of the midrise office building in Fig. 5.3.7 while the north face features flat panels with punched openings.

Solar control through the use of shading devices is most effective when designed specifically for each facade, since

---

Fig. 5.3.1 Panels used to shade glass

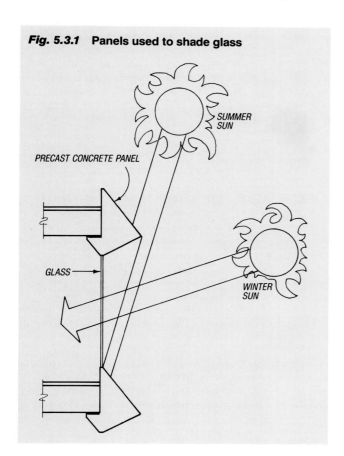

Fig. 5.3.2 Shading elements

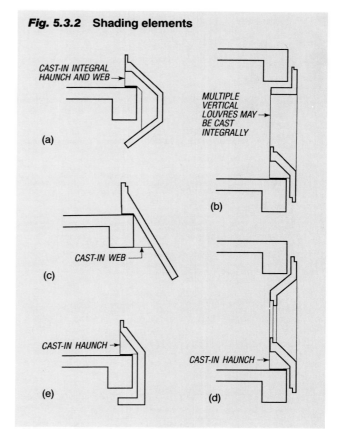

Fig. 5.3.3

Fig. 5.3.4

Fig. 5.3.5

Fig. 5.3.6

time and duration of solar radiation vary with the sun's altitude and azimuth. The designer can predict accurately the location and angles of the sun, designing overhangs or fins to shade exactly the area desired.* This type of envelope response can be seasonal (shade during certain times of the year) or daily (shade during certain hours of the day).

The versatility of precast concrete was used to change the window opening configuration with respect to wall orientation in order to maximize solar gains in the winter and minimize them in the summer, Fig. 5.3.8. Since the windows are small relative to the wall surface, the window units were splayed back on two different planes (at the sill and jamb) so that the windows could be recessed and shaded.

Fig. 5.3.7

Fig. 5.3.8

WEST AND NORTH PANEL   EAST AND SOUTH PANEL

East and west facing windows are more effectively shaded by vertical projecting planes, Fig. 5.3.9. Vertical projections from either side of the window narrow the peripheral view from the window. The further south a building is located, the more important shading east- and west-facing windows becomes, and the less important shading south-facing windows becomes. This is due to the high position of the summer sun in southern latitude with the resulting decrease in direct sunlight transmitted by the south-facing windows.

* American Institute of Architects. *Architectural Graphic Standards.* 8th ed. New York, John Wiley & Sons, 1988.

Fig. 5.3.9

Fig. 5.3.10

Fig. 5.3.11

Fig. 5.3.12

In Fig. 5.3.10 the top floor is cantilevered over the main floor to shade the windows. All second floor windows on the east and west sides are oriented directly south or north for sun control. The vertical wingwall shading devices completely shade the windows during the four summer months.

The use of three-dimensional precast concrete window wall units permits windows to be recessed within an enclosing concrete surround. The sides may be vertical or angled. Deeply recessed windows are particularly effective in minimizing solar heat gains without reducing natural light and view. Eggcrate shading works well on walls facing southeast, and is particularly effective for a southwest orientation. Because of its high shading efficiency, the eggcrate device (deeply recessed windows) is often used in hot climates. The deep, recessed window areas and massive overhangs in Fig 5.3.11 illustrate the total flexibility of design that precast concrete offers the architect.

Three foot deep "eyebrows," Fig. 5.3.12, was the shading device used to keep out the sun's rays in the summer and reduce cooling loads.

Precast concrete and inclined glass can work together for optimum use of daylighting. Direct sun strikes the glass at an angle and is reflected, reducing glare, while indirect sunlight reflects off the sill of the precast concrete panel and through the glass to provide soft natural light at the perimeter of the building, Fig. 5.3.13. By keeping the direct rays of the sun out of the building, cooling loads are considerably reduced. "Eyelid" or hooded shading devices and inclined glass can be very effective in controlling the penetration of the sun into a building by reducing the area of glass exposed to the sun, Fig. 5.3.14. This shading device softens the brightness contrast between the interior and exterior. Rounded jambs or deep window wells could also be used to soften brightness contrasts, Fig. 5.3.15.

The information and design criteria that follow are taken from or derived from the ASHRAE *Handbook of Fundamentals,* hereafter referred to as the ASHRAE Handbook, and from the ASHRAE Standard 90A, *Energy Conservation in New Building Design*, hereafter referred to as the ASHRAE Standard, except where noted. It is important to note that all design criteria are not given and the criteria used may change from time to time as the ASHRAE Standard and Handbook are revised. It is therefore essential to consult the applicable codes and revised references for the specific values and procedures that govern in a particular area when designing the energy conservation systems of a particular structure.

Thermal codes and standards specify the heat transmission requirements for buildings in many different ways. Prescriptive standards specify heat transmittance (U) or thermal resistance (R) values for each building component, whereas with performance standards, two buildings are equivalent if they use the same amount of energy, regardless of the U or R values of the components. This allows the designer to choose conservation strategies that provide the required performance at the least cost. It is important to have basic knowledge concerning heat loss

Fig. 5.3.13

Fig. 5.3.14

Fig. 5.3.15 Window forms

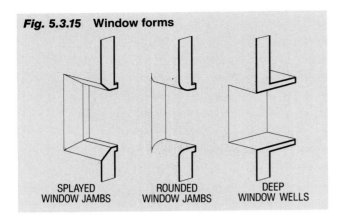

SPLAYED
WINDOW JAMBS   ROUNDED
WINDOW JAMBS   DEEP
WINDOW WELLS

and heat gain of many materials. The fundamentals and design aids that are necessary to properly analyze and compare the losses and heat gains through envelopes for code compliance are presented in this section.

5.3.3 THERMAL PROPERTIES OF MATERIALS, SURFACES AND AIR SPACES

Common thermal properties of materials and air spaces are based on steady state tests, which measure the heat that passes from the warm side to the cool side of the test specimen. The results of the tests determine the thermal conductivity, k, or, for non-homogeneous sections, compound sections and air spaces, the thermal conductance, C, for the total thickness. The values of k and C do not include film or surface conductances, (f_i and f_o), or resistances.

The overall thermal resistance of a wall section is the sum of the resistances of the wall layers, R (reciprocal of k, C, f_i and f_o). The R- values of most construction materials are not greatly influenced by the direction of heat flow. The R-values of surfaces and air spaces differ depending on whether they are vertical, sloping, or horizontal and, if horizontal, whether heat flow is up or down. Also, the R-values of surfaces are affected by the velocity of air at the surfaces and by their reflective properties.

Tables 5.3.1 and 5.3.2 give the thermal resistances of surfaces and 3-1/2 in. air spaces. Table 5.3.3 gives the thermal properties of most commonly used building materials.

Only U-values are given for glass because the surface resistances and air space between panes account for nearly all of the U-value. Table 5.3.4 gives the thermal properties of various weight concretes in the "normally dry" condition. Normally dry is the condition of concrete containing an equilibrium amount of free water after extended exposure to room temperature air at 35 to 50 percent relative humidity.

Thermal conductances and resistances of other building materials are usually reported for oven dry conditions. However, higher moisture content in concrete causes higher thermal conductance.

In years past, the U-factor was considered the most significant indication of heat gain, principally because laboratory tests have shown that thermal transmission is directly proportional to the U-factor during steady-state heat flow. However, the steady-state condition is rarely realized in actual practice.

External conditions (temperatures, position of the sun, presence of shadows, etc.) vary throughout a 24-hr. day, and heat gain is not instantaneous through most solid materials, resulting in the phenomenon of time lag (thermal inertia). As temperatures rise on one side of a wall, heat begins to flow toward the cooler side. Before heat transfer can be achieved, the wall must undergo a temperature increase. The Btu's of thermal energy necessary to achieve this increase are directly proportional to the specific heat and density of the wall.

Due to its density, concrete has the capacity to absorb and store large quantities of heat. This thermal mass allows concrete to react very slowly to changes in outside temperature. This characteristic of thermal mass reduces peak heating and cooling loads and delays the time at which these peak loads occur by several hours, Fig. 5.3.16(a). This delay improves the performance of heating and cooling equipment, since the peak cooling loads are delayed until the evening hours, when the outside temperature has dropped. Mass effect, glass area, air infiltration, ventilation, building orientation, exterior color, shading or reflections from adjacent structures, surrounding surfaces or vegetation, building aspect ratio, number of stories, wind direction and speed, all have an effect on insulation requirements.

Table 5.3.1 Thermal resistances, R$_f$, of surfaces

Position of surface	Direction of heat flow	Still air, R$_{fi}$			Moving air, R$_{fo}$	
		Non reflective surface	Reflective surface		Non reflective surface	
			Aluminum painted paper	Bright aluminum foil	15 mph Winter design	7½ mph Summer design
Vertical	Horizontal	0.68	1.35	1.70	0.17	0.25
Horizontal	Up	0.61	1.10	1.32	0.17	0.25
	Down	0.92	2.70	4.55	0.17	0.25

Table 5.3.2 Thermal resistances, R_a, of air spaces[1]

Position of air space	Direction of heat flow	Air space		Non reflective surfaces	Reflective surfaces		
		Mean temp. °F	Temp. diff. °F		One side[2]	One side[3]	Both sides[3]
Vertical	**Horizontal (walls)**	**Winter**					
		50	10	1.01	2.32	3.40	3.63
		50	30	0.91	1.89	2.55	2.67
	Horizontal (walls)	**Summer**					
		90	10	0.85	2.15	3.40	3.69
Horizontal	**Up (roofs)** **Down (floors)**	**Winter**					
		50	10	0.93	1.95	2.66	2.80
		50	30	0.84	1.58	2.01	2.09
		50	30	1.23	3.86	8.17	9.60
		50	10	1.24	4.09	9.27	11.15
	Down (roofs)	**Summer**					
		90	10	1.00	3.41	8.19	10.07

1. For 3½ in. air space thickness. The values, with the exception of those for reflective surfaces, heat flow down, will differ about 10% for air space thicknesses of ¾ in. to 6 in. Refer to Table 2, Chapter 23 of the ASHRAE Handbook for values of other thicknesses, reflective surfaces, heat flow directions, mean temperatures, and temperature differentials.
2. Aluminum painter paper.
3. Bright aluminum foil.

Table 5.3.3 Thermal properties of various building materials[1]

Material	Unit weight, pcf	Resistance, R per inch of thickness, 1/k	Transmittance, U	Specific heat, Btu/(lb.)(°F)
Insulation, rigid				
Cellular glass	8.5	2.86		0.18
Glass fiber, organic bonded	4.0 - 9.0	4.00		0.23
Mineral fiber, resin binder	15	3.45		0.17
Expanded polystyrene extruded smooth skin surface	1.8 - 3.5	5.00		0.29
Expanded polystyrene molded bead	1.0	3.85		—
	1.25	4.00		—
Cellular polyurethane/ polyisocyanurate (unfaced)	1.5	6.25-5.56		0.38
Miscellaneous				
Gypsum board	50	0.88		0.26
Particle board	50	1.06		0.31
Plaster				
cement, sand agg.	116	0.20		0.20
gyp., L.W. agg.	45	0.63		—
gyp., sand agg.	105	0.18		0.20
Wood, hard (maple, oak)	38 - 47	0.94 - 0.80		0.39
Wood, soft (pine, fir)	24 - 41	1.00 - 0.89		0.33
Plywood	34	1.25		0.29
Glass doors & windows; clear[2]				
Single, winter			1.10	
Single, summer			1.04	
Double, winter[3]			0.59	
Double, summer[3]			0.61	

1. See Table 5.3.4 for all concretes.
2. Does not incude correction for sash resistance. Refer to Chapter 27 of the ASHRAE Handbook for sash correction.
3. ¼ in. air space; coating on either glass surface facing air space.

Table 5.3.4 Thermal properties of concrete

Description	Concrete weight, pcf	Thickness, in.	Resistance, R Per inch of thickness, 1/k	Resistance, R For thickness shown, 1/C	Specific heat, Btu/(lb)(°F)
Concretes including normal weight, lightweight and lightweight insulating concretes	145				0.19
	140		0.10-0.05		
	130				
	120		0.18-0.09		
	110				
	100		0.27-0.17		
	90				
	80		0.40-0.29		
	70				
	60		0.63-0.56		
	50				
	40		1.08-0.90		
	30		1.33-1.10		
	20		1.59-1.20		
Normal weight solid panels	145	2		0.15	0.19
		3		0.23	
		4		0.30	
		5		0.38	
		6		0.45	
		8		0.60	
Structural lightweight solid panels	110	2		0.38	0.19
		3		0.57	
		4		0.76	
		5		0.95	
		6		1.14	
		8		1.52	

Table 5.3.5 Wall U-values: solid and sandwich panels; winter and summer conditions[1]

Concrete weight pcf	Type of wall panel	Thickness, t, and resistance, R of concrete t^2	R	Winter $R_{fo} = 0.17$, $R_{fi} = 0.68$ None	4	6	8	10	Summer $R_{fo} = 0.25$, $R_{fi} = 0.68$ None	4	6	8	10
145	Solid walls, and sandwich panels	2	0.15	1.00	.20	.14	.11	.09	.93	.20	.14	.11	.09
		3	0.23	.93	.20	.14	.11	.09	.86	.19	.14	.11	.09
		4	0.30	.87	.19	.14	.11	.09	.81	.19	.14	.11	.09
		5	0.38	.81	.19	.14	.11	.09	.76	.19	.14	.11	.09
		6	0.45	.77	.19	.14	.11	.09	.72	.19	.14	.11	.09
		8	0.60	.69	.18	.13	.11	.09	.65	.18	.13	.10	.09
110	Solid walls, and sandwich panels	2	0.38	.81	.19	.14	.11	.09	.76	.19	.14	.11	.09
		3	0.57	.70	.18	.13	.11	.09	.67	.18	.13	.11	.09
		4	0.76	.62	.18	.13	.10	.09	.59	.18	.13	.10	.09
		5	0.95	.56	.17	.13	.10	.09	.53	.17	.13	.10	.08
		6	1.14	.50	.17	.13	.10	.08	.48	.16	.12	.10	.08
		8	1.52	.42	.16	.12	.10	.08	.41	.16	.12	.10	.08

1. When insulations having other R-values are used, U-values can be interpolated with adequate accuracy, or U can be calculated as shown in Section 5.3.4. When a finish, air space or any other material layer is added, the new U-value is:

$$U = \cfrac{1}{\cfrac{1}{U \text{ from table}} + R \text{ of added finish, air space, or material}}$$

2. Thickness for sandwich panels, t is the sum of the thicknesses of the wythes.

Fig. 5.3.16   Heating and cooling load comparisons

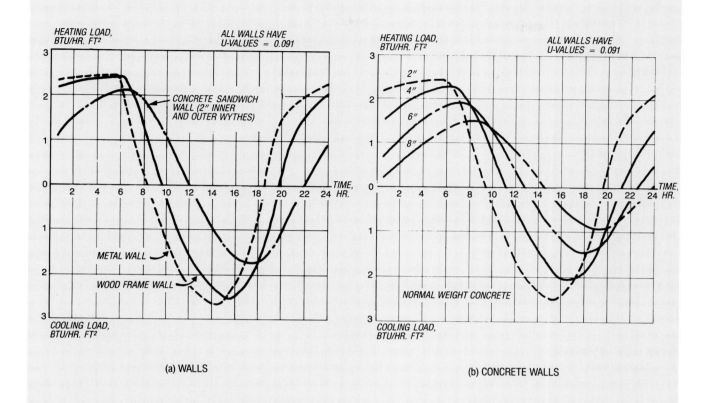

(a) WALLS

(b) CONCRETE WALLS

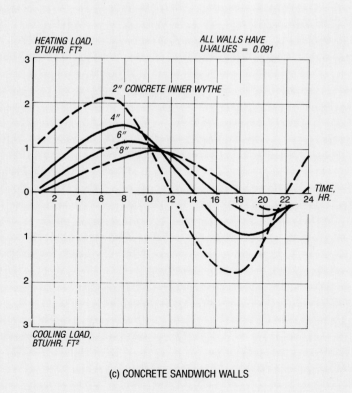

(c) CONCRETE SANDWICH WALLS

5.3.4 COMPUTATION OF THERMAL TRANSMITTANCE VALUES

The heat transmittance (U-values) of a building wall is computed by adding together the R-values of the materials in the section, the air film surfaces (R_{fi} and R_{fo}) and air spaces (R_a) within the section. The reciprocal of the sum of the R's is the U-value:

$$U = \frac{1}{R_{fi} + R_{materials} + R_a + R_{fo}} \qquad \text{(Eq. 5.3.1)}$$

where $R_{materials}$ is the sum of all opaque materials in the wall. A number of typical wall U-values are given in Table 5.3.5. These wall tables can be applied to sandwich type panels as well as single wythe panels insulated on one side. A design example showing how to calculate U values for a wall using R values taken from Tables 5.3.1 through 5.3.5 is presented below:

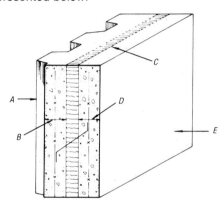

		R Winter	R Summer	Table
A.	Surface, outside	0.17	0.25	5.3.1
B.	Concrete, 2 in. (110 pcf)	0.38	0.38	5.3.5
C.	Polystyrene insulation (1.25 pcf), 1½ in.	6.00	6.00	5.3.3
D.	Concrete, 2½ in. (110 pcf)	0.48	0.48	5.3.5
E.	Surface, inside	0.68	0.68	5.3.1
	Total R =	7.71	7.79	
	U = 1/R	0.13	0.13	

5.3.5 THERMAL STORAGE EFFECTS

Analytical and experimental studies have shown that the use of heavy materials in buildings has the effect of reducing heating and cooling peak loads, and thus reducing equipment size, compared with lightweight materials, because of greater heat storage capacity. Small equipment that runs continuously uses less energy than large equipment that is run intermittently as it responds to peak loads. By lowering peak loads, energy is saved.

Unlike cooling design, the ASHRAE Standard does not prescribe a method to account for thermal storage effects for heating. However, it does permit a performance approach to heating design which can be used to take advantage of the benefits of mass.

Energy use differences between light and heavy materials are illustrated in the hour-by-hour computer analyses shown in Fig. 5.3.16.

Fig. 5.3.16a compares the heat flow through three walls having the same U-value, but made of different materials. The concrete wall consisted of a layer of insulation sandwiched between inner and outer wythes of 2-in. concrete and weighed 48.3 psf. The metal wall, weighing 3.3 psf, had insulation sandwiched between an exterior metal panel and 1/2 in. drywall. The wood frame wall weighed 7.0 psf and had wood siding on the outside, insulation between 2x4 studs, and 1/2-in. drywall on the inside. The walls were exposed to simulated outside temperatures that represented a typical spring day in a moderate climate. The massive concrete wall had lower peak loads by about 13% for heating and 30% for cooling than the less massive walls.

Concrete walls of various thicknesses that were exposed to the same simulated outside temperatures, are compared in Fig. 5.3.16b. The walls had a layer of insulation sandwiched between concrete on the outside and 1/2 in. drywall on the inside; U-values were the same. The figure shows that the more massive the wall the lower the peak loads and the more the peaks were delayed.

Fig. 5.3.16c compares concrete sandwich panels having an outer wythe of 2 in., various thicknesses of insulation, and various thicknesses of inner wythes. All walls had U-values of 0.091 and were exposed to the same simulated outside temperatures. The figure shows that by increasing the thickness of the inner concrete wythe, peak loads were reduced and delayed.

The energy saving benefits of thermal mass are most pronounced when the outside temperature fluctuates above and below the inside temperature causing a reversal of heat flow in the envelope. These conditions exist on a daily basis at all locations in the United States and Canada during at least some months of the year. Thermal mass is most effective in conserving energy in the sun-belt regions in the southern United States, because these temperature fluctuations occur throughout the year. Designs employing thermal mass for energy conservation should be given a high priority in these areas.

Another factor affecting the behavior of thermal mass is the availability of so called "free heat." This includes heat generated inside the building by lights, equipment, appliances, and people. It also includes heat from the sun entering through windows. Generally, during the heating season, benefits of thermal mass increase with the availability of "free heat," Tables 5.3.6 and 5.3.7. Thus office buildings which have high internal heat gains from lights, people, and large glass areas represent an ideal application for thermal mass designs. This is especially true if the glass has been located to take maximum advantage of the sun. Building codes and standards now provide for the benefits of thermal mass. In increasing numbers, they are beginning to acknowledge the effect of the greater heat storage capacity in buildings having high thermal mass.

Table 5.3.6 Design Considerations for Building with High Available Free Heat[1]

Climate Classification		Thermal Mass	Increase Insulation	External Fins[2]	Surface Color		Daylighting	Reduce Infiltration
					Light	Dark		
Winter								
Long Heating Season (6000 Degree Days or more)	With sun and wind[3,4]	1	2	2		2	1	3
	With sun without wind	1	2			2	1	3
	Without sun and wind		2			1		3
	Without sun with wind	1	2	2		1		3
Moderate Heating Season (3000-6000 Degree Days)	With sun and wind	2	2	1		1	2	2
	With sun without wind	2	2			1	2	2
	Without sun and wind	1	2					2
	Without sun with wind	1	2	1				2
Short Heating Season (3000 Degree Days or less)	With sun and wind	3	1				2	1
	With sun without wind	3	1				2	1
	Without sun and wind	2	1					1
	Without sun with wind	2	1					1
Summer								
Long Cooling Season (1500 hr @ 80°F)	Dry or humid	3		3	3		3	3
Moderate Cooling Season (600-1500 hr. @ 80°F)	Dry or humid	3		2	2		2	3
Short Cooling Season Less than 600 hr @ 80°F	Dry or humid	2		1	1		1	2

* Higher numbers indicate greater importance.
1. Includes office buildings, factories, and commercial buildings.
2. Provide shading and protection from direct wind.
3. With sun: Sunshine during at least 60% of daylight time.
4. With wind: Average wind velocity over 9 mph.

Table 5.3.7 Design Considerations for Building with Low Available Free Heat[1]

Climate Classification		Thermal Mass	Increase Insulation	External Fins[2]	Surface Color		Reduce Infiltration
					Light	Dark	
Winter							
Long Heating Season (6000 Degree Days or more)	With sun and wind[3,4]		3	2		3	3
	With sun without wind		3			3	3
	Without sun and wind		3			2	3
	Without sun with wind		3	2		2	3
Moderate Heating Season (3000-6000 Degree Days)	With sun and wind	1	2	1		2	3
	With sun without wind	1	2			2	3
	Without sun and wind		2			1	3
	Without sun with wind	1	2	1		1	3
Short Heating Season (3000 Degree Days or less)	With sun and wind	2	1			1	2
	With sun without wind	2	1			1	2
	Without sun and wind	1	1				2
	Without sun with wind	1	1				2
Summer							
Long Cooling Season (1500 hr @ 80°F)	Dry[5] or humid[6]	3		2	2		3
Moderate Cooling Season (600-1500 hr. @ 80°F)	Dry	2		1	1		2
	Humid	2		1	1		3
Short Cooling Season Less than 600 hr @ 80°F	Dry or humid	1					1

* Higher numbers indicate greater importance.
1. Includes low-rise residential buildings and some warehouses.
2. Provide shading and protection from direct wind.
3. With sun: Sunshine during at least 60% of daylight time.
4. With wind: Average wind velocity over 9 mph.
5. Dry: Daily average relative humidity less than 60% during summer.
6. Humid: Daily average relative humidity greater than 60% during summer.

5.3.6 CONDENSATION CONTROL

Moisture which condenses on the interior of a building is unsightly and can cause damage to the building and its contents. Even more undesirable is the condensation of moisture within a building wall where it is not readily noticed until damage has occurred. Moisture problems in colder climates are most likely in heated occupied buildings, which usually generate substantial amounts of water vapor. Moisture problems in warm humid climates are most likely to occur because of moisture accumulating near cooled surfaces. All air in buildings contains water vapor. In many buildings moisture is added to the air by industrial processes, cooking, laundering, or humidifiers. If the inside surface temperature of a wall is too cold, the air contacting this surface will be cooled below its dew-point temperature and water will condense on that surface. Condensation occurs first on the surface with the lowest temperature. Condensation on interior room surfaces can be controlled both by suitable construction and by precautions such as: (1) reducing the interior dew point temperature; (2) raising the temperatures of interior surfaces that are below the dew point, generally by use of insulation; and (3) using vapor retarders.

The interior air dew point temperature can be lowered by removing moisture from the air, either through ventilation or dehumidification. Adequate surface temperatures can be maintained during the winter by incorporating sufficient thermal insulation, using double glazing, circulating warm air over the surfaces, or directly heating the surfaces, and by paying proper attention during design to the prevention of thermal bridging.

Moisture can move into or across a wall assembly by means of vapor diffusion and air movement. If air, especially exfiltrating, hot, humid air, can leak into the enclosure, then this will be the major source of moisture. Air migration occurs from air pressure differentials independent of moisture pressure differentials.

Infiltration and exfiltration are air leakage into and out of a building through cracks or joints between infill components and structural elements, interstices around windows and doors, through floors and walls and at openings for building services. They are often a major source of energy loss in buildings.

Atmospheric air pressure differences between the inside and outside of a building envelope exist because of the action of wind, the density difference between outside cold heavy air and inside warm light air creating a "stack effect" and the operation of equipment such as fans. The pressure differences will tend to equalize, and the air will flow through holes or cracks in the building envelope carrying with it the water vapor it contains.

A thorough analysis of air leakage is very complex, involving many parameters, including wall construction, building height and orientation.

Many condensation-related problems in building enclosures are caused by exfiltration and subsequent condensation within the enclosure assembly. Condensation due to air movement is usually much greater than that due to vapor diffusion for most buildings. However, when air leakage is controlled or avoided, the contribution from vapor diffusion can still be significant. In a well designed wall, attention must therefore be paid to the control of air flow and vapor diffusion.

An air barrier and vapor retarder are both needed, and in many instances a single material can be used to provide both of these as well as other functions. The principal function of the air barrier is to stop outside air from entering the building through the walls, windows or roof, and inside air from exfiltrating through the building envelope to the outside. This applies whether the air is humid or dry, since air leakage can result in problems other than the deposition of moisture in cavities. Exfiltrating air carries away heating and cooling energy, while incoming air may bring in pollution as well as reduce the effectiveness of a rain screen wall system.

Materials and the method of assembly chosen to build an air barrier must meet several requirements if they are to perform the air leakage control function successfully.

1. There must be continuity throughout the building envelope. The low air permeability materials of the wall must be continuous with the low air barrier materials of the roof (e.g., the roofing membrane) and must be connected to the air barrier material of the window frame, etc.

2. Each membrane or assembly of materials intended to support a differential air pressure load must be designed and constructed to carry that load, inward or outward, or it must receive the necessary support from other elements of the wall. If the air barrier system is made of flexible materials then it must be supported on both sides by materials capable of resisting the peak air pressure loads; or it must be made of self-supporting materials, such as board products adequately fastened to the structure. If an air pressure difference cannot move air, it will act to displace the materials that prevent the air from flowing.

3. The air barrier system must be virtually air-impermeable. A value for maximum allowable air permeability has not yet been determined. However, materials such as polyethylene, gypsum board, precast concrete panels, metal sheeting or glass qualify as low air impermeable materials when joints are properly sealed, whereas concrete block, acoustic insulation, open cell polystyrene insulation or fiberboard would not, see Fig. 5.3.17.

The metal and glass curtain wall industry, notably in the U.S., has adopted a value of 0.06 CFM/ft² (0.3 l/m²·s) at 1.57 lbs/ft² (75 Pa) as the maximum allowable air leakage rate for these types of wall construction. This value however is considered high for buildings in Canada and some Canadian manufacturers of metal and glass curtain wall systems claim that their system will meet 0.1 l/m²·s or better for the same air pressure difference. But, it is the total assembled air barrier system (main areas plus joints) which must exhibit practically zero leakage.

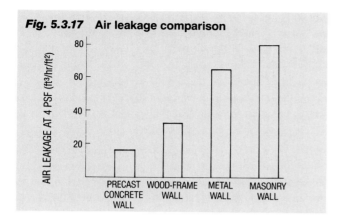

Fig. 5.3.17 Air leakage comparison

AIR LEAKAGE AT 4 PSF (ft³/hr/ft²)

PRECAST CONCRETE WALL | WOOD-FRAME WALL | METAL WALL | MASONRY WALL

4. The air barrier assembly must be durable in the same sense that the building is durable, and be made of materials that are known to have a long service life or be positioned so that it may be serviced from time to time.

In climates where the heating season dominates, it is strongly recommended that the visible interior surface of a building envelope be installed and treated as the primary air barrier and vapor retarder. Where floors and cross walls are of solid concrete, it is necessary to seal only the joints, as floors and walls themselves do not constitute air paths. Where hollow partitions, such as steel stud or hollow masonry units are used, the interior finish of the envelope should first be made continuous. Where this is impractical, polyethylene film should be installed across these junctions and later sealed to the interior finish material. An interior finish of gypsum wallboard or plaster painted with two coats of enamel paint will provide a satisfactory air barrier/vapor retarder in many instances if the floor/wall and ceiling/wall joints are tightly fitted and sealed with caulking.

A recent development is an air barrier and vapor retarder system consisting of panel joints sealed from the inside with a foam backer rod and sealant, plus a thermal fusible membrane (TFM) seal around the panel, covering the gap between the structure and the panel. Surfaces should be clean, as dry as possible, smooth, and free of foreign matter which may impede adhesion. The bond between concrete and the membrane may be improved by priming the concrete before fusing the membrane to it.

While it is preferable that the air barrier system be placed on the warm side of an insulated assembly, where thermal stresses will be at a minimum, it is not an essential requirement. (This does not necessarily mean on the inside surface of the wall.) The position of the air barrier in a wall is more a matter of suitable construction practice and the type of materials to be used. However, if this barrier is positioned on the outside of the insulation, consideration must be given to its water vapor permeability in case it should also act as a barrier to vapor which is on its way out from inside the wall assembly. This situation may be prevented by choosing an air barrier material that is ten to twenty times more permeable to water vapor diffusion than the vapor barrier material.

In the case of construction assemblies which do not lend themselves to the sealing of interior surfaces, or where it is desirable to limit condensation to very small amounts, such as sandwich wall panel construction (which may have no air leaks through the panels themselves or any air space which can ensure venting and drying out in summer), the use of a separate vapor retarder must be considered. In such cases, the insulation material itself, if it is of a rigid closed-cell type, can be installed on a complete bed of adhesive applied to the interior of the inner wythe of the wall with joints fully sealed with adhesive, to provide a complete barrier to both air and vapor movement.

While the discussion above has been concerned with the flat areas of walls, the joints between them may well present the most important design and construction problems. There are many kinds of joints and the following are considered the most critical: the roof/wall connection, the wall/foundation connection, the wall/window or door connection, soffit connections, corner details, and connections between different types of exterior wall systems, such as brick and precast concrete, or curtain wall and precast concrete.

Water vapor diffusion, another way in which indoor water can move through a building envelope to condense in the colder zones, occurs when water vapor molecules diffuse through solid interior materials at a rate dependent on the permeability of the materials, the vapor pressure and temperature differentials. Generally, the colder the outside temperature the greater the pressure of the water vapor in the warm inside air to reach the cooler, drier outside air.

The principal function of a vapor retarder made of low permeability materials is to stop or, more accurately, to retard the passage of moisture as it diffuses through the assembly of materials in a wall. Vapor diffusion control is simple to achieve and is primarily a function of the water vapor diffusion resistance of the chosen materials and their position within the building envelope assembly. The vapor retarder should be clearly identified by the designer and be clearly identifiable by the general contractor.

In temperate climates, vapor retarders should be applied on or near the warm side (inner surface) of assemblies. Vapor retarders may be structural, or in the form of thin sheets, or as coatings. Vapor retarders may also be positioned part way into the insulation but, to avoid condensation, they should be no further in than the point at which the temperature drops to its dew point.

In climates with high humidities and high temperatures, especially where air-conditioning is virtually continuous, the ingress of moisture may be minimized by a vapor retarder system in the building envelope near the outer surface. For air-conditioned buildings in hot and humid climates without extended cold periods, it may be more economical to use only adequate air infiltration retarding systems rather than vapor retarders since the interior temperature is very rarely below the dew point of the outside air.

Where warm and cold sides may reverse, with resulting reversal of vapor flow, careful analysis of the condition is recommended rather than to ignore the problem and omit any vapor retarders. The designer should refer to the *ASHRAE Handbook of Fundamentals* or ASTM C755, *Selection of Vapor Retarders for Thermal Insulation*. In general, a vapor retarder should not be placed at both the inside and outside of wall assemblies. The warm side of a wall should have a vapor retarder with a resistance to permeability at least 5 times greater than the cold side.

High thermal-conductance paths reaching inward from or near the colder surfaces may cause condensation within the construction. High-conductance paths may occur at junctions of floors and walls, walls and ceilings, and walls and roofs; around wall or roof openings; at perimeters of slabs on the ground; and at connections.

Fittings installed in outer walls, such as electrical boxes without holes and conduits, should be completely sealed against moisture and air passage, and they should be installed on the warm side of unbroken vapor retarders or air barriers that are completely sealed.

Condensation Upon a Wall should be prevented. The U-value of a wall must be such that the surface temperature will not fall below the dew-point temperature of the room air in order to prevent condensation on the interior surface of a wall. Fig. 5.3.18 gives U-values for any combination of outside temperatures and inside relative humidities above which condensation will occur on the interior surfaces. For example, if a building were located in an area with an outdoor design temperature of -10 deg. F and it was desired to maintain a relative humidity within the building of 25%, the wall must be designed so that all components have a U-value less than 0.78, otherwise there is likely to be a problem with condensation. In many designs the desire to conserve energy will dictate the use of lower U-values than those required to avoid the condensation problem. However, individual components may still have U-values high enough to allow condensation to occur.

The degree of wall heat transmission resistance that must be provided to avoid condensation may be determined from the following relationship:

$$R_t = \frac{R_{fi}\,(t_i - t_o)}{(t_i - t_s)}$$ (Eq. 5.3.2)

Fig. 5.3.18 **Relative humidity at which visible condensation occurs on inside surfaces. Inside temperature, 70°F**

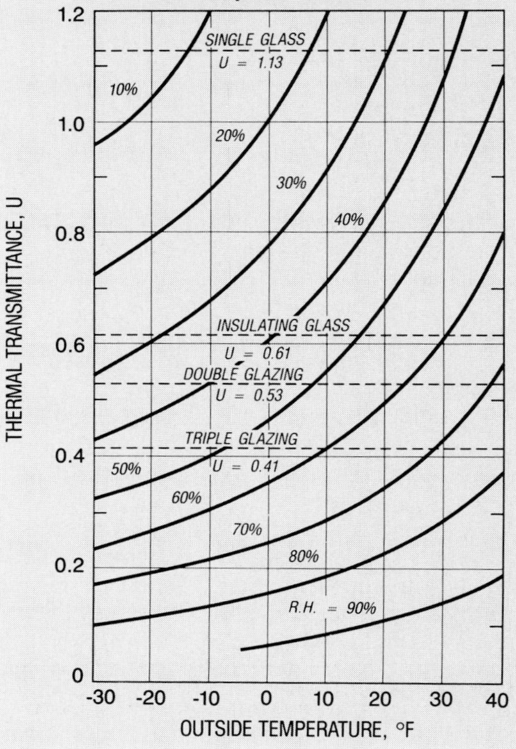

Dew-point temperatures to the nearest deg. F for various values of t_i and relative humidity are shown in Table 5.3.8.

Determine R_t when the room temperature and relative humidity to be maintained are 70 deg. F and 40%, and the outside temperature is -10 deg. F. From Table 5.3.8 the dew-point temperature t_s is 45 deg. F, and from Table 5.3.1 $R_{fi} = 0.68$

$$R_t = \frac{0.68\,[70 - (-10)]}{(70 - 45)} = 2.18$$

$$U = 0.46$$

Table 5.3.8 Dew-point temperatures, °F[1]

Dry bulb or room temperature, °F	Relative humidity, %									
	10	20	30	40	50	60	70	80	90	100
40	−7	6	14	19	24	28	31	34	37	40
45	−3	9	18	23	28	32	36	39	42	45
50	−1	13	21	27	32	37	41	44	47	50
55	5	17	26	32	37	41	45	49	52	55
60	7	21	30	36	42	46	50	54	57	60
65	11	24	33	40	46	51	55	59	62	65
70	14	27	38	45	51	56	60	63	67	70
75	17	32	42	49	55	60	64	69	72	75
80	21	36	46	54	60	65	69	73	77	80
85	23	40	50	58	64	70	74	78	82	85
90	27	44	55	63	69	74	79	83	85	90

1. Temperatures are based on barometric pressure of 29.92 in. Hg.

Table 5.3.9 Typical permeance (M) and permeability (μ) values[1]

Material	M perms	μ perm-in.
Concrete (1:2:4 mix)[2]	—	3.2
Wood (sugar pine)	—	0.4 - 5.4
Expanded polystyrene (extruded)	—	1.2
Paint - 2 coats		
Asphalt paint on plywood	0.4	
Enamels on smooth plaster	0.5 - 1.5	
Various primers plus 1 coat flat oil paint on plaster	1.6 - 3.0	
Expanded polystyrene (bead)	—	2.0 - 5.8
Plaster on gypsum lath (with studs)	20.00	
Gypsum wallboard, 0.375 in.	50.00	
Polyethylene, 2 mil	0.16	
Polyethylene, 10 mil	0.03	
Aluminum foil, 0.35 mil	0.05	
Aluminum foil, 1 mil	0.00	
Built-up roofing (hot mopped)	0.00	
Duplex sheet, asphalt laminated aluminum foil one side	0.002[3]	

1. ASHRAE Handbook, Chapter 21, Table 2.
2. Permeances for concrete vary depending on the concrete's water-cement ratio and other factors.
3. Dry-cup.

Condensation Within a Wall should be prevented. The passage of water vapor through material is in itself generally not harmful. It becomes of consequence when, at some point along the vapor flow path, a temperature level is encountered that is below the dew-point temperature and condensation results.

Building materials have water vapor permeances from very low to very high, Table 5.3.9. When properly used, low permeance materials keep moisture from entering a wall assembly, and materials with higher permeance allow construction moisture and moisture which enters inadvertently or by design to escape.

When a material such as plaster or gypsum board has a permeance which is too high for the intended use, one or two coats of paint is frequently sufficient to lower the permeance to an acceptable level, or a vapor retarder can be used directly behind such products. Polyethylene sheet, aluminum foil and roofing materials are commonly used. Proprietary vapor retarders, usually combinations of foil and polyethylene or asphalt, are frequently used in freezer and cold storage construction.

Concrete is a relatively good vapor retarder. A minimum thickness of 4 in. of concrete is normally regarded as a satisfactory vapor retarder provided it remains crack-free. Permeance is a function of the water-cement ratio of the concrete. A low water-cement ratio, such as that used in most precast concrete members, results in concrete with low permeance. Where climatic conditions demand insula-

tion, a vapor retarder is generally necessary in order to prevent condensation. A closed cell insulation, if properly applied, will serve as its own vapor retarder.

DESIGN EXAMPLE:

Representative exterior wall panel

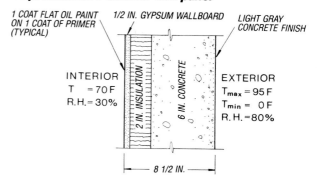

The wall construction shown will be investigated for possible development of water vapor condensation.

Step 1
The overall vapor pressure differential through the wall section may be determined from saturated vapor pressures listed in Table 5.3.10 and the assumed temperatures and relative humidities:

	RH	SVP	
room vapor pressure	.30 x .739	= 0.23 in. Hg	
outdoor vapor pressure	.80 x .038	= 0.03 in. Hg	
overall vapor pressure differential		= 0.20 in. Hg	

This vapor differential must be distributed among the components of the wall section according to their respective vapor transfer resistances.

Step 2
Determine the heat flow properties of the wall section from data listed in Tables 5.3.1 through 5.3.4:

Layer	Thermal Resistance R
Interior surface, R_{fi}	0.68
1/2 in. gypsum wallboard	0.45
2 in. expanded polystyrene insulation	8.00
6 in. normal weight concrete, 145 pcf	0.45
Outside surface, R_{fo}, 15 mph	0.17
Total resistance, R_t	9.75
U	0.10

Step 3
The individual heat flow resistances involved in the wall section together with the sum of these resistances are necessary to the establishment of the temperature gradient through the wall section and the location of possible vapor condensation points. The total temperature drop through the wall in this case is 70 deg. F and can be distrib-

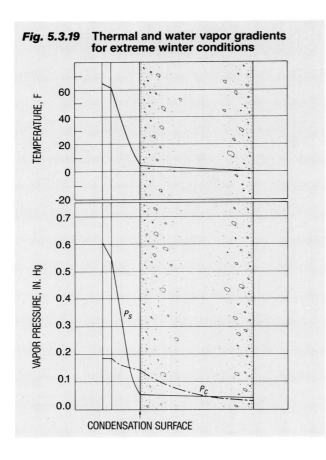

Fig. 5.3.19 Thermal and water vapor gradients for extreme winter conditions

CONDENSATION SURFACE

uted among the individual components in proportion to their resistances. The interface temperatures can then be determined and the temperature gradient plotted as shown in Fig. 5.3.19 with the wall section drawn to physical scale.

Layer	Temperature Drop, deg. F
Inside air to inside surface	(0.68 ÷ 9.75) 70 = 4.88
1/2 in. gypsum wallboard	(0.45 ÷ 9.75) 70 = 3.23
2 in. expanded polystyrene insulation	(8.00 ÷ 9.75) 70 = 57.44
6 in. normal weight concrete, 145 pcf	(0.45 ÷ 9.75) 70 = 3.23
Outside air (15 mph wind)	(0.17 ÷ 9.75) 70 = 1.22
Total	= 70

As an alternate procedure to the arithmetic method for determination of the temperature gradient under steady-state parallel heat flow, a graphical method may be used. In the graphical method, a cross section of the wall is drawn wherein the thickness shown for each component is proportional to its thermal resistance. Then by plotting a temperature scale on the cross section and a straight line joining the inside and outside temperatures (representing the temperature gradient) the temperature at any point in the construction can be read.

The assumed steady-state conditions are seldom reached owing to fluctuations in the temperatures to which the envelope is exposed and to the heat storage capacities of the concrete, see Section 5.3.5 for a discussion on thermal storage. Unless simplified procedures are followed, the solution of practical cases of heat flow through walls and roofs can become very complicated. Some inaccuracies may be introduced by these simplifications, but the results obtained provide a valuable guide for design of walls and roofs. The determination of the interface temperatures to a precision greater than 1 or 2 deg. F is, however, unwarranted. Paths of high conductivity, called thermal bridges, do produce inaccuracies that often require special consideration. Thermal bridges are discussed later in this section.

The temperature existing at any point in a wall under any given exterior and interior temperature conditions is of great significance in designing problem-free building enclosures. An ability to calculate the thermal gradient permits the designer to forecast the magnitude of the movements caused by external temperature changes, to predict the location of condensation and freezing planes in the wall, and to assess the suitability of any construction. The temperature gradient will not, in itself, give the designer all the information required to select and assemble building components, but it is an essential first step.

The selection of appropriate outside air temperatures requires considerable judgment. The effects of heat storage in materials must be recognized, as must the fact that wall or roof surface temperatures can be higher than air temperature because of solar radiation, and colder than air temperature because of clear sky radiation. These temperature modifications vary with the color, texture, thickness, weight and orientation of the surface materials and with the intensity of the radiation.

Step 4
From Table 5.3.10 the saturated vapor pressures at various surfaces and interfaces within the wall section may be obtained from temperatures at these locations, using the previously calculated temperature drops:

Vapor Pressures at Saturation

Location (Fig. 5.3.19)	Temp., deg. F	SVP, in. Hg
Room air	70	0.739
Inside surface (70 − 4.88)	= 65.12	0.625
Interface-wallboard and insulation (65.12 − 3.23)	= 61.89	0.558
Interface-insulation and concrete (61.89 − 57.44)	= 4.45	0.047
Outside surface (4.45 − 3.23)	= 1.22	0.040
Outside air	0	0.038

These saturated vapor pressures are plotted in Fig. 5.3.19 to form the SVP gradient, P_s, through the wall section.

Step 5
To check the location where condensation is likely to take place, the vapor pressure gradient necessary for vapor

Table 5.3.10 Water vapor pressures at saturation for various dry bulb temperatures

Temp., °F	SVP, in. Hg	Temp., °F	SVP, in. Hg	Temp., °F	SVP, in. Hg	Temp., °F	SVP, in. Hg
− 30	.007	+ 17	.089	+ 38	.229	+ 59	.503
− 20	.0126	18	.093	39	.238	60	.522
− 10	.022	19	.098	40	.248	61	.540
− 5	.029	20	.103	41	.257	62	.560
0	.038	21	.108	42	.268	63	.580
+ 1	.040	22	.113	43	.278	64	.601
2	.042	23	.119	44	.289	65	.622
3	.044	24	.124	45	.300	66	.644
4	.046	25	.130	46	.312	67	.667
5	.049	26	.137	47	.324	68	.690
6	.051	27	.143	48	.336	69	.714
7	.054	28	.150	49	.349	70	.739
8	.057	29	.157	50	.362	71	.765
9	.060	30	.165	51	.376	72	.791
10	.063	31	.172	52	.390	73	.818
11	.066	32	.180	53	.404	74	.846
12	.069	33	.188	54	.420	75	.875
13	.073	34	.195	55	.436	76	.905
14	.077	35	.203	56	.452	77	.935
15	.081	36	.212	57	.468	78	.967
16	.085	37	.220	58	.486	79	.999
						80	1.032

Note: 1 in. Hg = 0.491 psi   Actual V.P. = SVP (RH)

transfer continuity, P_c, is plotted as shown in Fig. 5.3.19. The vapor pressure gradient, P_c, is obtained by a calculation procedure similar to that used to determine the temperature gradient, described in Step 3. It is based upon the total vapor pressure drop (0.23 -0.03 = 0.20 in. Hg) and the respective vapor transfer resistances of the different components of the section from Table 5.3.10, thus:

Wall Components Continuity	Vapor Transfer Resistance, $\frac{1}{M}$ or $\frac{1}{\mu}$	V.P. Drop for Continuity	V.P. for Continuity
Inside surface film		= 0	0.230
1 coat of primer and paint	0.50	(0.50 ÷ 3.06) × 0.20 = 0.033	0.197
1/2 in. gypsum wallboard	0.02	(0.02 ÷ 3.06) × 0.20 = 0.001	0.196
2 in. insulation	0.67	(0.67 ÷ 3.06) × 0.20 = 0.044	0.152
6 in. concrete	1.87	(1.87 ÷ 3.06) × 0.20 = 0.122	0.030
Outside surface film	0	0 = 0	0.030
Total R = 3.06		Total drop = 0.20	

The actual vapor pressure drop, P_a, from the inside surface of the wall to any material interface may be taken as the difference between the vapor pressure at the inside wall surface and the saturated vapor pressure at the material interface.

Continuous vapor flow conditions are preserved provided this vapor pressure does not exceed the saturation vapor pressure. If P_c does cross P_s, condensation will occur, usually at the nearest outer interface or surface. For discontinuous vapor flow, the vapor flow to and away from the condensation surface must be recalculated. The difference will be equal to the condensation rate. The vapor flow to or from a point is equal to the actual vapor pressure difference divided by the vapor resistance to or from that point.

To ensure that condensation does not take place, the fundamental requirement is that, at all points through the thickness of the building enclosure, the vapor pressure set by the condition of continuous flow (P_c) must be less than the maximum permissible vapor pressure set by the saturation vapor pressure (P_s) corresponding to the temperature at that point. This condition can be achieved either by changing the various vapor flow resistances to reduce the values of P_c (i.e., by adding a vapor barrier), by changing the various thermal resistances to raise the temperature and thus the values of P_s, or by a combination of both methods.

Thermal Bridges are high conductivity metal ties or solid concrete sections that penetrate the insulation layer of precast concrete sandwich cladding as described in Section 5.3.8. These bridges may lead to localized cold areas where surface condensation can occur, particularly where the interior relative humidity is maintained at high levels, which may cause annoying or damaging wet streaks.

The effect of these "bridges" on the heat transmittance can be calculated with reasonable accuracy by the zone method described in ASHRAE Handbook. The net effect of metal ties is to increase the U-value by 10 or 15 percent, depending on type, size and spacing. For example, a wall as shown in Fig. 5.3.20 would have a U-value of 0.13 if the effect of the ties is neglected. If the effect of 1/4 in. diameter ties at 16 in. on center is included, U = 0.16; at 24 in. spacing, U = 0.15.

5.3.7 APPLICATION OF INSULATION

Where insulation is required on a precast concrete wall, it may be applied to the panel normally to the back or it may be fully incorporated in the panel, resulting in a sandwich wall panel.

There are basically several approaches for the application of insulation to large flat surfaces:

1. Supplementary framing (e.g. steel studs) can be added to provide cavities for the installation of batts or rigid insulation and to support subsequent components of the assembly. There should be a gap between the framing and the panels to avoid a thermal bridge.

2. Rigid insulation can be fastened to concrete surfaces with adhesives, by impaling it on adhered pins ("stick clips"), and with various types of furring and mechanical fasteners.

Adhesives: This is the most obvious method of fastening anything to a large flat surface and there are a number of adhesives available for this use. Selection of the proper adhesive is important. It should be compatible with the type of insulation being used. The vehicles or thinners in some adhesives will attack foam plastic insulation. Also, some protein-based adhesives can provide nutrition for fungi and other micro-organisms unless they have preservatives included in their make-up.

The adhesive should not be applied in daubs. The use of daubs of adhesive creates an air space between the surface and the insulation. If the insulation is on the inner surface of the assembly, warm moist air circulating in this space will cause condensation. If the insulation is on the outer surface of the assembly, cold air circulating in this space will "short circuit" the insulation.

It is better to apply a full bed of adhesive or a grid of beads of adhesive, Fig. 5.3.21. A full adhesive bed is the preferable method from an adhesion point of view but where it is on the cold side of the insulation (e.g., applying insulation to the interior surface) it may act as a vapor trap preventing drying of any moisture which penetrates the interior air/vapor barrier. In this situation therefore the grid approach should be used.

Stick Clips: These are thin metal or plastic pins with a large perforated flat head at one end. The head is fastened to the concrete surface with a high quality adhesive which keys into the perforations. The clips are applied in a grid pattern, then the insulation is impaled on the pins and secured in

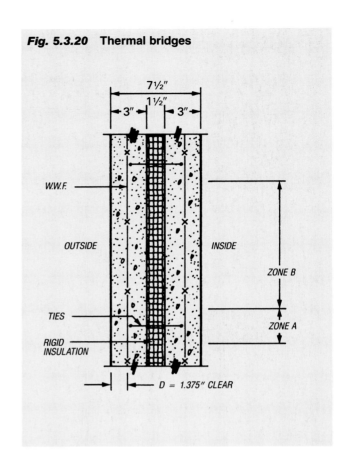

Fig. 5.3.20 Thermal bridges

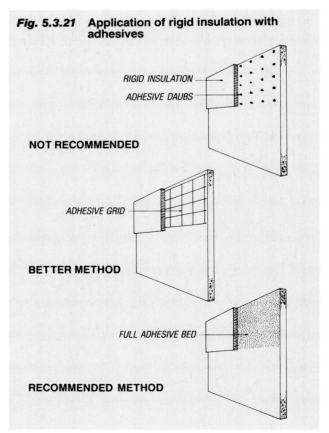

Fig. 5.3.21 Application of rigid insulation with adhesives

RIGID INSULATION
ADHESIVE DAUBS

NOT RECOMMENDED

ADHESIVE GRID

BETTER METHOD

FULL ADHESIVE BED

RECOMMENDED METHOD

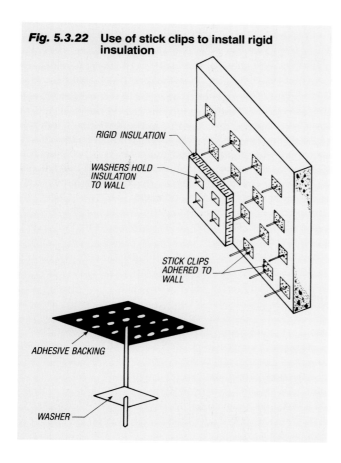

Fig. 5.3.22 Use of stick clips to install rigid insulation

RIGID INSULATION

WASHERS HOLD INSULATION TO WALL

STICK CLIPS ADHERED TO WALL

ADHESIVE BACKING

WASHER

place with a type of spring washer which is simply pushed over the end of the pin against the insulation. Sharp "teeth" on the washer grip the pin, Fig. 5.3.22. Although this method also relies on adhesive, the entire surface does not have to be covered, thus making it easier to clean the surface and permitting the use of high performance (and hence costly) adhesives.

Furring Systems: There are a number of types of plastic, wood or metal furring which can be applied over the insulation and fastened, through it, to the concrete surface. Fig. 5.3.23 illustrates some of the approaches. The furring is usually applied along the joint between two insulation boards so that one piece of furring contributes to the support of two insulation boards. This may require special preparation of the edges of the insulation such as grooving to accept the legs of the U-channel in Fig. 5.3.23b. Depending on the size of the insulation boards and the amount of support required by any subsequent finish, furring may also be applied in the middle of the insulation boards. This may also require special preparation of the insulation.

The insulation may be held in place temporarily prior to application of the furring by light daubs of adhesive. These should be very light to avoid holding the insulation away from the surface as mentioned under adhesives.

The furring can be fastened with powder-driven fasteners or a special type of concrete nail which is driven into a pre-drilled hole. The available length of fasteners usually limits

Fig. 5.3.23 Furring systems

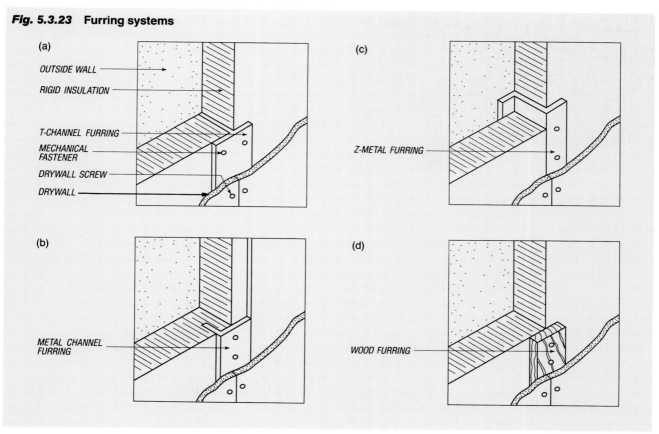

(a)

OUTSIDE WALL

RIGID INSULATION

T-CHANNEL FURRING

MECHANICAL FASTENER

DRYWALL SCREW

DRYWALL

(b)

METAL CHANNEL FURRING

(c)

Z-METAL FURRING

(d)

WOOD FURRING

the thickness of insulation to about 4 in.

Where this method is used to apply insulation to the inside of a wall, Fig. 5.3.23d, the interior finish is applied by screwing or nailing it to the furring members.

Insulation may be plant or jobsite applied:

1. Mechanical; Most commonly performed at jobsite. If done in precast concrete plant, see note below.

2. Adhesive; As above.

3. Spraying; Normally accomplished at jobsite after installation. If done in precast concrete plant, see note below.

4. Poured; Face to be insulated must be face-up during casting. Bulkheads permit simple application of insulation following concrete casting and initial curing. Very lightweight concrete mixes should be checked for variation due to shrinkage to avoid possible delamination. For soft insulations the note below is also valid.

5. Wet Application; Insulation should have a bondable surface. Shear ties should be used between concrete and insulation.

Note: For all insulation applied in the precast concrete plant, by whatever method, the initial cost saving in application should be weighed against the cost of added protection during handling and transportation and possible protection against inclement weather. The latter will depend mostly on the type of insulation used.

Where precast concrete cladding is applied over a previously erected wall, as would be the case with a load-bearing masonry wall, (Figs. 5.3.24, 5.3.25 and 5.3.26) or a concrete end shear wall, it is necessary to leave holes in such walls for access to the connection points for the precast concrete panels. Care must be taken in filling these holes after the precast concrete panels are installed in order to maintain the integrity of the envelope's airtightness and thermal resistance. The thermal consideration is especially true where the insulation is installed on the outer surface of the inner wall prior to erection of the precast concrete panels, Fig. 5.3.25, as is recommended in order to avoid thermal bridges at the slabs. One solution is to fill around the panel connections with pre-packaged foam-in-place urethane. The effect of these holes on the envelope's airtightness will be less of a concern where the approach of treating the interior finish as the primary air barrier is adopted. This is not to suggest that the holes should not be properly sealed when this approach is adopted. They also represent weaknesses in the wall's secondary line of defense against rain penetration.

Access to the back of the panels for sealing the joints is not a problem where the inner wall is erected after the precast concrete panels, as in the examples in Figs. 5.3.27 and 5.3.28 where the inner wall is a steel stud type, or where there is no inner wall as in Fig. 5.3.29. This, of course, assumes the panel joints are offset from the slabs and cross-walls or exterior columns.

The steel stud approach is shown in Fig. 5.3.27 with a layer

Fig. 5.3.24 **Precast concrete cladding over masonry bearing wall — insulated in interior**

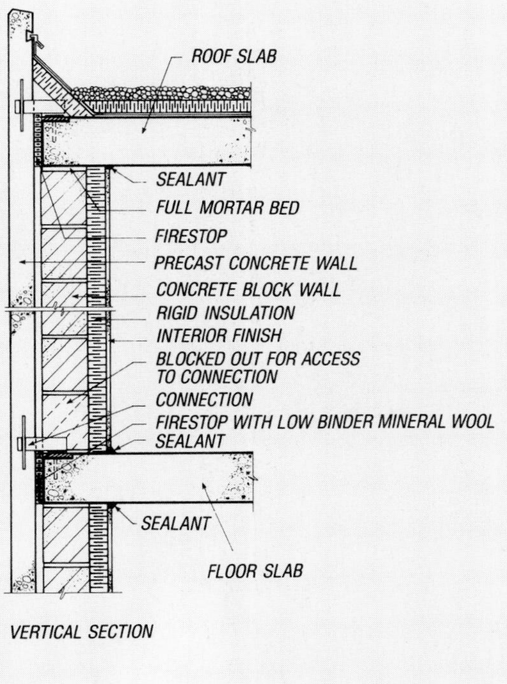

ROOF SLAB
SEALANT
FULL MORTAR BED
FIRESTOP
PRECAST CONCRETE WALL
CONCRETE BLOCK WALL
RIGID INSULATION
INTERIOR FINISH
BLOCKED OUT FOR ACCESS TO CONNECTION
CONNECTION
FIRESTOP WITH LOW BINDER MINERAL WOOL
SEALANT
SEALANT
FLOOR SLAB

VERTICAL SECTION

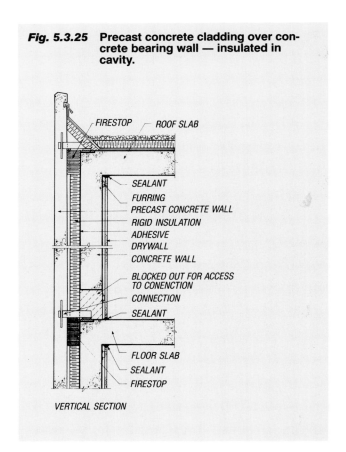

Fig. 5.3.25 **Precast concrete cladding over concrete bearing wall — insulated in cavity.**

FIRESTOP   ROOF SLAB
SEALANT
FURRING
PRECAST CONCRETE WALL
RIGID INSULATION
ADHESIVE
DRYWALL
CONCRETE WALL
BLOCKED OUT FOR ACCESS TO CONENCTION
CONNECTION
SEALANT
FLOOR SLAB
SEALANT
FIRESTOP

VERTICAL SECTION

Fig. 5.3.26 Details of precast concrete cladding over masonry bearing wall

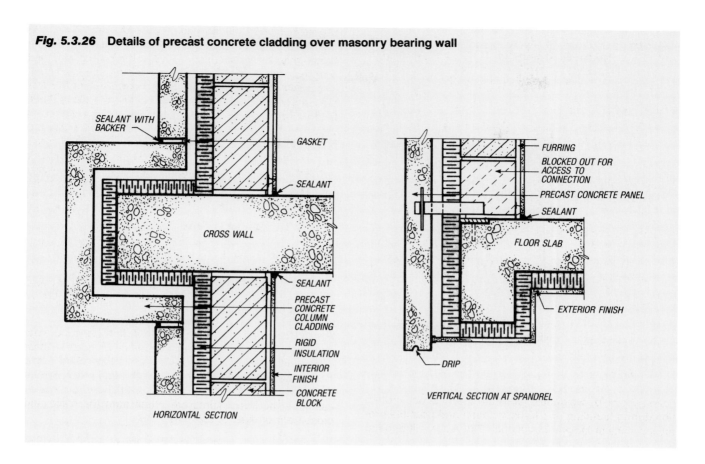

SEALANT WITH BACKER

GASKET

SEALANT

CROSS WALL

SEALANT

PRECAST CONCRETE COLUMN CLADDING

RIGID INSULATION

INTERIOR FINISH

CONCRETE BLOCK

HORIZONTAL SECTION

FURRING

BLOCKED OUT FOR ACCESS TO CONNECTION

PRECAST CONCRETE PANEL

SEALANT

FLOOR SLAB

EXTERIOR FINISH

DRIP

VERTICAL SECTION AT SPANDREL

Fig. 5.3.27 Precast wall with steel stud backing

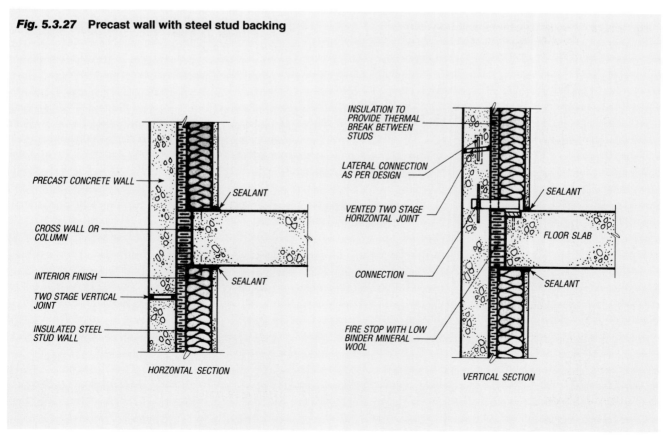

PRECAST CONCRETE WALL

CROSS WALL OR COLUMN

INTERIOR FINISH

TWO STAGE VERTICAL JOINT

INSULATED STEEL STUD WALL

SEALANT

SEALANT

HORZONTAL SECTION

INSULATION TO PROVIDE THERMAL BREAK BETWEEN STUDS

LATERAL CONNECTION AS PER DESIGN

VENTED TWO STAGE HORIZONTAL JOINT

CONNECTION

FIRE STOP WITH LOW BINDER MINERAL WOOL

SEALANT

FLOOR SLAB

SEALANT

VERTICAL SECTION

Fig. 5.3.28 Precast concrete spandrel panel with steel stud backing

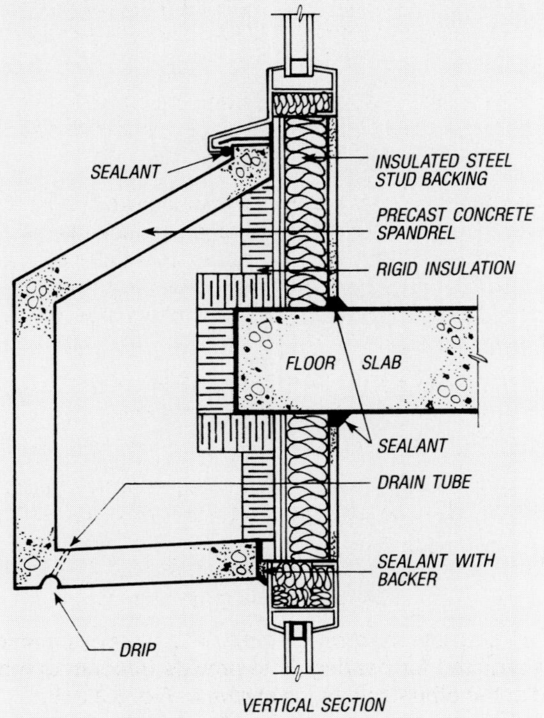

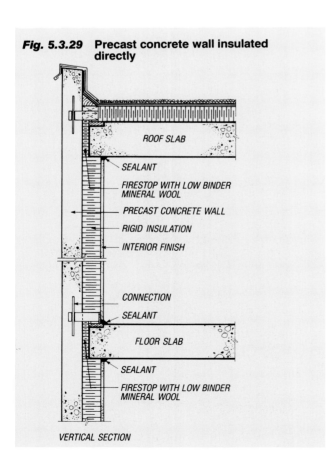

SEALANT

INSULATED STEEL STUD BACKING

PRECAST CONCRETE SPANDREL

RIGID INSULATION

FLOOR SLAB

SEALANT

DRAIN TUBE

SEALANT WITH BACKER

DRIP

VERTICAL SECTION

Fig. 5.3.29 Precast concrete wall insulated directly

ROOF SLAB

SEALANT

FIRESTOP WITH LOW BINDER MINERAL WOOL

PRECAST CONCRETE WALL

RIGID INSULATION

INTERIOR FINISH

CONNECTION

SEALANT

FLOOR SLAB

SEALANT

FIRESTOP WITH LOW BINDER MINERAL WOOL

VERTICAL SECTION

of rigid insulation between the precast concrete panels and the studs in addition to the batt insulation between the studs. This both allows higher R values to be achieved and reduces the effect of thermal bridging through the studs. The rigid insulation would be applied to the inner surface of the precast concrete panels after sealing the panel joints and prior to erecting the steel studs.

5.3.8 PRECAST CONCRETE SANDWICH PANELS

Precast concrete sandwich wall panels are ideally suited for energy conservation. Precast concrete sandwich wall panels offer a weatherproof exterior, effective insulation and a finished interior surface in one unit. Also, the danger of toxic fumes caused by the burning of cellular plastics is practically eliminated with insulation encased in the panel. Sandwich panels can run directly from the footing and serve as an insulated foundation wall as do the 8-in. exterior bearing walls in Fig. 5.3.30. This is very important in reducing heat losses to the ground, especially where deep frost lines prevail. The interior wall energy requirements were satisfied by placing insulation between the exposed aggregate surfaces. The main advantage of insulated precast concrete sandwich panels is that they have a low steady state U-value, yet retain the thermal mass associated with concrete. A range of U-values can be obtained by using different insulation thicknesses and varying the concrete unit weights.

Sandwich panels are composed of two concrete wythes (layers) separated by a layer of insulation and, depending on the design, an air space between the exterior wythe and the insulation. If either of the concrete wythes is replaced by a material other than concrete, it may still be classified as a sandwich panel, provided that the remaining concrete layer is the major (structural) wythe and the product lends itself to production in a precast concrete plant.

In place, sandwich panels provide a means of transferring load, and insulating the structure. They may be cladding panels, in which case they support only their own weight and transfer lateral loads to the supporting structure or they may act as beams, as bearing panels or shear walls, transferring loads imposed from remote parts of the structure.

In the past, both composite and non-composite panels

Fig. 5.3.30

have been used, Fig. 5.3.31. Composite panels, Fig. 5.3.31b, are those in which inner and outer wythes are interconnected through the insulation by means of rigid ties, or regions of solid concrete, that restrict relative movement between the wythes. The two wythes act together to resist externally imposed loads. Depending on the rigidity of the connector system (ties or ribs) wythe interaction may be total or partial. Non-composite panels, Fig. 5.3.31a, are those in which one wythe is supported from the other by relatively flexible ties, and/or hangers, allowing differential movement of the wythes with changing temperatures and humidity conditions. Non-composite panels with an air space allow for ventilation of the outer wythe and pressure equalization. For non-composite panels, one wythe is usually assumed to be "structural" and all loads are carried by that wythe. The structural wythe is normally thicker and stiffer than the non-structural wythe, and is usually located on the interior (warm) side of the panel to reduce thermal stresses due to temperature variation. Occasionally it may be the exterior wythe, particularly in the case of sculptured panels such as ribbed panels, that serves the structural function.

For equal overall thickness of panel, a composite element (Fig. 5.3.31b) will demonstrate greater lateral stiffness. However, because the deformation of the outer wythe will affect the inner wythe, experience indicates that the lateral bowing of composite panels with mild steel reinforcement is less predictable than that of non-composite panels. While the introduction of prestress in both wythes of a com-

posite panel has no effect on thermal bowing, it can be used to induce a negative bow to counteract the tendency of the panel to bow outwards, thus improving the behavior of the element. While this is difficult to calculate, it is a workable solution used successfully by experienced precasters.

Panels with perimeter or interior ribs of concrete, or openings with surrounding ribs, are not recommended because:

1. The ribs act as restraints between the two concrete layers, each of which is subjected to significantly differing deformations, thus developing forces which may lead to cracking. This is true of any composite panel. It should be noted that the degree of composite action cannot be established accurately by analysis.

2. The ribs act as significant heat sinks, and will reduce the insulating effectiveness of the panel, as well as possibly causing local condensation and discoloration.

Some precasters have reported successful use of panels with ribs top and bottom only, and with no ribs on the sides. Such an arrangement provides less restraint than surrounding ribs; however, it is suggested that this be used with caution and based on previous experience.

A more satisfactory arrangement, where a solid concrete rib is required for bearing, is to provide freedom of movement at the other 3 sides, as shown in Fig. 5.3.31b.

Precasters who advocate the use of non-composite sand-

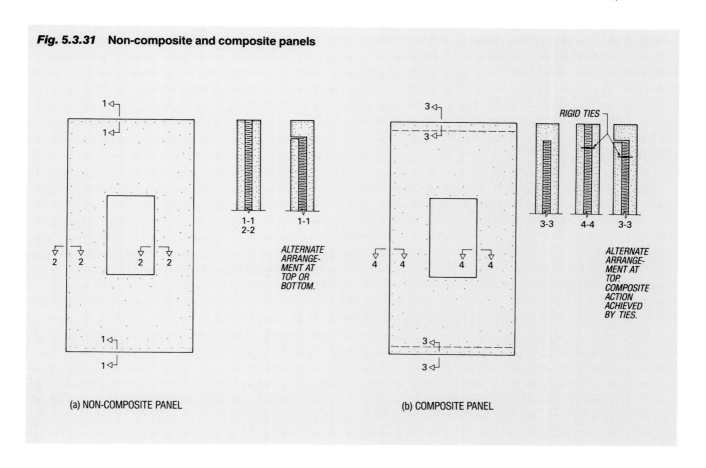

Fig. 5.3.31 Non-composite and composite panels

(a) NON-COMPOSITE PANEL

(b) COMPOSITE PANEL

wich panels emphasize the advantage of the structural wythe being protected by the insulation from extremes of temperature, thus minimizing the bowing of the structural wythe and eliminating thermal stresses in the structure. The exterior wythe is free to expand and contract with variations in temperature and thus will remain crack-free.

Precasters who advocate the use of composite panels point out that the structural wythe of a non-composite panel must carry all of the applied loads in addition to the weight of the other wythe thus increasing overall dimensions, weight and expense, whereas a composite panel can be designed to share the loads between the wythes thus reducing dimensions, weight and expense.

Insulation and concrete thermal properties are discussed earlier in Section 5.3.2. The insulation should have low absorption or a water-repellent coating should be used to minimize absorption of water from the fresh concrete, as this can have an adverse effect on the performance of the insulation.

The thickness of the insulation will be determined by the thermal characteristics of the material and the design temperatures of the structure. A minimum thickness of 1 in. is recommended. The deflection characteristics of the inter-wythe connectors should be considered in relation to the insulation thickness. Although one does not necessarily limit the other, the two must be designed to be compatible.

Openings in the insulation around connectors should be packed with insulation to avoid forming concrete thermal bridges between wythes.

The maximum sizes of insulation commercially available, consistent with the shape of the panel, are recommended. This will minimize joints and the resulting cold sinks. Lapped abutting ends of single layer insulation, or staggered joints with double layer insulation, will effectively remove cold sinks at joints, Fig. 5.3.32.

The insulation itself is capable of transferring a certain amount of shear between the wythes, the value being dependent upon the thickness and properties of the insulation. It may be necessary to break the bond between the insulation and the concrete wythes of non-composite panels by physical or chemical methods to eliminate unintended restraint. This will allow relatively free movement between the wythes for the dissipation of temperature and other volume change stresses. While such bond may be destroyed in time, it is strongest at the initial stages of casting, when the concrete has its least tensile strength.

Panels may be manufactured by incorporating bondbreakers of polyethylene or reinforced paper sheets, or by applying form release oils to the insulation or by using two layers of insulation with staggered joints which will allow movement between the two insulation sheets. This movement may be prevented if the layers are not placed in a level plane. Similarly, the use of sheeting as a bondbreaker can be nullified by the unevenness of the bottom layer of concrete and hence the insulation. Under certain conditions, air gaps may be utilized between the insulation and the outer wythe. This also ensures prevention of shear transfer.

Fig. 5.3.32 Preferred installation of insulation sheets

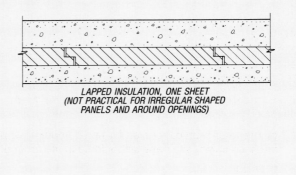

STAGGERED INSULATION, TWO SHEETS

LAPPED INSULATION, ONE SHEET
(NOT PRACTICAL FOR IRREGULAR SHAPED
PANELS AND AROUND OPENINGS)

The use of tape or sheeting with a single layer of insulation serves to bridge insulation joints, thus minimizing the danger of concrete bridges between the wythes. Polyethylene sheeting on the warm side of the insulation also serves as a vapor retarder. In this case, it is necessary to seal around mechanical ties between the wythes to provide continuity of the vapor retarder. It should be noted that a 4 in. minimum thickness of the inner structural concrete wythe is normally regarded as a satisfactory vapor retarder, provided that it remains crack-free.

Wythe minimum thickness is dependent upon structural requirements, finish, reinforcement protection, handling considerations and past experience.

In order to minimize differential temperature across the thickness of the non-structural wythe (non-composite panels), it should be as thin as architectural details will permit. The following limitations are applicable:

1. At the thinnest point, thickness should not be less than 2 in. but preferably a minimum of 2-1/2 in.

2. Thickness should be sufficient to provide proper reinforcement cover.

3. Thickness should be sufficient to provide required anchorage of the wythe connector devices.

4. At no point should the thickness be less than three times the size of the maximum aggregate.

The thickness of the structural wythe should be determined by structural analysis, and by the need to accommodate architectural details. The wythe should not be less than 3 in. thick, although a thinner wythe may be successfully used with rather high quantities of reinforcement and with a higher risk of cracking and bowing. If the wythes are prestressed, the wythe should not be less than 2-1/2 in. thick.

The other limitations listed above for the non-structural wythe also should be considered. Loadbearing structural wythes are, in most cases, supported at the bottom edge. They may have a lateral tie near the top and a mid-height connection to the adjacent panels to prevent differential bowing. Non-loadbearing composite or non-composite panels can be supported by hanging from suitably designed connections. It is worth noting that top hung panels eccentrically supported will bow outwards less than bottom supported units.

Panel size will be primarily determined by architectural considerations. Additionally, the following restrictions are applicable:

1. The maximum dimension and weight should be determined based on transportation and handling requirements.

2. a) The maximum dimension for non-composite or prestressed composite panels should be in the order of:

    L = 48t

    where:

    L = maximum dimension, in.

    t = overall thickness of panel excluding any ribs, in.

    b) The maximum dimension for non-prestressed composite panels should be in the order of:

    L = 48c

    where:

    L = maximum dimension, in.

    c = overall thickness of panel minus thickness of insulation, in.

    In addition, the area of a composite panel should not exceed about 40 sq ft. The suggested restrictions on length and area are empirical, and are based on experience that composite non-prestressed panels tend to bow to a greater degree than non-composite panels. Where bowing is not a serious consideration, these figures could be exceeded by an experienced precaster.

3. The maximum dimension from that point on a composite panel where the two wythes are rigidly connected to the furthest free edge, should not exceed approximately 15 ft, unless the panel is prestressed, or experience indicates otherwise.

Special procedures which will reduce the differential shrinkage rate, or differential temperature rate will, in turn, permit larger panels. Such procedures include: (1) use of low shrinkage concrete, and (2) jointing of the non-structural wythe. Any joints should preferably be complete all the way to the insulation and should be provided at corners of large openings in the panels.

Wythe Connectors. When one wythe is non-structural, its weight must be transferred to the structural wythe. This is generally accomplished by using shear connectors and metallic tension/compression ties passing through the insulation at regularly spaced intervals. An alternative to the use of shear connectors is a solid rib at the bottom of the panel to serve as a support and to still allow the panel to move without restraint that may cause cracking. The solid rib is located at the foundation to minimize the effect on the insulation value of the panel.

Shear connectors may be bent reinforcing bars, sleeve anchors, expanded metal, or welded wire trusses, Fig. 5.3.33.

Galvanized mild steel bars approximately 1/4 in. in diameter are bent into configurations shown in Figs. 5.3.33a, b, and c. The inclined legs of the bent bars carry the vertical shear load. These bars are usually placed along the horizontal axis of the panel. Sometimes, placement along the vertical axis is added because of panel rotation, torsional forces or special shear requirements.

Sleeve anchors up to 20 in. in diameter, may be used to connect the two concrete wythes, Fig. 5.3.33d. Since movements due to temperature differentials radiate away from the sleeve anchor, no restraint to the movement of the suspended wythe exists, permitting free floating of the

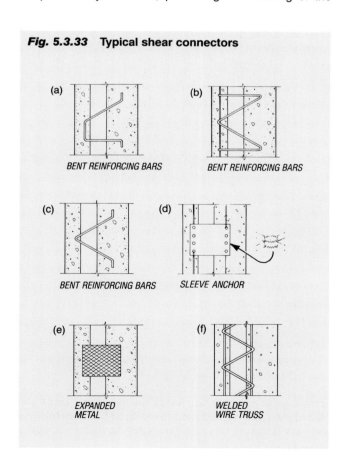

Fig. 5.3.33 **Typical shear connectors**

(a) BENT REINFORCING BARS

(b) BENT REINFORCING BARS

(c) BENT REINFORCING BARS

(d) SLEEVE ANCHOR

(e) EXPANDED METAL

(f) WELDED WIRE TRUSS

non-loadbearing wythe. The metal sleeve, in conjunction with anchorage bars, carries all applied shear loads from any direction.

Expanded metal can be installed or tied to the reinforcement to connect the concrete wythes, Fig. 5.3.33e. Cut sections (approximately 8 in. long) or continuous sections may be used to provide a one-directional shear capacity. Simple tie rods are also used as tension/compression ties to resist wind loads and individual layer separation.

For ribbed insulated panels, it is best to position the shear connector in the rib area of the panel, Fig. 5.3.34. By so doing, it will be easier to position and assure proper embedment depths for an effective connection between the two concrete wythes.

Due to expansion and contraction, it is preferable to have only one anchoring center. In a panel with two ribs, a shear connector can be positioned in either one of the ribs and, in the other rib, a flat anchor can be positioned vertically. It should have the same vertical shear capacity as the shear connector, and should be located on the same horizontal axis. The flat anchor has little or no horizontal shear capacity and therefore restraint of the exterior concrete face is minimized, Fig. 5.3.34. In a multi-ribbed panel, the shear connector should be positioned in the rib closest to the center, and flat anchors used in the other ribs. To complete the connection, metal tension/compression ties passing through the insulation are spaced at regular intervals to prevent the wythes from separating. The spacing of ties should be approximately 2 ft on centers, but not more than 4 ft, or at least 2 ties per 10 sq ft of panel area. Around the perimeter of the panel, and the perimeter of openings larger than 2 ft, tie spacing should not exceed 2 ft on centers, Fig. 5.3.35. Experience indicates that these spacings will assure proper performance and contain wythe bowing to acceptable levels.

Tension/compression ties should be flexible enough so as not to provide significant resistance to temperature and shrinkage stresses in the direction parallel to the panel surface. They should have sufficient anchorage in each wythe to safely transfer applied loads. This can be accomplished through hooking around or tying to the reinforcement, or by bending or deforming the ends of the ties. Typical tension/compression ties are shown in Fig. 5.3.36.

Fig. 5.3.34   Anchorage for ribbed panels

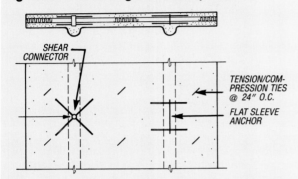

Fig. 5.3.35   Arrangement of connectors between wythes

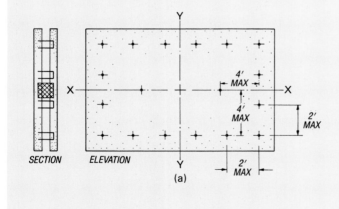

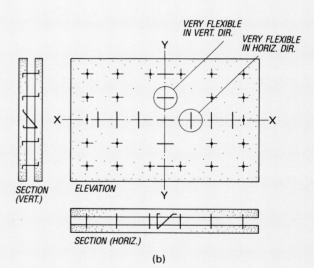

One device transfers weight of non-structural wyth to structural wythe, as well as any racking shear.

Ties spaced in the field of the panel transfer direct wind forces and stripping loads from the non-structural wythe to the structural wythe.

Devices spaced along the x-x axis transfer weight of non-structural wythe to the structural wythe. Similar devices along the Y-Y axis transfer racking shear from non-structural wythe to structural wythe.

Ties spaced in the field of the panel transfer direct wind loads and stripping forces from the non-structural wythe to the structural wythe.

Fig. 5.3.36 Tension/compression ties

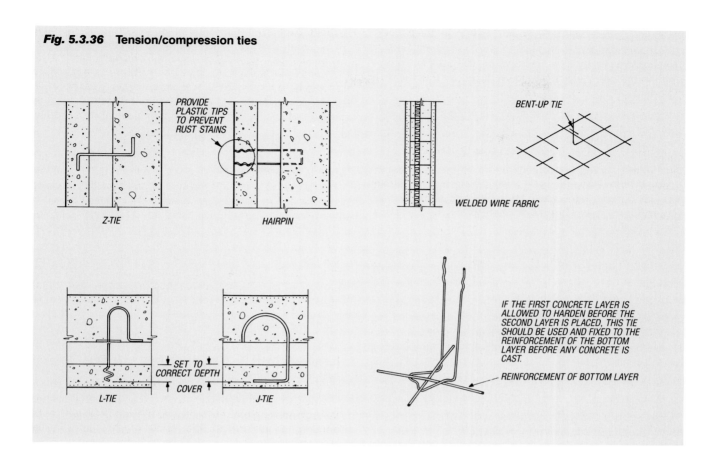

Wire tie connectors are usually 12 to 14 gauge, and preferably of stainless steel, Type 304 or 316. Galvanized metal or plastic ties may also be acceptable. Ties of welded wire fabric and reinforcing bars are sometimes used. Shaped, crimped, or bent ties should be cold bent. Ties should be arranged or coated so that galvanic reaction between the tie and reinforcement will not occur. Connectors which are intended to accommodate differential movement between wythes should be sufficiently ductile to withstand 5000 cycles of reversal at an amplitude of 1/8 in. In buildings with high relative humidities, over 60%, it may be desirable to use plastic ties to avoid condensation at tie locations. Consideration may have to be given to the effect of the plastic tie on the fire resistance of the wall.

General Architectural Design Considerations of precast concrete sandwich panels are similar to the design of single wythe architectural precast concrete panels. However, there are some special considerations for precast concrete sandwich wall panels.

A recognized problem with composite precast concrete sandwich panels is the tendency of longer panels to bow outwards under prolonged exposure to the hot sun. (In non-composite sandwich panels, there is less tendency to develop thermal bowing. The inner wythe of a temperature-controlled building is kept at a constant temperature with virtually no thermal gradient. Although the exterior wythe experiences extremes of temperature, the temperature gradient through the wythe is minimal leaving it unaffected by a steep thermal gradient through the wall system. This

practically eliminates or minimizes thermal bowing, and is the main advantage of non-composite sandwich panels.) Even where the bowing problem is recognized and considered in design, the possible failure of sealant between adjacent panels at corners or where abrupt changes in building shape occur, remains a major design and construction consideration for successful performance. The tendency of panels to bow is influenced chiefly by panel size, degree of composite action, the rigidity of connection between the wythes, and daily temperature variations on the exterior face. Differential movement between adjacent panels is not normally a problem and through good design and detailing practice, the effect of thermal bowing can be satisfactorily accommodated.

For panels with large openings, joints in the outer wythes at the corners of such opening are desirable. These joints should preferably be completely through to the insulation layer and may subsequently be caulked or treated architecturally, in the same manner as the joints between panels.

Control joints should be provided in large panels to break the outer wythe into units which will not craze or crack due to extreme temperature changes, or shrinkage and creep of the concrete. The pattern for such control joints becomes an important architectural feature and aligning such joints with adjacent panels must be done carefully. These can be minimized by having the real panel joint expressed as a recess, but this may not be possible if the outer wythe is already of minimum thickness. Alternatively, the pattern may be varied and only maintained in alternate panels, so that a small misalignment will not be noticeable. The problem of crazing or cracking and the need for control joints in the outer wythe can be reduced by prestressing the panels.

Good corner details are essential and should be carefully detailed. Mitered corners should be restrained and the panels adequately prestressed or reinforced to resist the restraint forces. Panels are not easy to weatherseal even with returns as the bowing will be in different planes. In addition, the panel with even a small return will be stiffer than its neighbor, and both joints on either side of a corner may suffer. A special corner unit, which is not necessarily flush with the adjacent panels, can be effectively used to camouflage bowing in the two different planes, Fig. 5.3.37. It should be noted that some manufacturers have successfully developed miter corner details to suit their own product.

If other materials are incorporated in a wall with precast concrete sandwich wall panels, no attempt should be made to make this material flush with the concrete surface, as it is unlikely that this material will act and bow exactly like the concrete panels. If it is essential that they are in the same theoretical plane, it is suggested that they be framed around with material which is not flush with the walls, similar to suggestions for corner columns. A door or window frame can be attached to the inside wythe since the movement is confined to the exterior wythe, Fig. 5.3.37. Window frames should have thermal breaks between the exterior frame and the interior frame. Although extruded aluminum window frames are more commonly used in precast concrete cladding, wood windows covered with aluminum will experience less heat loss through the frame. A substantial part of the total heat loss through a window can occur through its frame.

5.4 ACOUSTICAL PROPERTIES

5.4.1 GLOSSARY

Airborne sound — Sound that reaches the point of interest by propagation through air.

Background level — The ambient sound pressure level existing in a space.

Decibel (dB) — A logarithmic unit of measure of sound pressure or sound power. Zero on the decibel scale corresponds to a standardized reference pressure ($20\mu Pa$) or sound power (10^{-12} watt).

Flanking transmission — Transmission of sound by indirect paths other than through the primary barrier.

Frequency (Hz) — The number of complete vibration cycles per second.

Noise — Unwanted sound.

Noise Criteria (NC) — A series of curves, used as design goals to specify satisfactory backgound sound levels as they relate to particular use functions.

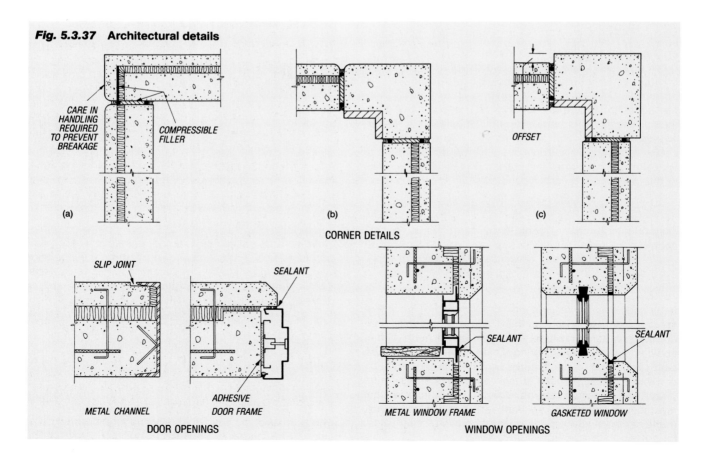

Fig. 5.3.37 Architectural details

CARE IN HANDLING REQUIRED TO PREVENT BREAKAGE

COMPRESSIBLE FILLER

(a)

OFFSET

(b)

(c)

CORNER DETAILS

SLIP JOINT

SEALANT

METAL CHANNEL

ADHESIVE DOOR FRAME

DOOR OPENINGS

SEALANT

METAL WINDOW FRAME

SEALANT

GASKETED WINDOW

WINDOW OPENINGS

Noise Reduction (NR) — The difference in decibels between the space-time average sound pressure levels produced in two enclosed spaces by one or more sound sources in one of them.

Noise Reduction Coefficient (NRC) — The arithmetic average of the sound absorption coefficients at 250, 500, 1000, and 2000 Hz expressed to the nearest multiple of 0.05 (ASTM C423).

RC curves — A revision of the NC curves based on empirical studies of background sounds.

Reverberation — The persistence of sound in an enclosed or partially enclosed space after the source of sound has stopped.

Sabin — The unit of measure of sound absorption (ASTM C423).

Sound absorption coefficient (a) — The fraction of randomly incident sound energy absorbed or otherwise not reflected off a surface (ASTM C423).

Sound Pressure Level (SPL) — Ten times the common logarithm of the ratio of the square of the sound pressure to the square of the standard reference pressure of $20\mu Pa$. Commonly measured with a sound level meter and microphone, this quantity is expressed in decibels.

Sound Transmission Class (STC) — The single number rating system used to give a preliminary estimate of the sound insulation properties of a partition system. This rating is derived from measured values of transmission loss (ASTM E413).

Sound Transmission Loss (TL) — Ten times the common logarithm of the ratio, expressed in decibels, of the airborne sound power incident on the partition that is transmitted by the partition and radiated on the other side (ASTM E90).

Structure-borne sound — Sound that reaches the point of interest over at least part of its path by vibration of a solid structure.

5.4.2 GENERAL

The basic purpose of architectural acoustics is to provide a satisfactory environment in which the desired sounds are clearly heard by the intended listeners and the unwanted sounds (noise) are isolated or absorbed.

Under most conditions, the architect can design the building to satisfy the acoustical needs of the tenant. Good acoustical design utilizes reflective and absorptive surfaces, sound barriers, and vibration isolators. Some surfaces must reflect sound so that the loudness will be adequate in all areas where listeners are located. Other surfaces must absorb sound to avoid echoes, sound distortion, and long reverberation times. Sound is isolated from rooms where it is not wanted by selected wall and floor/ceiling constructions. Vibrations generated by mechanical equipment are isolated from the structural frame of the building by means of mechanical isolators or compressible materials.

Most acoustical situations can be described in terms of: (1) sound source, strength and path; (2) sound transmission path; and (3) sound receiver.

A satisfactory acoustical environment is one in which the character and magnitude of all sounds are compatible with the intended space function. People are highly adaptable to the sensations of heat, light, odor, and sound, with sensitivities varying widely. The human ear can detect a sound the intensity of rustling leaves, 10 dB, and can tolerate, although briefly, the powerful exhaust of a jet engine at 120 dB, 10^{12} times the intensity of the rustling leaves sound.

5.4.3 SOUND LEVELS

The problems of sound insulation are usually considerably more complicated than those of sound absorption. Sound insulation involves greater reductions in sound level than can be achieved by absorption. These large reductions can only be achieved by continuous, impervious barriers. If the problem also involves structure-borne sound, it may be necessary to introduce resilient layers or discontinuities into the barrier.

Sound absorbing materials and sound insulating materials are used for two different purposes. There is not much sound absorption from an 8-in. concrete wall; similarly, high sound insulation is not available from a porous, light-weight material that may be applied to room surfaces for sound absorption. It is important to recognize that the basic mechanisms of sound absorption and sound insulation are quite different.

5.4.4 SOUND TRANSMISSION LOSS

Sound transmission loss measurements are made at 16 frequencies at one-third octave intervals covering the range from 125 to 4000 Hz. The testing procedure is described in ASTM Specification E90, *Laboratory Measurement of Airborne Sound Transmission Loss of Building Partitions*. To simplify specification of desired performance characteristics the single number Sound Transmission Class (STC) was developed.

Airborne sound reaching a wall, floor or ceiling produces vibrations in the wall which are radiated with reduced intensity on the other side. Airborne sound transmission loss in wall assemblies is a function of their weight, stiffness, and vibration damping characteristics.

Weight is concrete's greatest asset when it is used as a sound insulator. For sections of similar design, but different weights, the STC increases approximately 6 units for each doubling of weight, Fig. 5.4.1. This figure describes sound transmission class as a function of weight based on experimental data. Precast concrete walls usually do not need additional treatments in order to provide adequate sound insulation. If desired, greater sound insulation can be obtained by using resiliently attached layer(s) of gypsum board or other building materials. The increased transmission loss occurs because the energy flow path is increased to include a dissipative air column and additional mass.

The acoustical test results of airborne sound transmission loss of 4, 6, and 8 in. solid flat panels are shown in Fig.

Fig. 5.4.1 Sound transmission class as a function of wall weight

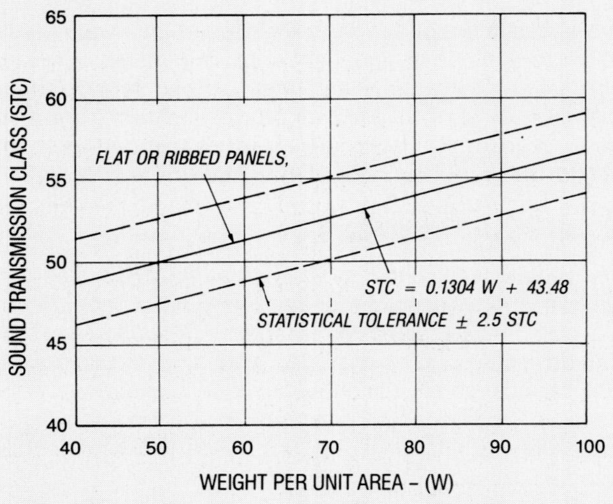

STC = 0.1304 W + 43.48
STATISTICAL TOLERANCE ± 2.5 STC
FLAT OR RIBBED PANELS,

(graph: SOUND TRANSMISSION CLASS (STC) vs WEIGHT PER UNIT AREA – (W))

Fig. 5.4.2 Acoustical test data of solid flat concrete panels — normal weight concrete

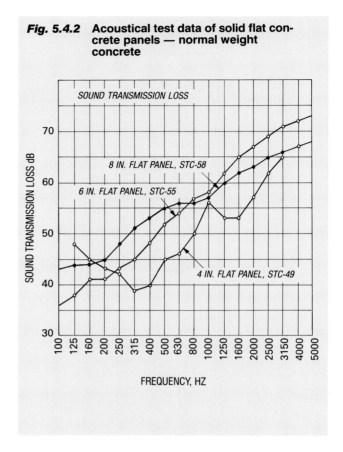

SOUND TRANSMISSION LOSS
8 IN. FLAT PANEL, STC-58
6 IN. FLAT PANEL, STC-55
4 IN. FLAT PANEL, STC-49

(graph: SOUND TRANSMISSION LOSS dB vs FREQUENCY, HZ)

5.4.2. Table 5.4.1 presents the ratings for various precast concrete assemblies. The effects of various assembly treatments on sound transmission can also be predicted from results of previous tests shown in Table 5.4.2. The improvements are additive, but in some cases the total effect may be slightly less than the sum.

5.4.5 ABSORPTION OF SOUND

A sound wave always loses part of its energy as it is reflected by a surface. This loss of energy is termed sound absorption. It appears as a decrease in sound pressure of the reflected wave. The sound absorption coefficient is the fraction of energy incident but not reflected per unit of surface area. Sound absorption can be specified at individual frequencies or as an average of absorption coefficients (NRC). A dense non-porous concrete surface typically absorbs 1 to 2% of incident sound and has an NRC of 0.015. In cases where additional sound absorption is desired, a coating of acoustical material can be spray applied, acoustical tile can be applied with adhesive, or an acoustical ceiling can be suspended. Most of the spray applied fire retardant materials used to increase the fire resistance of precast concrete and other floor-ceiling systems can also be used to absorb sound. The NRC of the

Table 5.4.1 Airborne sound transmission class ratings from tests of precast concrete assemblies

Assembly No.	Description	STC
1	4 in. flat panel, 54 psf	49
2	5 in. flat panel, 60 psf	52*
3	6 in. flat panel, 75 psf	55
4	Assembly 2 with "Z" furring channels, 1 in. insulation and ½ in. gypsum board, 75.5 psf	62
5	Assembly 2 with wood furring, 1½ in. insulation and ½ in. gypsum board, 73 psf	63
6	Assembly 2 with ½ in. space, 1⅝ in. metal stud row, 1½ in. insulation and ½ in. gypsum board	63*
7	8 in. flat panel, 95 psf	58
8	10 in. flat panel, 120 psf	59*

*Estimated values

Table 5.4.2 Typical improvements for wall treatments used with precast concrete elements

Treatment	Increased Airborne STC
Wall furring, ¾ in. insulation & ½ in. gypsum board attached to concrete wall	3
Separate metal stud system, 1½ in. insulation in stud cavity & ½ in. gypsum board attached to concrete wall	5 to 10
Plaster direct to concrete	0

sprayed fiber types range from 0.25 to 0.75. Most cementitious types have an NRC from 0.25 to 0.50.

The design of the auditorium in Fig. 5.4.3 required selected areas of high resolution and reflectivity which was achieved by using 54 ft high smooth surface precast concrete panels.

5.4.6 ACCEPTABLE NOISE CRITERIA

As a rule, a certain amount of continuous sound can be tolerated before it becomes noise. An "acceptable" level neither disturbs room occupants nor interferes with the communication of wanted sound.

Fig. 5.4.3

The most generally accepted used noise criteria today are expressed as the Noise Criteria (NC) curves, Fig. 5.4.4 and Table 5.4.3. The figures in Table 5.4.4 represent general acoustical goals. They can also be compared with anticipated noise levels in specific rooms to assist in evaluating noise reduction problems.

The main criticism of NC curves is that they are too permissive when the control of low or high frequency noise is of concern. For this reason, Room Criteria (RC) curves were developed, Fig. 5.4.5. RC curves are the result of extensive studies based on the human response to both sound pressure level and frequency and take into account the requirements for speech intelligibility.

A low background level obviously is necessary where listening and speech intelligibility is important. Conversely, higher ambient levels can persist in large business offices or factories where speech communication is limited to short distances. Often, the minimum target levels are just as important as the maximum permissible levels listed in Table 5.4.3. In an office or residence, it is desirable to have a certain ambient sound level to assure adequate acoustical privacy between spaces and minimize the transmission loss requirements of unwanted sound (noise).

These undesirable sounds may be from exterior sources such as automobiles and aircraft, or they may be generated as speech in an adjacent classroom or music in an adjacent apartment. They may also be direct impact-induced sound such as foot-falls on the floor above, rain on a

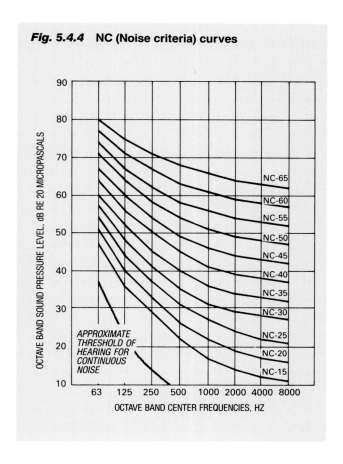

Fig. 5.4.4 NC (Noise criteria) curves

Table 5.4.3 Noise criteria (NC) curves

Noise criteria curves	Octave band center frequency, Hz							
	63	125	250	500	1000	2000	4000	8000
NC-15*	47	36	29	22	17	14	12	11
NC-20*	51	40	33	26	22	19	17	16
NC-25*	54	44	37	31	27	24	22	21
NC-30	57	48	41	35	31	29	28	27
NC-35	60	52	45	40	36	34	33	32
NC-40	64	56	50	45	41	39	38	37
NC-45	67	60	54	49	46	44	43	42
NC-50	71	64	58	54	51	49	48	47
NC-55	74	67	62	58	56	54	53	52
NC-60	77	71	67	63	61	59	58	57
NC-65	80	75	71	68	66	64	63	62

*The applications requiring background levels less than NC-25 are special purpose spaces in which an acoustical consultant should set the criteria.

lightweight roof construction or vibrating mechanical equipment. Thus, the designer must always be ready to accept the task of analyzing the many potential sources of intruding sound as related to their frequency characteristics and the rates at which they occur. The level of toleration that is to be expected by those who will occupy the space must also be established. Figs. 5.4.6 and 5.4.7 are the spectral characteristics of common noise sources.

With these criteria, the problem of sound isolation now must be solved, namely the reduction process between the high unwanted noise source and the desired ambient level. Once the objectives are established, the designer then should refer to available data, e.g. Fig. 5.4.1 or Table 5.4.1 and select the system which best meets these requirements. In this respect, precast concrete systems have superior properties and can with minimal effort comply with these criteria. When the insulation value has not been specified, selection of the necessary barrier can be determined analytically by (a) identifying exterior and /or interior noise sources, and (b) by establishing acceptable interior noise criteria.

Fig. 5.4.5   RC (Room criteria) curves

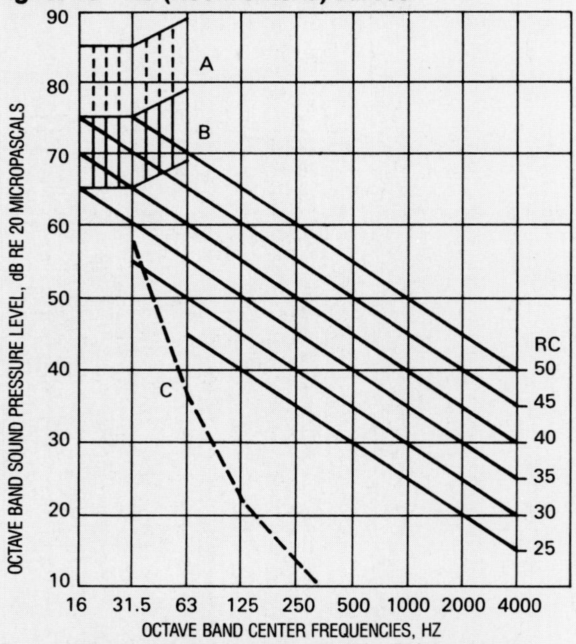

Region A: High probability that noise-induced vibration levels in lightweight wall/ceiling constructions will be clearly feelable; anticipate audible rattles in light fixtures, doors, windows, etc.
Region B: Noise-induced vibration levels in lightweight wall/ceiling constructions may be moderately feelable; slight possibility of rattles in light fixtures, doors, windows, etc.
Region C: Below threshold of hearing for continuous noise.

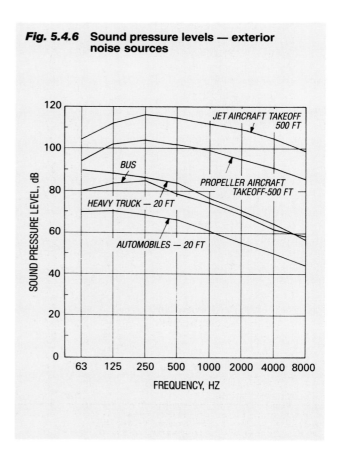

Fig. 5.4.6   Sound pressure levels — exterior noise sources

Example: Sound insulation criteria

Assume a precast, prestressed concrete office building is to be erected adjacent to a major highway. Private and semiprivate offices will run along the perimeter of the structure. The first step is to determine the degree of insulation required of the exterior wall system.

The 500 Hz requirement, 38 dB, can be used as the first approximation of the wall STC category. However, if win-

Sound pressure level — (dB)								
Frequency (Hz)	63	125	250	500	1000	2000	4000	8000
Bus traffic source noise (Fig. 5.4.6)	80	83	85	78	74	68	62	58
Private office noise criteria—NC 35 (Fig. 5.4.4)	60	52	45	40	36	34	33	32
Required Insulation	20	31	40	38	38	34	29	26

Table 5.4.4 Recommended category classification and suggested noise criteria range for steady background noise as heard in various indoor functional activity areas.*

Type of space	NC or RC curve
1. Private residences	25 to 30
2. Apartments	30 to 35
3. Hotels/motels	
a. Individual rooms or suites	30 to 35
b. Meeting/banquet rooms	30 to 35
c. Halls, corridors, lobbies	35 to 40
d. Services/support areas	40 to 45
4. Offices	
a. Executive	25 to 30
b. Conference rooms	25 to 30
c. Private	30 to 35
d. Open-plan areas	35 to 40
e. Computer/business machine areas	40 to 45
f. Public circulation	40 to 45
5. Hospitals and clinics	
a. Private rooms	25 to 30
b. Wards	30 to 35
c. Operating rooms	25 to 30
d. Laboratories	30 to 35
e. Corridors	30 to 35
f. Public areas	35 to 40
6. Churches	25 to 30**
7. Schools	
a. Lecture and classrooms	25 to 30
b. Open-plan classrooms	30 to 35**
8. Libraries	30 to 35
9. Concert Halls	**
10. Legitimate theatres	**
11. Recording studios	**
12. Movie theatres	30 to 35

*Design goals can be increased by 5dB when dictated by budget constraints or when noise intrusion from other sources represents a limiting condition.
** An acoustical expert should be consulted for guidance on these critical spaces.

dows are planned for the wall, a system of about 50-55 STC should be selected (see following composite wall discussion). Individual transmission loss performance values of this system are then compared to the calculated need.

Sound pressure level — (dB)						
Frequency (Hz)	125	250	500	1000	2000	4000
Required insulation	31	40	38	38	34	29
6 in. Precast solid concrete wall (Fig. 5.4.2)	38	43	52	59	67	72
Deficiencies	—	—	—	—	—	—

The selected wall should meet or exceed the insulation needs at all frequencies. However, to achieve the most efficient design conditions, certain limited deficiencies can be tolerated. Experience has shown that the maximum deficiencies are 3 dB at two frequencies or 5 dB on one frequency point.

5.4.7 COMPOSITE WALL CONSIDERATIONS

Windows and doors are often the weak link in an otherwise effective sound barrier. Minimal effects on sound transmission loss will be achieved in most cases by proper selection of glass, Table 5.4.5. The control of sound transmission through windows requires large cavities between layers (multiple glazing), heavy layers (thicker glass) and reduc-

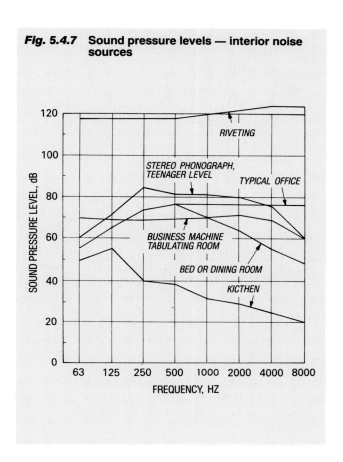

Fig. 5.4.7 Sound pressure levels — interior noise sources

Table 5.4.5 Acoustical properties of glass

Sound transmission class (STC)				
Type and overall thickness	**Inside lite**	**Construction space**	**Outside lite**	**STC**
⅛″ Plate or float	—	—	⅛″	23
¼″ Plate or float	—	—	¼″	28
½″ Plate or float	—	—	½″	31
1″ Insulated glass	¼″	½″ Air space	¼″	31
¼″ Laminated	⅛″	0.030 Vinyl	⅛″	34
1½″ Insulated glass	¼″	1″ Air space	¼″	35
¾″ Plate or float	—	—	¾″	36
1″ Insulated glass	¼″	½″ Air space	¼″ Laminated	38
1″ Plate or float	—	—	1″	37
2¾″ Insulated glass	¼″	2″ Air space	½″	39
4¾″ Insulated glass	¼″	4″ Air space	½″	40
6¾″ Insulated glass	¼″	6″ Air space	¼″ Laminated	42

Transmission loss (dB)															
Frequency (Hz)															
125	160	200	250	315	400	500	630	800	1000	1250	1600	2000	2500	3150	4000
¼ inch plate glass - 28 STC															
24	22	24	24	21	23	21	23	26	27	33	36	37	39	40	40
1 inch insulating glass with ½ inch air space - 31 STC															
25	25	22	20	24	27	27	30	32	33	35	34	29	31	33	36
1 inch insulating glass laminated with ½ inch air space - 38 STC															
30	29	26	28	31	34	35	37	37	38	38	40	41	40	41	44

tion of the structural connection between layers (separate frames and sashes for inner and outer layers). They certainly have to be as airtight as possible: usually fixed windows provide much better sound transmission control than operable windows. The sound transmission loss through a door depends upon the material and construction of the door and the effectiveness of the seal between the door and its frame. There is a mass law dependence of STC on weight (psf) for both wood and steel doors. The approximate relationships are:

For steel doors: STC = 15 + 27 log W

For wood doors: STC = 12 + 32 log W

where W = weight of the door, psf.

These relationships are purely empirical and a large deviation can be expected for any given door.

For best results, the distances between adjacent door and/or window openings should be maximized, staggered when possible and held to a minimum area. Minimizing openings allows the wall to retain the acoustical properties of the precast concrete. The design characteristics of the door or window systems must be analyzed prior to specification. Such qualities as frame design, door construction, and glazing thickness are vital performance criteria. Installation procedures must be exact and care given to the framing of each opening. Gaskets, weatherstripping, and raised thresholds serve as both thermal and acoustical seals and are recommended.

Fig. 5.4.8 can be used to calculate the effective acoustic

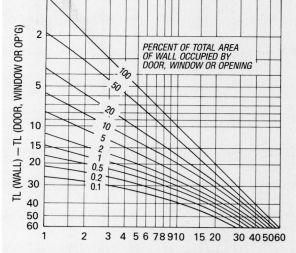

Fig. 5.4.8 Chart for calculating the effective transmission loss of a composite barrier. (For purposes of approximation STC values can be used in place of TL values.)

DECIBELS TO BE SUBTRACTED FROM TL OF WALL FOR EFFECTIVE TL OF COMPOSITE BARRIER

isolation of a wall system which contains a composite of elements, each with known individual transmission loss data. (Note: for purposes of approximation, STC values can be used in place of TL values).

Example: Composite wall insulation criteria

To complete the office building wall acoustical design from Section 5.4.6 assume the following:

1. The glazing area represents 10% of the exterior wall area.

2. The windows will be double glazed with a 38 STC acoustical insulation rating.

The problem now becomes the test of determining the combined effect of the concrete-glass combination and a redetermination of criteria compliance.

Sound pressure level — (dB)						
Frequency (Hz)	125	250	500	1000	2000	4000
6 in. Precast solid concrete wall (Fig. 5.4.2)	38	43	52	59	67	72
Double glazed windows (Table 5.4.5)	30	28	35	38	41	44
Correction (Fig. 5.4.8)	− 2	− 6	− 7	−11	−15	−19
Combined transmission loss	36	37	45	48	52	53
Insulation requirements	31	40	38	38	34	29
Deficiencies	—	3	—	—	—	—

The maximum deficiency is 3 dB and occurs at only one frequency point. The 6 in. precast concrete wall with double glazed windows will provide the required acoustical insulation.

Floor-ceiling assembly acoustical insulation requirements are determined in the same manner as walls by using Figs. 5.4.2 and 5.4.8.

5.4.8 LEAKS AND FLANKING

Performance of a building section with an otherwise adequate STC can be seriously reduced by a relatively small hole (or any other path) which allows sound to bypass the acoustical barrier. All noise which reaches a space by paths other than through the primary barrier is called flanking noise. Common flanking paths are openings around doors or windows, electrical outlets, telephone and television connections, and pipe and duct penetrations. Suspended ceilings in rooms where walls do not extend from the ceiling to the roof or floor above also allow sound to travel to adjacent rooms by flanking.

Anticipation and prevention of leaks begins at the design stage. Flanking paths (gaps) at the perimeters of interior precast walls and floors are generally sealed during construction with grout or drypack. All openings around penetrations through walls or floors should be as small as possible and must be sealed airtight. The higher the required STC of the barrier, the greater the importance of sealing all openings, see Fig. 5.4.8.

Perimeter leakage commonly occurs at the intersection between an exterior cladding panel and a floor slab. It is of vital importance to seal this gap to retain the acoustical integrity of the system and provide the required fire stop

between floors. One way to seal the gap is to place a 4 pcf density mineral wool blanket between the floor slab and the exterior wall. Fig. 5.4.9 demonstrates the acoustical isolation effects of this treatment.

Flanking paths can be minimized by:

1. Interrupting the continuous flow of energy with dissimilar materials, i.e., expansion or control joints or air gaps.

2. Increasing the resistance to energy flow with floating floor systems, full height and/or double partitions, and suspended ceilings.

3. Using primary barriers which are less subject to the creation of flanking paths. Although not easily quantified, an inverse relationship exists between the performance of an element as a primary barrier and its propensity to transmit flanking sound. In other words, the probability of existing flanking paths in a concrete structure is much less than in one of steel or wood framing.

If the acoustical design is balanced, the maximum amount of acoustic energy reaching a space via flanking should not equal the energy transmitted through the primary barriers. In exterior walls, the proper application of sealant and backup materials in the joints between units will not allow sound to flank the wall.

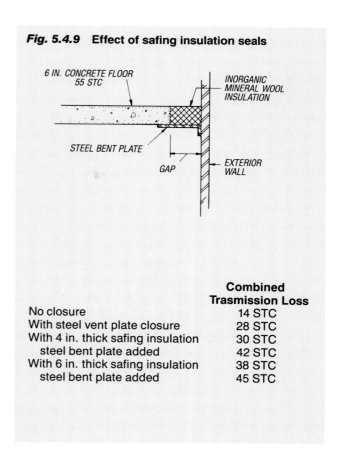

Fig. 5.4.9 Effect of safing insulation seals

	Combined Trasmission Loss
No closure	14 STC
With steel vent plate closure	28 STC
With 4 in. thick safing insulation steel bent plate added	30 STC
	42 STC
With 6 in. thick safing insulation steel bent plate added	38 STC
	45 STC

5.5 FIRE RESISTANCE

5.5.1 GENERAL

In the interest of life safety and property protection, building codes require that resistance to fire be considered in the design of buildings. The degree of fire resistance required is dependent on the type of occupancy, the size of the building, its location (proximity to property lines and within established fire zones), and the amount and type of fire detection and suppression equipment available in the structure.

In Fig. 5.5.1 the dimensional constraints imposed by the site required building to the property lines. A precast concrete panel system was selected over a unit masonry wall system to cost-effectively solve the problem of the required 4-hour fire-rated exterior property-line walls that are architecturally consistent with street elevations. These large walls had to be considered as "temporary" with the prospect of being concealed by adjacent buildings some time in the future.

Although life safety is of paramount importance, casualty insurance companies and owners are also concerned with the damage that might be inflicted on the building and its contents during a fire. This means that fire containment must be considered as well as fire resistance. Insurance rates are often substantially lower for buildings with higher fire resistance ratings and containment designs.

Fig. 5.5.1

In the past, fire resistance ratings were assigned on the basis of results of standard fire tests. In recent years, there has been a trend toward calculating the fire endurance of building components, rather than relying entirely upon fire tests. To facilitate this trend, much research work has been conducted on the behavior of materials and building components in fires. This section summarizes the available information on the behavior of architectural precast concrete under fire conditions. See Section 4.5.7 for a discussion on fire protection of connections and Section 4.7.8 for fire resistance of joints.

Fire resistance ratings of building components are measured and specified in accordance with a common standard, ASTM E119. Fire endurance is defined as the period of resistance to the standard fire exposure which elapses at the prior hour or half-hour before an "end point" is reached. The major "end points" used to evaluate performance in a fire test include:

1. Collapse of loadbearing specimens (structural end point).

2. Formation of holes, cracks, or fissures through which flames or gases hot enough to ignite cotton waste may pass (flame passage end point).

3. Temperature increase of the unexposed surface of floors, roofs, or walls reaching an average of 250 deg. F or a maximum of 325 deg. F at any one point (heat transmission end point).

4. Collapse of walls and partitions during a hose stream test or inability to support twice the superimposed load following the hose stream test.

A fire resistance rating (sometimes called a fire rating, a fire resistance classification, or an hourly rating) is a legal term defined in building codes, usually based on fire endurance. Building codes specify required fire resistance ratings for various types of construction and occupancy. Performance is defined by the authorities (regulatory and insurance) as the maximum time for which each component would survive if it were subjected to a standard test. The standard tests provide arbitrary fire exposure, arbitrary load, and arbitrary restraint.

5.5.2 FIRE ENDURANCE OF WALLS

The fire endurances of precast concrete walls, as determined by fire tests, are almost universally governed by the ASTM E119 criteria for heat transmission (temperature rise of the unexposed surface) rather than by structural behavior during fire tests. This is probably due to the low level of stresses, even in concrete bearing walls, and the fact that reinforcement generally does not perform a primary structural function. Thus, the thickness of concrete cover on the reinforcement is seldom a factor in determining the fire endurance of concrete wall panels.

Most of the information on heat transmission was derived from fire tests of assemblies tested in a horizontal position simulating floors or roofs. The data are slightly conservative for assemblies tested vertically, i.e., as walls. Nevertheless, it is suggested that no correction be made unless more specific data derived from fire tests of walls are used.

For concrete wall panels, the temperature rise of the unexposed surface depends mainly on the thickness and aggregate type of the concrete. Other less important factors include unit weight, moisture condition, air content, and maximum aggregate size. Within the usual ranges, water-cement ratio, strength, and age have only insignificant effects.

From information that has been developed from fire tests, it is possible to estimate accurately the thickness of many

Table 5.5.1 Fire endurances for single-course concrete panels

Aggregate	Thickness in inches for fire endurance of			
	1 hr	2 hr	3 hr	4 hr
All lightweight	2.47	3.56	4.35	5.10
Sand-lightweight	2.63	3.76	4.62	5.37
Carbonate	3.25	4.67	5.75	6.63
Siliceous	3.48	5.00	6.15	7.05

Table 5.5.2 Thickness of inside wythes (in inches) to provide various fire endurances for two-course panels

Fire endur-ance	Inside wythe material (fire-exposed side)	Siliceous aggregate concrete* (outside wythe)			Sand-lightweight concrete (outside wythe)		
		1½ in.	2 in.	3 in.	1½ in.	2 in.	3 in.
1 hr	Carbonate aggregate concrete	1.9	1.4	0.45	1.7	1.0	0
1 hr	Siliceous aggregate concrete	2.0	1.48	0.48	1.7	1.0	0
1 hr	Lightweight aggregate concrete	1.5	1.2	0.25	1.13	0.63	0
1 hr	Cellular concrete (30 pcf)	0.7	0.5	0.2	0.5	0.3	0
1 hr	Perlite concrete (30 pcf)	0.8	0.6	0.2	0.7	0.4	0
1 hr	Vermiculite concrete (30 pcf)	0.9	0.6	0.2	0.7	0.4	0
1 hr	Sprayed mineral fiber	0.4	0.25	0.1	0.4	0.2	0
1 hr	Sprayed vermiculite cementitous material	0.4	0.25	0.1	0.4	0.2	0
2 hr	Carbonate aggregate concrete	3.25	2.8	1.9	3.2	2.6	1.25
2 hr	Siliceous aggregate concrete	3.5	3.0	2.0	3.3	2.7	1.3
2 hr	Lightweight aggregate concrete	2.5	2.1	1.4	2.26	1.76	0.76
2 hr	Cellular concrete (30 pcf)	1.2	1.0	0.6	1.2	0.9	0.4
2 hr	Perlite concrete (30 pcf)	1.4	1.1	0.7	1.3	0.9	0.4
2 hr	Vermiculite concrete (30 pcf)	1.6	1.3	0.8	1.4	1.1	0.4
2 hr	Sprayed mineral fiber	1.1	0.8	0.5	1.0	0.8	0.3
2 hr	Sprayed vermiculite cementitous material	1.0	0.8	0.5	1.0	0.75	0.3
3 hr	Carbonate aggregate concrete	4.4	3.9	3.0	4.2	3.7	2.4
3 hr	Siliceous aggregate concrete	4.65	4.15	3.15	4.4	3.8	2.5
3 hr	Lightweight aggregate concrete	3.4	3.1	2.4	3.12	2.62	1.62
3 hr	Cellular concrete (30 pcf)	1.6	1.3	0.9	1.6	1.3	0.8
3 hr	Perlite concrete (30 pcf)	1.9	1.6	1.1	1.8	1.4	0.8
3 hr	Vermiculite concrete (30 pcf)	2.2	1.8	1.3	2.0	1.6	1.0
3 hr	Sprayed mineral fiber	N.A.	1.4	0.9	N.A.	1.3	0.85
3 hr	Sprayed vermiculite cementitous material	1.6	1.35	0.85	1.6	1.3	0.8
4 hr	Carbonate aggregate concrete	5.15	4.8	3.85	5.2	4.7	3.5
4 hr	Siliceous aggregate concrete	5.55	5.05	4.05	5.5	4.9	3.7
4 hr	Lightweight aggregate concrete	4.2	3.8	3.0	3.87	3.37	2.37
4 hr	Cellular concrete (30 pcf)	2.1	1.9	1.4	2.0	1.7	1.1
4 hr	Perlite concrete (30 pcf)	2.3	2.0	1.5	2.3	1.9	1.3
4 hr	Vermiculite concrete (30 pcf)	2.7	2.3	1.7	2.6	2.2	1.5
4 hr	Sprayed mineral fiber	N.A.	N.A.	1.4	N.A.	N.A.	1.4
4 hr	Sprayed vermiculite cementitous material	N.A.	1.8	1.3	N.A.	1.75	1.25

*Tabulated values for thickness of inside wythe are conservative for carbonate aggregate concrete.
N.A. means not applicable, i.e., a thicker outside wythe is needed.

types of one-course and multi-course walls that will provide fire endurances of 1, 2, 3, or 4 hr based on the temperature rise of the unexposed surface. Based on fire test data, the thicknesses shown in Fig. 5.5.2 and Tables 5.5.1 and 5.5.2 can be expected to provide the fire endurances indicated for single-course and two-course walls. Fig. 5.5.2 shows the fire endurance (heat transmission) of concrete as influenced by aggregate type and thickness. Interpolation of varying concrete unit weights is acceptable in this figure. Table 5.5.1 provides the thickness (in inches) of solid concrete wall panels for various fire endurances, while Table 5.5.2 provides the same for two-course panels.

As used in this section, concrete aggregates are designated as lightweight, sand-lightweight, carbonate, or siliceous.

1. Lightweight aggregates include expanded clay, shale, slate, and fly ash. These materials produce concretes having unit weights of about 95 to 105 pcf without sand replacement.

2. Lightweight concretes in which sand is used as part or all of the fine aggregate, and unit weight is less than 120 pcf, are designated as sand-lightweight.

3. Carbonate aggregates include limestone and dolomite (i.e., minerals consisting mainly of calcium carbonate and/or magnesium carbonate).

4. Siliceous aggregates include quartzite, granite, basalt, and most hard rocks other than limestone and dolomite.

Table 5.5.3 shows the thicknesses of concrete wall panels required to provide fire endurances of 2 and 3 hrs when the fire-exposed surface is covered with 5/8 in. Type X gypsum wallboard. Mechanical fastening devices must be used to attach the wallboard to the wall panels rather than adhesives.

Ribbed Panel Heat Transmission is influenced by both the thinnest portion of the panel and the panel's "equivalent thickness." Here, equivalent thickness is defined as the net cross sectional area of the panel divided by the width of the cross section. In calculating the net cross sectional area of a ribbed panel, portions of the ribs that project beyond twice the minimum thickness should be neglected, Fig. 5.5.3.

The fire endurance (as defined by the heat transmission end point) can be governed by the thinnest section, the

average thickness, or a combination of the two. The following rule-of-thumb expressions describe the conditions under which each set of criteria governs.

Let t = minimum thickness, in.

t_e = equivalent thickness of panel, in.

s = rib spacing, in.

If $t < s/4$, fire endurance R is governed by t and is equal to R_t.

If $t > s/2$, fire endurance R is governed by t_e and is equal to R_{te}.

If $s/2 > t > s/4$:

$$R = R_t + (4t/s - 1)(R_{te} - R_t) \qquad \text{(Eq 5.5.1)}$$

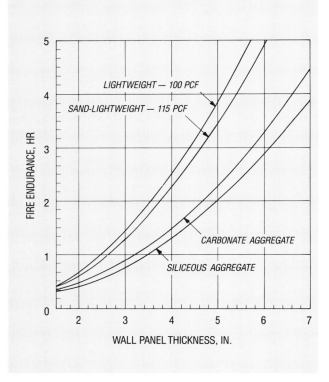

Fig. 5.5.2 **Fire endurance (heat transmission) as a function of panel thickness**

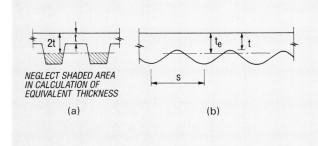

Fig. 5.5.3 **Cross sections of ribbed wall panels**

NEGLECT SHADED AREA IN CALCULATION OF EQUIVALENT THICKNESS

(a)   (b)

Table 5.5.3 Use of 5/8 in. Type X gypsum wallboard

| Aggregate | Thickness (in.) of concrete panel for fire endurance of | | | |
| | With 7/8 in. air space | | With 6-in. air space | |
	2 hr	3 hr	2 hr	3 hr
Sand-lightweight	2.5	3.6	1.2	2.4
Carbonate	2.8	4.0	1.3	2.7
Siliceous	2.9	4.2	1.3	2.8

where R is the fire endurance of a concrete panel and subscripts t and te relate the corresponding R values to concrete slab thicknesses t and t_e, respectively. These expressions apply to many ribbed and corrugated panels, but they give excessively low results for panels with widely spaced grooves or rustications. Consequently, engineering judgment must be used when applying the above expressions.

Sandwich Panels have insulating materials between the two wythes of concrete, see Section 5.3.8. Several building codes require that, where non-combustible construction is specified, combustible elements in walls shall be limited to thermal and sound insulation having a flame spread classification of not more than 75, when the insulation is sandwiched between two layers of non-combustible material such as concrete. When insulation is not installed in this manner, it is required to have a flame spread of not more than 25. Data on flame spread classification is available from insulation manufacturers.

It should be noted that the cellular plastics melt and are consumed at about 400 to 600 deg. F. Thus, thickness greater than 1.0 in. or changes in composition probably have only a minor effect on the fire endurance of sandwich panels. The danger of toxic fumes caused by the burning cellular plastics is practically eliminated when the plastics are completely encased within concrete sandwich panels.

It is possible to calculate the thicknesses of various materials in a sandwich panel required to achieve a given fire rating using Equation 5.5.2.

$$R^{0.59} = (R_1^{0.59} + R_2^{0.59} \dots R_n^{0.59})^{1.7} \qquad \text{(Eq. 5.5.2)}$$

where R = the fire endurance of the composite assembly in minutes and R_1, R_2, and R_n = the fire endurance of each of the individual courses in minutes.

Table 5.5.4 lists fire endurances for insulated precast concrete sandwich panels with either cellular plastic, glass fiber board, or insulating concrete used as the insulating material. The values were obtained using Eq. 5.5.2. A design graph for solving this equation is provided, Fig. 5.5.4.

Window Walls that are required to be fire resistive have limits imposed on the area of openings by various building codes. These limits are based on the construction type, occupancy, spatial separation (distance between a building and its neighbor or property line), and fire zone. For example, the 1988 Uniform Building Code permits no openings in exterior walls of office buildings when the spatial separation is less than 5 ft, and requires the walls in Type I and Type II-FR, to be of 4-hr fire resistive construction. In Types II-1 hr, II-N and V construction, 1 hr fire resistive classification is required for spatial separations of less than 20 ft.

Protected openings are required in exterior walls if the spatial separation is between 5 ft and 20 ft. for Types I, II-FR, III and IV construction; in Types II-1 hr, II-N and V, protected openings are required for spatial separations between 5 ft and 10 ft. Where openings are required to be protected,

Fig. 5.5.4   Design aid for use in solving Eq.5.5.2

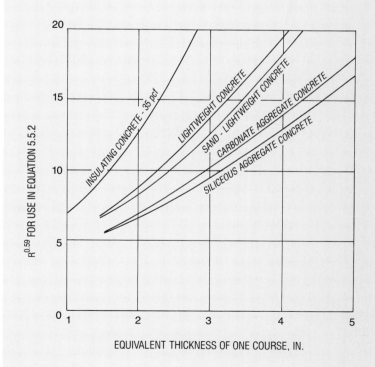

EQUIVALENT THICKNESS OF ONE COURSE, IN.

R, MINUTES	$R^{0.59}$
60	11.20
120	16.85
180	21.41
240	25.37

MATERIAL	$R^{0.59}$
cellular plastic (1 in. or thicker)	2.5
¾ in. glass fiber board	4.0
1½ in. glass fiber board	8.5
continuous air space	3.33
two continuous air spaces	6.67
2 in. foam glass	10.6

Table 5.5.4 Fire endurance of sandwich panels

Inside wythe	Insulation	Outside wythe	Fire endurance, hr:min.
1½ in. Sil	1 in. CP	1½ in. Sil	1:23
1½ in. Carb.	1 in. CP.	1½ in. Carb.	1:23
1½ in. SLW	1 in. CP	1½ in. SLW	1:45
2 in. Sil.	1 in. CP	2 in. Sil	1:50
2 in Carb	1 in. CP	2 in. Carb	2:00
2 in. SLW	1 in. CP	2 in. SLW	2:32
3 in. Sil	1 in. CP	3 in. Sil	3:07
1½ in. Sil	¾ in. GFB	1½ in. Sil	1:39
2 in. Sil	¾ in. GFB	2 in. Sil	2:07
2 in. SLW	¾ in. GFB	2 in. SLW	2:52
2 in. Sil	¾ in. GFB	3 in. SLW	3:10
1½ in. Sil	1½ in. GFB	1½ in. Sil	2:35
2 in. Sil	1½ in. GFB	2 in. Sil	3:08
2 in. SLW	1½ in. GFB	2 in. SLW	4:00
1½ in. Sil	1 in. IC	1½ in. Sil	2:12
1½ in. SLW	1 in. IC	1½ in. SLW	2:39
2 in. Carb	1 in. IC	2 in. Carb	2:56
2 in. SLW	1 in. IC	2 in. SLW	3:33
1½ in. Sil	1½ in. IC	1½ in. Sil	2:54
1½ in. SLW	1½ in. IC	1½ in. SLW	3:24
2 in. Sil	1½ in. IC	3 in. Sil	4:16
2 in. Sil	2 in. IC	2 in. Sil	4:25
1½ in. SLW	2 in. IC	1½ in. SLW	4:19

Notes:
Carb = carbonate aggregate concrete
Sil = siliceous aggregate concrete
SLW = sand-lightweight concrete (115 pcf maximum)
CP = cellular plastic (polystyrene or polyurethane)
GFB = glass fiber board
IC = lightweight insulating concrete (35 pcf maximum)

the area of such openings is limited to half the total area of the wall in each story, and the openings must be protected by fire assemblies having 3/4 hour fire-protection ratings.

The above pertains to the 1988 Uniform Building Code requirements for office buildings. Requirements for other occupancies differ somewhat but generally follow the same pattern and certain exceptions often apply. Requirements in other codes also differ. Perhaps the most comprehensive requirements are those in the National Building Code of Canada, which relate spatial separation and maximum area of unprotected openings to the area and height-length ratio of the exposed building face. Percentages of unprotected opening areas are tabulated in the code for various combinations of area of building face, height-length ratio, and spatial separation.

The percentage of openings permitted increases: (1) as the spatial separation increases; (2) as the area of the exposing building face decreases; and (3) as the ratio of either height-length (H/L) or length-height (L/H) increases, i.e., a greater percentage is permitted for H/L or L/H of 10:1 than for H/L or L/H of 3:1.

As an example, an exposed face of an office building having an area of 3500 sq ft, a L/H = 2:1, and a limiting distance of 23 ft can have a maximum of 18 percent of unprotected openings. If the ratio of L/H or H/L were 10:1 or more, the area of unprotected openings could be increased to 30 percent, or if the spatial separation were 40 ft and L/H were 10:1, the area of unprotected openings permitted is 59 percent of the exposed face.

The National Building Code of Canada also permits a higher limit on the unexposed surface temperature if the area of unprotected openings is less than the maximum allowed.

An equivalent opening factor, F_{eo}, is then applied in a formula to determine the corrected area of openings:

$$A_c = A + A_f F_{eo} \qquad \text{(Eq. 5.5.3)}$$

where

A_c = corrected area of unprotected openings including actual and equivalent openings.

A = actual area of unprotected openings

A_f = area of exterior surface of the exposing building face exclusive of openings, on which the temperature limitation of the standard fire test is exceeded.

Fig. 5.5.5 shows the relation between F_{eo} (as defined in the National Building Code of Canada) and panel thickness for three types of concrete.

To illustrate the use of Fig. 5.5.5, suppose that for a particular building face, a 2-hr fire resistance rating is required and the area of unprotected openings permitted is 57 percent. Suppose also that the actual area of unprotected openings is 49 percent and that the window wall panels are made of carbonate aggregate concrete (referred to as Type N in NBC of Canada). Determine the minimum thickness of the panel.

In this case A_c = 57%, A = 49%, A_f = 100-49 = 51%, hence:

$$F_{eo} = \frac{A_c - A}{A_f} = \frac{57 - 49}{51} = 0.16$$

From Fig. 5.5.5, for F_{eo} = 0.16 at 2 hr, the minimum panel thickness is 2.3 in. Thus, if the panel is 2.3 in. thick or thicker, the code requirements will be satisfied.

5.5.3 DETAILING OF FIRE BARRIERS

One of the purposes of code provisions for fire resistive construction is to limit the involvement of a fire to the room or compartment where the fire originates. Thus, the floors, walls, and roof surrounding the compartment must serve as fire barriers.

Most codes require that fire walls start at the foundation and extend continuously through all stories to and above the roof, except where the roof is of fire resistive construction, in which case the wall must be tightly fitted against the underside of the roof. If the roof and walls are of combustible construction, fire walls must extend not only through the roof, but must extend through the sides of the building beyond eaves or other combustible projections.

Fig. 5.5.5 Equivalent opening factor, F_{eo}

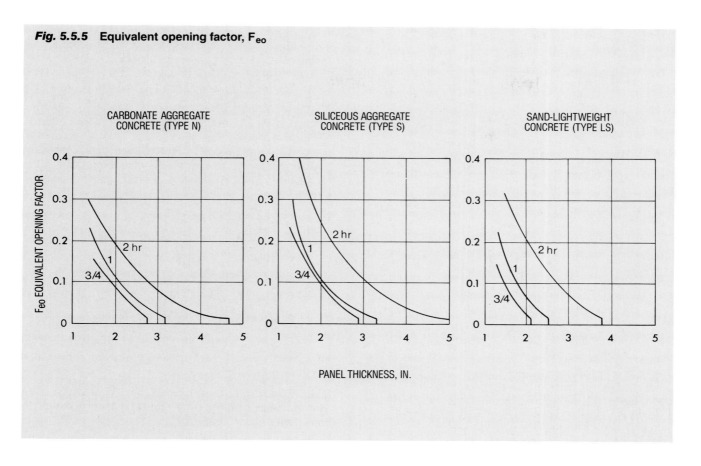

CARBONATE AGGREGATE CONCRETE (TYPE N)

SILICEOUS AGGREGATE CONCRETE (TYPE S)

SAND-LIGHTWEIGHT CONCRETE (TYPE LS)

F_{eo} EQUIVALENT OPENING FACTOR

PANEL THICKNESS, IN.

When protected openings are required in walls, coverings for such openings must be fire resistive. Most codes require that fire doors have fire resistive classifications of three-fourths of the classification required for the wall. Glazed openings in fire doors or fire windows are limited in area by code provisions, and the glass must be reinforced with wire mesh. Fire dampers must be used in ducts, unless fire tests show that they are not needed.

When precast concrete wall panels are designed and installed in such a manner that no space exists between the wall panel and floor, a fire below the floor cannot pass through the joint between the floor and wall. However, some wall panels are designed in such a manner that a space exists between the floor and wall. This space is referred to as a "safe-off" area.

Fig. 5.5.6 shows two methods of fire stopping such safe-off areas. The sketch shows safing supported on a metal plate, while the safing in the photo is supported by impaling pins (not shown). Safing insulation is available in the form of mineral fiber mats of varying dimensions and densities. Care must be taken during installation to be sure that the entire safe-off area is sealed. The safing insulation provides an adequate firestop and accommodates differential movement between the wall panel and the floor. See Section 5.4.8 for a discussion on the acoustical isolation effects of this treatment.

Fig. 5.5.6 Methods of installing safing insulation

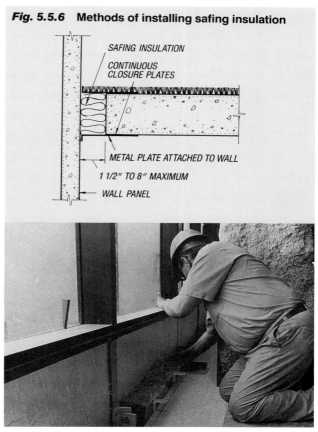

SAFING INSULATION

CONTINUOUS CLOSURE PLATES

METAL PLATE ATTACHED TO WALL

1 1/2" TO 8" MAXIMUM

WALL PANEL

5.5.4 COLUMNS AND COLUMN COVERS

Reinforced Concrete Columns have for many years served as the standard for fire resistive construction. Indeed, the performance of concrete columns in actual fires has been excellent.

The inherent fire resistance of concrete columns results from three factors:
1. Minimum size of a structural column is generally such that the inner core of the column retains much of its strength even after long periods of fire exposure.
2. Concrete cover to the reinforcing bars is generally 1-7/8 in. or more, thus providing considerable fire protection for the reinforcement.
3. Ties or spirals contain the concrete within the core.

Table 5.5.5 shows typical building code requirements for reinforced concrete columns. The values shown in Table 5.5.5 apply to both precast and cast-in-place concrete columns. In addition, they apply to cast-in-place concrete columns clad with precast concrete column covers, whether the covers serve merely as cladding or as forms for the cast-in-place column.

Precast Concrete Column Covers are often used to clad steel columns for architectural reasons. Such covers also provide fire protection for the columns. Fig. 5.5.7 shows the relationship between the thickness of a concrete column cover and the fire endurance for various steel column sections. The fire endurances shown are based on an empirical relationship. It was also found that the air space between the steel core and the column cover has only a minor effect on the fire endurance. An air space will probably increase the fire endurance but only by an insignificant amount.

Most precast concrete column covers are 3 in. or more in thickness, but some are as thin as 2-1/2 in. From Fig. 5.5.7, it can be seen that such column covers provide fire endurances of at least 2-1/2 hr. For steel column sections other than those shown, including shapes other than wide flange beams, interpolation between the curves on the basis of weight per foot will generally give reasonable results.

For example, the fire endurance afforded by a 3 in. thick column cover of normal weight concrete for a 8 x 8 x 1/2 in. steel tube column will be about 3 hr 20 min (the weight of the section is 47.35 lb per ft).

Table 5.5.5 Minimum sizes of concrete columns*

Types of Concrete	Minimum Column Dimension, In., for Fire Resistance Rating of				
	1 hr	1½ hr	2 hr	3 hr	4 hr
Siliceous	8	8	10	12	14
Carbonate	8	8	10	12	14
Sand-Lightweight	8	8	9	10.5	12

*The minimum cover to the main reinforcement in columns for fire resistance ratings of 1 hr., 1½ hr., 2 hr. and 3 hr. shall be 1½ in.; for 4 hr. the minimum cover to the main reinforcement shall be 2 in. for siliceous aggregate concrete and 1½ in. for carbonate aggregate concrete or sand-lightweight concrete.

Fig. 5.5.7 Fire endurance of steel columns afforded protection by concrete column covers

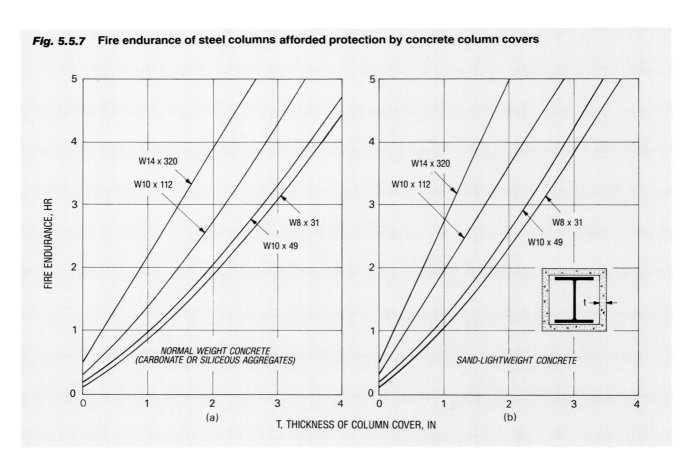

T, THICKNESS OF COLUMN COVER, IN

Fig. 5.5.8 displays some of the various shapes that precast concrete column covers are made, including (a) four flat panels with butt or mitered joints that fit together to enclose the steel column, (b) four L-shaped units, (c) two L-shaped units, (d) two U-shaped units, and (e) and (f) U-shaped units in combination with flat closure panels. There are, of course, many other combinations which may be used to accommodate isolated columns, corner columns, and columns in walls.

Fig. 5.5.8a would probably be most vulnerable to bowing during fire exposure, while Fig. 5.5.8f would probably be the least vulnerable.

To be fully effective, the column covers must remain in place without severe distortion. Many types of connections are used to hold the column covers in place. Some connections consist of bolted or welded clip angles attached to the tops and bottoms of the covers. Others consist of steel plates embedded in the covers that are welded to angles, plates, or other shapes which are, in turn, welded or bolted to the steel column. In any case, these connections are used primarily to position the column covers. As such, they are not highly stressed. Thus, temperature limits need not be applied to the steel in most column cover connections.

If restrained, either partially or fully, concrete panels tend to deflect or bow when exposed to fire. For example, for a steel column that is clad with four flat panels attached top and bottom, the column covers will tend to bulge at midheight, thus tending to open gaps along the sides. The

gap size decreases as the panel thickness increases.

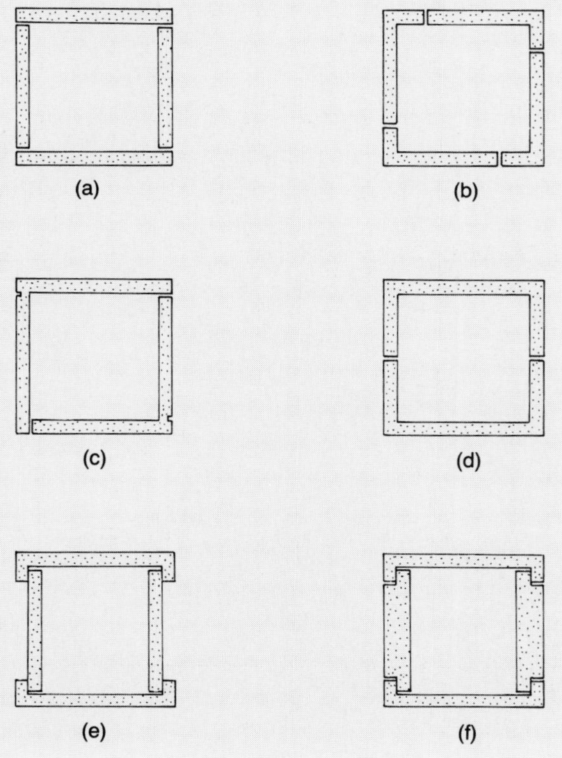

With L, C, or U-shaped panels, the gap size is further reduced. The gap size can be further minimized by connections installed at midheight. In some cases, shiplap joints can be used to minimize the effects of joint openings.

Joints should be sealed to prevent the passage of flame to the steel column. A non-combustible material such as sand-cement mortar or ceramic fiber blanket, can be used to seal the joint and then caulking applied.

As the covers are heated, they tend to expand. Connections should accommodate such expansion without subjecting the cover to additional loads. Thus, the precast concrete column covers should not be restrained vertically. Fire resistive compressible materials, such as mineral fiber safing, can be used to seal the tops or bases of the column covers, thus permitting the column covers to expand without restraint. Similarly the connections between the covers and columns should be flexible (or soft) enough to accommodate thermal expansion without inducing much stress into the covers.

5.6 ROOFING

The most vulnerable parts of any roof system are the joints between the horizontal roof deck and vertical surfaces. Most roof leaks occur at these locations. Therefore, designers should carefully consider the design of flashing details at vertical wall to roof intersections. For further information, refer to the NRCA Construction Details in the NRCA Roofing and Waterproofing Manual.

Three major considerations govern the detailing of roofing adjoining precast concrete units:

1. Relative movement between the roof and the precast concrete panels should be assessed, and allowances made to safely accommodate this movement with flashing, counterflashing, and expansion strips.

2. Hardware required in the precast concrete should be detailed and located to tolerances stated on the working drawings and selected to suit the particular conditions.

3. Details should reflect involvement of the minimum number of trades, assuring that the work of each trade can be completed independently of the others.

The details in this section depict jobsite fabricated construction. Many roofing material manufacturers now offer pre-fabricated flashing pieces or permit the use of materials for flashing purposes other than those that are shown here. Specifics on these proprietary designs vary greatly. The individual roofing material manufacturers specifications should be consulted when proprietary designs are used. Fig. 5.6.1 shows four of the most common roofing systems in use today. Illustrations indicate only the edges of built-up roofs and cant strips, and refer only to materials directly affecting the precast concrete details. Built-up roofing when improperly applied can split and pull away from a precast concrete panel, allowing moisture to enter. Proper application for a built-up roof is shown in Fig. 5.6.1a and a modified bitumen roofing detail is shown in Fig. 5.6.1b.

Fig. 5.5.8 **Types of column covers**

(a)

(b)

(c)

(d)

(e)

(f)

Fig. 5.6.1 Counterflashing for concrete walls or parapets

(a) BUILT-UP ROOFING

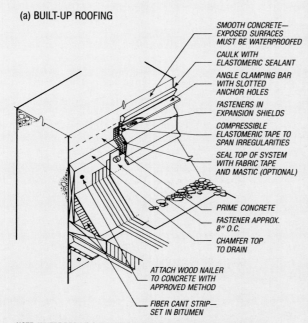

SMOOTH CONCRETE—
EXPOSED SURFACES
MUST BE WATERPROOFED

CAULK WITH
ELASTOMERIC SEALANT

ANGLE CLAMPING BAR
WITH SLOTTED
ANCHOR HOLES

FASTENERS IN
EXPANSION SHIELDS

COMPRESSIBLE
ELASTOMERIC TAPE TO
SPAN IRREGULARITIES

SEAL TOP OF SYSTEM
WITH FABRIC TAPE
AND MASTIC (OPTIONAL)

PRIME CONCRETE

FASTENER APPROX.
8" O.C.

CHAMFER TOP
TO DRAIN

ATTACH WOOD NAILER
TO CONCRETE WITH
APPROVED METHOD

FIBER CANT STRIP—
SET IN BITUMEN

NOTE: WHERE DECK IS SUPPORTED BY AND FASTENED TO THE CONCRETE WALL, VERTICAL WOOD NAILER SHOULD BE SECURED TO THE WALL WITH SUITABLE FASTENERS.

(b) MODIFIED BITUMEN

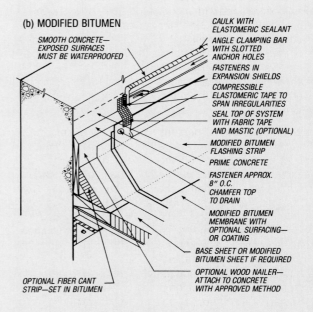

CAULK WITH
ELASTOMERIC SEALANT

ANGLE CLAMPING BAR
WITH SLOTTED
ANCHOR HOLES

FASTENERS IN
EXPANSION SHIELDS

COMPRESSIBLE
ELASTOMERIC TAPE TO
SPAN IRREGULARITIES

SEAL TOP OF SYSTEM
WITH FABRIC TAPE
AND MASTIC (OPTIONAL)

MODIFIED BITUMEN
FLASHING STRIP

PRIME CONCRETE

FASTENER APPROX.
8" O.C.

CHAMFER TOP
TO DRAIN

MODIFIED BITUMEN
MEMBRANE WITH
OPTIONAL SURFACING—
OR COATING

BASE SHEET OR MODIFIED
BITUMEN SHEET IF REQUIRED

OPTIONAL WOOD NAILER—
ATTACH TO CONCRETE
WITH APPROVED METHOD

SMOOTH CONCRETE—
EXPOSED SURFACES
MUST BE WATERPROOFED

OPTIONAL FIBER CANT
STRIP—SET IN BITUMEN

NOTE: WHERE DECK IS SUPPORTED BY AND FASTENED TO THE CONCRETE WALL, VERTICAL WOOD NAILER SHOULD BE SECURED TO THE WALL WITH SUITABLE FASTENERS.

(c) PVC SINGLE-PLY

SMOOTH CONCRETE—
EXPOSED SURFACES
MUST BE WATERPROOFED

CAULK WITH
ELASTOMERIC SEALANT

ANGLE CLAMPING BAR
WITH SLOTTED
ANCHOR HOLES

FASTENERS IN
EXPANSION SHIELDS

COMPRESSIBLE
ELASTOMERIC TAPE TO
SPAN IRREGULARITIES

PVC FLASHING MEMBRANE
ADHERED TO WALL AND WELDED
TO MEMBRANE

PRIME CONCRETE

PVC MEMBRANE

FASTENING STRIP APPLIED TO
DECK OR WALL—FASTENED
12" O.C.

SLIP SHEET IF REQUIRED

ALTERNATE FLASHING ARRANGEMENT

PVC COATED
METAL FLASHING

PVC MEMBRANE WELDED
TO PVC COATED METAL
FLASHING

FASTENED AS
REQUIRED

SLIP SHEET
IF REQUIRED

(d) EPDM SINGLE-PLY

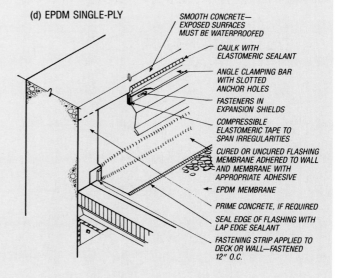

SMOOTH CONCRETE—
EXPOSED SURFACES
MUST BE WATERPROOFED

CAULK WITH
ELASTOMERIC SEALANT

ANGLE CLAMPING BAR
WITH SLOTTED
ANCHOR HOLES

FASTENERS IN
EXPANSION SHIELDS

COMPRESSIBLE
ELASTOMERIC TAPE TO
SPAN IRREGULARITIES

CURED OR UNCURED FLASHING
MEMBRANE ADHERED TO WALL
AND MEMBRANE WITH
APPROPRIATE ADHESIVE

EPDM MEMBRANE

PRIME CONCRETE, IF REQUIRED

SEAL EDGE OF FLASHING WITH
LAP EDGE SEALANT

FASTENING STRIP APPLIED TO
DECK OR WALL—FASTENED
12" O.C.

ALTERNATE BASE FLASHING ARRANGEMENT

FLASHING STRIP

EPDM MEMBRANE

Single ply polymeric membrane systems have evolved along several lines. The thermoplastic sheets such as PVC, Fig. 5.6.1c, are weldable. Many of these sheets are internally reinforced with a scrim or fabric and are seamed by hot air welding. EPDM membranes' lower price and improved performance have helped them become a leader in the industry, Fig. 5.6.1d. When single ply membranes are returned vertically on the back of precast concrete panels to act as flashing, the back surface must be given a smooth trowel finish. Surface irregularities may puncture the roofing membrane.

The bending radius of present composition roofing materials except for single plies is generally limited to 45 deg. To allow for this bending radius, all vertical surfaces must have cant strips installed between the roof and the vertical surface. The base flashing should extend vertically, from the horizontal plane of the roof, at least 8 in. but no more than 14 in. above the finished roof surface. Walls requiring flashings higher than 14 in. should receive special moisture-proofing. A wood nailer strip or suitable detail allowing mechanical fastening of the base flashing at the top must be provided. Since metals have a high coefficient of expansion, metal flashings and panel connections must be isolated from the roof membrane wherever possible to prevent thermal movements of the metal from splitting the membrane. Flashing details that require metal flanges to be sandwiched into the roof membrane should be avoided.

For all walls and projections that receive base flashing, metal counterflashings should be installed in the wall above the base flashing. The design of this detail should be two-piece (reglet and counterflashing), allowing installation of the counterflashing after the base flashing. Single-piece installations cannot be flashed properly. Also, it is difficult to perform re-roofing and roofing maintenance without deforming the metal when single-piece installations are used. Sheet metal should NEVER be used as a base flashing.

Roofing details that may be used where substantial movements are expected, such as a lower roof abutting a tower structure, are illustrated in Fig. 5.6.2a. Some proprietary roofing materials now on the market allow direct fastening to precast concrete, even when moderate movements are expected. Their pre-formed shapes are designed to accommodate such movements. If these special products are used, details for the precast concrete would be similar to Fig. 5.6.2b. An alternate method of attaching the counterflashing is shown in Fig. 5.6.2c.

The major requirement for a reglet is strict adherence to close tolerances in placing. This may be overlooked by the precaster unless clearly stated in the contract documents. Fig. 5.6.3 outlines the normal requirements. Also, see tolerances given in Section 4.6.2. Cast-in reglets are difficult to align properly. When not properly aligned, they can hinder the proper installation of counterflashing. Therefore, the use of cast-in-reglets is generally not recommended.

The method of fastening the roofing to the precast con-

Fig. 5.6.2 Typical roof flashing details

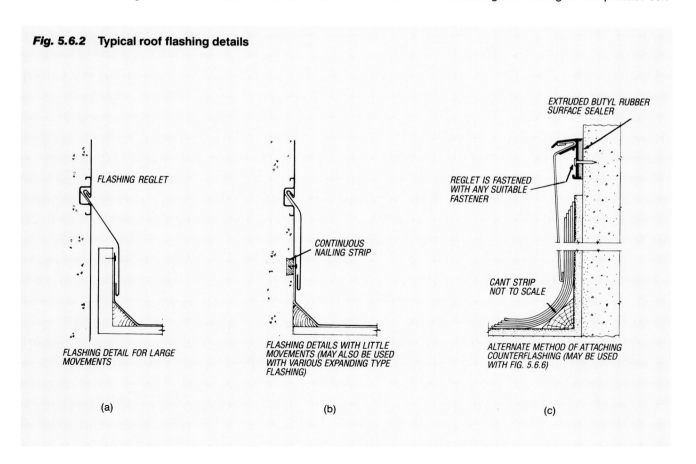

FLASHING REGLET

FLASHING DETAIL FOR LARGE MOVEMENTS

(a)

CONTINUOUS NAILING STRIP

FLASHING DETAILS WITH LITTLE MOVEMENTS (MAY ALSO BE USED WITH VARIOUS EXPANDING TYPE FLASHING)

(b)

EXTRUDED BUTYL RUBBER SURFACE SEALER

REGLET IS FASTENED WITH ANY SUITABLE FASTENER

CANT STRIP NOT TO SCALE

ALTERNATE METHOD OF ATTACHING COUNTERFLASHING (MAY BE USED WITH FIG. 5.6.6)

(c)

Fig. 5.6.3 Requirements for locating roof flashing reglets

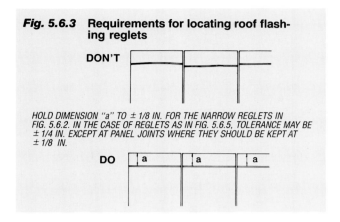

DON'T

HOLD DIMENSION "a" TO ± 1/8 IN. FOR THE NARROW REGLETS IN FIG. 5.6.2. IN THE CASE OF REGLETS AS IN FIG. 5.6.5, TOLERANCE MAY BE ± 1/4 IN. EXCEPT AT PANEL JOINTS WHERE THEY SHOULD BE KEPT AT ± 1/8 IN.

DO

crete units varies with local practices. Most roofers favor the continuous wood nailer, properly impregnated, as shown in Fig. 5.6.2b. The extra cost to the precaster should be reflected in equal or greater saving in the roofing contract. If the precaster has to provide tooling for accurate positioning of the reglet, it should be simple to include the wood nailer in such tooling.

For low parapet walls, the base flashing should be fastened to a vertical wood upright whose horizontal base is attached to the deck only. After the base flashing has been attached to the wood upright, the metal wall cap flashing may be installed. Then the counterflashing may be attached to the wall cap, extending down over the top of the base flashing. This method allows lateral movements of the wall without damage to the base flashing.

Where gravel stops are used, they should be raised above the roof surface using tapered cants and wood blocking, Fig. 5.6.4a. When this is not possible, the metal flanges for low-profile gravel stops should be set in mastic on top of the completed roof membrane and nailed at close intervals to the wood nailer. The metal flange should then be primed with asphalt primer and felt flashing strips applied. Interior drainage is recommended, and edges should be raised whenever possible. In the single-ply detail, Fig. 5.6.4b, prefabricated metal is used in place of the angular wood cant. The metal is fastened to a flat 2- by 6-in. nailer, and the membrane is fastened to it.

The waterproofing of parapet joints is important because they are exposed to weathering from all directions. Figs. 5.6.5a and b illustrate solutions using a one-stage joint gasket. Although the alternate parapet flashing indicated in Fig. 5.6.5c is likely to improve the maintenance-free life of the sealant, parapet joints require regular checking for performance. Figs. 5.6.5a and c show precast concrete parapets with guarded roof flashing while 5.6.5b is for a precast concrete parapet with flush roof flashing. Fig. 5.6.6 shows another possible solution for joint and roofing details involving a precast concrete parapet joint and cap flashing. Fig. 5.6.7a has a concrete cap with metal flashing only at the concrete cap joints and Fig. 5.6.7b has a continuous metal cap. The metal cap flashing should have at least a 4 in. exposed apron extending down over the base flashing. In most cases the metal counterflashing is

Fig. 5.6.4 Low parapet details

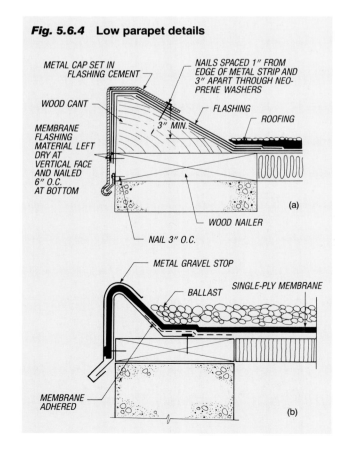

Fig. 5.6.5 Typical parapet and roofing details

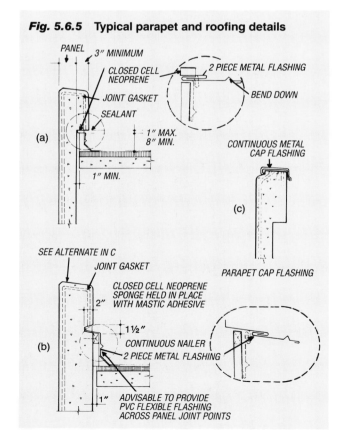

Fig. 5.6.6 Alternate parapet and roofing details

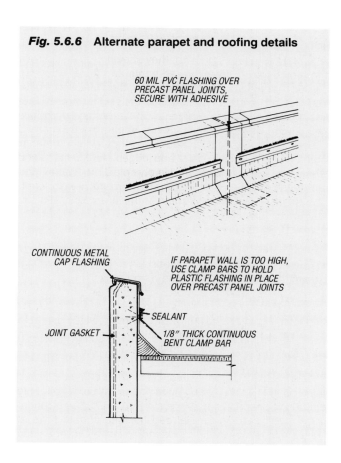

60 MIL PVC FLASHING OVER
PRECAST PANEL JOINTS,
SECURE WITH ADHESIVE

CONTINUOUS METAL
CAP FLASHING

IF PARAPET WALL IS TOO HIGH,
USE CLAMP BARS TO HOLD
PLASTIC FLASHING IN PLACE
OVER PRECAST PANEL JOINTS

SEALANT

JOINT GASKET

1/8" THICK CONTINUOUS
BENT CLAMP BAR

Fig. 5.6.7 Cap flashing

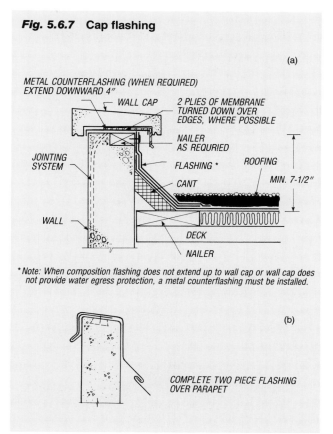

(a)

METAL COUNTERFLASHING (WHEN REQUIRED)
EXTEND DOWNWARD 4"

WALL CAP

2 PLIES OF MEMBRANE
TURNED DOWN OVER
EDGES, WHERE POSSIBLE

NAILER
AS REQURIED

JOINTING
SYSTEM

FLASHING *

ROOFING

CANT

MIN. 7-1/2"

WALL

DECK

NAILER

* Note: When composition flashing does not extend up to wall cap or wall cap does
not provide water egress protection, a metal counterflashing must be installed.

(b)

COMPLETE TWO PIECE FLASHING
OVER PARAPET

attached to the inside face of the metal wall cap flashing
with sheet metal screws. The choice of flashing material
and/or its treatment against corrosion should be based on
preventing potential staining of the precast concrete surface, see Section 3.6.7.

If a two-stage joint is used, the airseal, preferably a closed
cell neoprene sponge under pressure, is continued up the
back of the parapet, along the top, and down the front to
overlap the rain-barrier. The top and back of the parapet
normally receive a field-molded sealant to form a flush joint
as shown in Fig. 5.6.8. Fig. 5.6.9 shows a variation of the
coping detail where the airseal is protected by a small precast concrete slab grouted in after erection of the panels.
This is suitable only for narrow precast concrete products

Fig. 5.6.8 Parapet joint (two-stage solution)

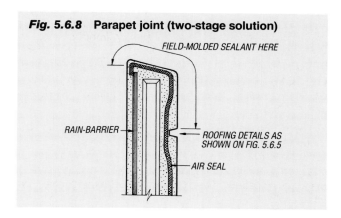

FIELD-MOLDED SEALANT HERE

RAIN-BARRIER

ROOFING DETAILS AS
SHOWN ON FIG. 5.6.5

AIR SEAL

Fig. 5.6.9 Parapet joint suitable for narrow units (where horizontal movement is slight)

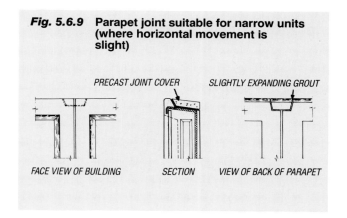

PRECAST JOINT COVER

SLIGHTLY EXPANDING GROUT

FACE VIEW OF BUILDING

SECTION

VIEW OF BACK OF PARAPET

Fig. 5.6.10 Inverted roof details

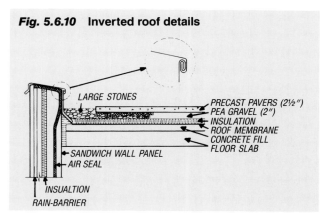

LARGE STONES

PRECAST PAVERS (2½")
PEA GRAVEL (2")
INSULATION
ROOF MEMBRANE
CONCRETE FILL
FLOOR SLAB

SANDWICH WALL PANEL
AIR SEAL

INSUALTION

RAIN-BARRIER

where horizontal movement is slight.

In any solution using two-stage joints, it is important that the airseal closely abuts the roof seal to complete the building envelope.

The application of precast concrete in connection with inverted roof membranes or a protected membrane roof is shown in Fig. 5.6.10. This concept provides a complete airseal envelope of the building with the insulation outside the envelope.

Prior to laying the insulation, the flashing must be installed. Installation techniques depend heavily on the type of membrane and where the flashing is being placed, Fig. 5.6.11. Check the membrane manufacturer's specifications concerning the membrane's perimeter securement, required length of flashing where it laps the membrane, and proper method of sealing the flashing to the membrane.

The tasks of the ballast include keeping the protective mat, insulation, and membrane from being blown off the roof; protecting these components from ultraviolet radiation; and preventing storm damage. Ballasts are usually crushed stone, gravel, concrete pavers, or a combination of these materials. The type of ballast is selected according to expected wind velocity, required roof drainage, and projected foot traffic.

When pavers are installed, they should be designed for durability, ease of removal, and ability to resist wind uplift. The last two criteria are functions of weight and installation. Manufacturers recommend that pavers be staggered, be no larger than two feet square, weigh no more than 60 pounds per paver, and no less than 22 psf. The manner in which the pavers are installed over the insulation can vary. Some manufacturers recommend that pavers be elevated above the insulation by means of pressure-treated wood blocks, paver pedestals manufactured specifically for this, Fig. 5.6.12, or a layer of permeable gravel or crushed stone.

Securing the pavers provides additional protection against wind uplift. Metal strapping is usually used with insulation/ballast board systems, but can also be used with concrete pavers, particularly in areas prone to high winds. Strapping fastens the individual pavers to one another so they work as a unit. The system is especially effective around penetrations. Figs. 5.6.13 and 5.6.14 illustrate how additional straps can be used to tie the perimeter pavers. Strapping should be corrosion-resistant and flexible enough to allow for flotation of the raft.

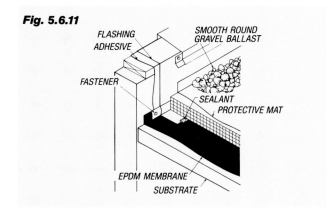

Fig. 5.6.11

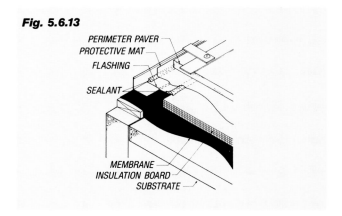

Fig. 5.6.13

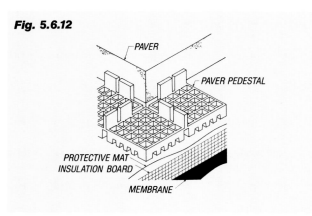

Fig. 5.6.12

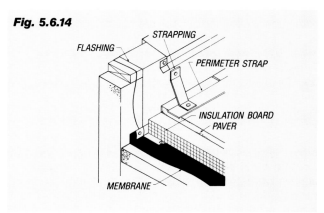

Fig. 5.6.14

SPECIFICATION CONSIDERATIONS

6.1 GENERAL

This chapter provides guidelines and basic information essential to proper specification of in-plant fabrication and field erection of architectural precast concrete with a variety of textures and finishes. It is not intended to be used for field-fabricated precast concrete panels, or precast structural concrete. It does not include dampproofing, special coatings applied to the panels, caulking around the panels, or loose attaching hardware.

Section 6.4 contains a Guide Specification. Each sub-section introduces the specification provisions which represent the minimum items which have to be considered for a typical architectural precast concrete specification. The Guide Specification provisions are on the left, and the Notes to Specifiers are on the right and discuss additional items or alternatives which may be required for specific applications of precast concrete or for particular job conditions.

Where appropriate, each sub-section also provides a short description of the rationale for the specification provisions, or a reference to the section in the Manual where such description has already been presented. In providing this notation of possible additional requirements, the final choice or adaption can be made in relation to the specific job conditions and specification format.

Specifiers should realize the shortcomings inherent in any attempt to answer all requirements or special conditions in a description of this nature due to the variety of precast concrete applications.

Architectural and structural drawings should show the locations, sections and dimensions necessary to define the size and shape of the architectural precast concrete. These drawings should also indicate the locations of joints, both functional and esthetic, and illustrate details between units. When more than one type of panel material or finish is used, the drawings should indicate the extent and location of each type. They should also illustrate the details to be used at corners of the structure and interfaces with other materials. Whether sizes and locations of steel reinforcement and details and locations of typical and special connection items and inserts are shown may be determined by local practices. If reinforcement and connections are not detailed, the drawings should identify the require-

ments for design, design loads, and indicate load support points and space allowed for connections.

Specifications supplement working drawings to adequately define the end results expected from the architectural precast concrete, as well as all other components affecting this work.

Specifications should describe the type and quality of the materials to be incorporated into the units, design strength of the concrete, finishes, and the tolerances for casting and erection. The methods and techniques used to achieve the specified results will vary with individual precast concrete manufacturers. Specifying the results desired without specifically defining manufacturing procedures will ensure a concise and accurate interpretation, and in turn encourage the best competitive bidding.

The key to preparing a specification for optimum quality of architectural precast concrete is to seek the counsel of the local producer representative.

The primary function of the precaster's **shop drawings (erection and production drawings)** is the translation of project drawings into usable information for accurate and efficient manufacture, handling and erection of the precast concrete units. In addition, the shop drawings provide the architect with a means of checking interfacing with adjacent materials.

Generally shop drawings are submitted to the general contractor who, after checking them and making notations, submits them to the architect/engineer for checking and review. Final approved shop drawings are returned to the precast concrete manufacturer by way of the general contractor. The general contractor should be responsible for project schedule, dimensions, and coordination with all construction phases. Fabrication should not commence until final approval — or "approved-as-noted" — has been received from both the architect and the general contractor. If shape drawings are submitted separately, approval would allow fabrication of molds and tooling. Alternatively, shop drawings may be approved initially for mold production and subsequently for panel production.

"Approved" or "Approved-as-noted" shop drawings should mean that the general contractor and architect/engineer have verified dimensions to be correct and final for the following: overall building dimensions, column centerlines, floor elevations, floor thicknesses, column and beam sizes, foundation elevations, the location of mechanical openings and other items pertinent to architectural precast concrete. The precast concrete manufacturer should be responsible for the precast concrete units fitting the building dimensions and conditions as given on the final approved drawings.

Some precasters prepare in-house shop (production) tickets from shop drawings listing schedules of precast concrete units. Others produce separate drawings of each individual and different unit by means of reproducibles from typical master mold units. All of these need not be submitted for approval. If any are required, the number should be stated in the specifications. Only the different architectural units and the typical units to be produced in each type of mold need be submitted for approval.

When erection drawings contain all information sufficient for design approval, production drawings, except for shape drawings, need not be submitted for approval, except in special cases. However, record copies are frequently requested. Guidelines for the preparation of shop drawings are given in the *PCI Drafting Handbook - Precast and Prestressed Concrete.*

Coordination and responsibility for supply of the items to be placed on or in the structure to allow placement of the precast concrete units depends on the type of structure and varies with local practice. The specifications must clearly specify responsibility for supply and installation of hardware. If not supplied by the precast concrete manufacturer, the supplier should be listed and requirements included in related trade specifications. See also Section 4.5.9 in this Manual. When the building frame is structural steel, erection hardware is normally supplied and installed as part of the structural steel. When the building frame is cast-in-place concrete, hardware, if not pre-designed or shown on drawings, may be supplied by the precast concrete manufacturer, general contractor, or be part of the miscellaneous steel subcontract, and placed by the general contractor according to a hardware layout prepared by the precast concrete supplier.

Assurance that the type and quantity of hardware items required to be cast into the precast concrete units for the use of other trades are specified by the architect and not duplicated, is of greater importance than the supplier. Such items may be inserts for windows, fastenings for other trades, dovetails for flashing or masonry, etc. Specialty items, however, should be supplied in a timely manner from the trade requiring them.

The architect should verify that the materials specified in the section on flashings are galvanically compatible with the reglets or counterflashing receivers to be installed. A check should also be made to ensure that concrete coatings, adhesives, and sealants specified in other sections are compatible with each other and with the mold release agents or surfaces to which they are to be applied. Establishing the compatibility of specified materials is the responsibility of the architect.

Supply and Installation by Others. Several items mentioned in the Guide Specification as possible supply and/or installation by others should be mentioned in the specifications covering the specific trades.

Such items may be:
- Cost of inspection by an independent testing laboratory.
- Hardware for interfacing with other trades (window, door, flashing and roofing items).
- Placing of contractor's hardware cast into or attached to the structure, including tolerances for such placing.
- Joint treatment for joints between precast concrete and other materials.
- Access to building and floors.
- Power supply.
- Cleaning.
- Water repellent coatings.
- Natural stone.

6.2 PERFORMANCE SPECIFICATIONS

Performance specifications are not necessarily the best solution to all building problems. They may be employed for specific items with good results as long as the architect identifies the purpose to be served and includes appropriate safeguards such as pre-qualification of precasters, pre-bid approval of materials, careful review of shop drawings, and architect's approval of initial production units.

Conventional specifications often resort to stringent requirements in order to protect the architect and the client.

Because of this, they do not always result in the best price within the desired or acceptable quality range. The principal advantage of performance specifications is that they can combine economy and optimum quality, utilizing established tooling and production techniques.

Performance specifications tend to create additional work for the architect at the design stage, because the end result must be clearly defined and several different proposals must be assessed. The accepted proposals will eventually become the standards for manufacturing. This additional work in the early stages is generally offset by time saved later in detailing in the architect's office.

Use of performance specifications can place the architect in a position of being legally responsible for the quality of the finished product because all materials, methods, equipment, and so forth are detailed in the specification. For this reason, performance specifications should be used with care.

Partial performance specifications may be used to the architect's advantage when the safeguards described in this section are included.

Properly prepared performance specifications should conform to the following criteria:

1. They should clearly state all limiting factors such as minimum or maximum thickness, depth, weight, and any other limiting dimensions. Acceptable limits of other requirements not detailed should be clearly provided. These limits may cover insulation (thermal and acoustical), interaction with other materials, services and appearance.

2. They should be written so that the scope of each subcontract is clearly defined. All subcontracts must be properly related to each other so that they combine to produce an integrated project.

3. If a method other than simultaneous competitive bidding through a general contractor is contemplated, the scope and the nature of the precast concrete work in relation to other trades should be carefully weighed in the final assessment of the precast concrete solution.

4. The architect should request samples, design and detail submissions from prospective bidders and make pre-bid approval of such submissions a prerequisite for bidding.

5. To the degree that such requests for pre-bid approvals form a part of the specifications, the architect should adhere to the following requirements:

   a. Sufficient time must be allowed for the bidder to submit samples or information for approval by the architect. Approval should be conveyed to the manufacturer in writing with sufficient time to allow completion of estimate and submittal of bid.

   b. Any proprietary pre-bid submittal should be treated in confidence and the individual producer's original solutions or techniques protected both before and after bidding.

6.3 INSPECTION CHECKLIST

This section contains a listing of general items to be examined during a routine inspection of the architectural precast concrete operations by the architect. Other items of particular importance to a specific project will also have to be addressed during the architect's office, plant, and site visits. The lists are divided into four time phases: during manufacture, before erection, during erection, and after erection.

6.3.1 DURING MANUFACTURE

1. Check qualifications of testing agencies.

2. Where in-plant testing is to be performed, verify that manufacturer's testing equipment is calibrated by testing agency personnel, and that manufacturer's testing personnel are qualified to perform the work.

3. Check casting and after-casting tolerances. Inspectors should be provided with quality instruments for measuring tolerances.

4. Verify that production panels match texture and color of accepted job mockup or panel samples.

5. Examine concrete test results.

6.3.2 BEFORE ERECTION

1. Check precast concrete panel erector's experience and qualification.

2. Check welder's qualifications.

3. Ensure that precast concrete panels are stored on site with resilient and stain resistant spacers.

4. Verify that panel identification marks are easily discernible.

5. Check field dimensions affecting erection.

6.3.3 DURING ERECTION

1. Check erection tolerances.

2. See that walls are clean and exposed metal is spot painted where welding has damaged the protective coating.

3. Check that proper handling points are used as per erection drawings.

6.3.4 AFTER ERECTION

1. Inspect repairs for accurate color and texture match to surrounding concrete.

2. Inspect panels after cleaning to see that they are properly prepared to receive joint sealant and water repellent coating.

6.4 GUIDE SPECIFICATION FOR ARCHITECTURAL PRECAST CONCRETE

This Guide Specification is intended to be used as a basis for the development of an office master specification or as an aid in the preparation of specifications for a particular project. In either case, this Guide Specification must be edited to fit the conditions of use.

Particular attention should be given to the deletion of inapplicable provisions. Necessary items related to a particular project should be included. Also, appropriate requirements should be added where blank spaces have been provided. Coordinate the specifications with the information shown on the contract drawings to avoid duplication.

The Guide Specifications are on the left. *Notes to Specifiers are on the right.*

GUIDE SPECIFICATIONS	NOTES TO SPECIFIERS

1. GENERAL

1.01 Description

A. Scope: This section establishes general criteria for materials, production, erection and evaluation of precast concrete as required for subsequent related sections of these Specifications. The work to be performed under this section of the specifications shall include all labor, material, equipment, related services, and supervision required for the manufacture and erection of the architectural precast concrete units shown on the contract drawings and schedules. All work covered in this section is subject to Instructions to Bidders, Special Conditions and General Conditions. The precast concrete manufacturer shall be responsible for, and be governed by all the requirements thereunder.

1.01.A   Local standard practice may indicate that responsibility for erection may not be included. Refer to Manual, Section 4.1.4.

B. Related work specified elsewhere:

1. Concrete reinforcement: Section _____ .

   1.01.B.1   Architectural precast concrete reinforcing steel requirements are different from cast-in-place reinforcement and should be specified in this section. Refer to Manual, Section 4.4.

2. Cast-in-place concrete: Section _____ .

   1.01.B.2   Placement of anchorage devices in cast-in-place concrete for precast concrete panels.

3. Precast, prestressed concrete: Section _____ .

   1.01.B.3   Precast concrete floor and roof slabs, beams, columns and other structural elements. Some items, such as prestressed wall panels on industrial buildings, could be included in either specification, depending on the desired finish.

4. Structural steel framing: Section _____ .

   1.01.B.4   Steel supporting structure, attachment of anchorage devices on steel for precast concrete panels, and sometimes loose anchors.

5. Water repellent coatings: Section _____ .

   1.01.B.5   For exposed face of panels.

6. Insulation: Section _____ .

   1.01.B.6   Insulation job-applied to precast concrete panels. Insulation cast in precast concrete panels during manufacture should be specified in this section.

7. Flashing and sheet metal: Section _____ .

   1.01.B.7   Counterflashing inserts and receivers, unless included in this section.

8. Sealants and caulking: Section _____ .

   1.01.B.8   Panel joint caulking and sealing.

9. Painting: Section _____ .

   1.01.B.9   Field touch-up painting. Delete when specified in this section.

GUIDE SPECIFICATIONS

10. Glass and glazing: Section _____ .

11. Glazing accessories: Section _____ .

C. Work installed but furnished by others:

1. Counterflashing receivers or reglets: Section _____ .

2. Inserts or attachments for _____ : Section _____ .

D. Testing agency provided by owner.

1.02 Applicable Specifications, Standards and Codes

The following specifications, standards, and codes shall govern the fabrication unless modified by this specification.

A. American Concrete Institute
ACI 318—*Building Code Requirements for Reinforced Concrete.*

B. American Welding Society
AWS D1.1—*Structural Welding Code—Steel.*
AWS D1.4—*Structural Welding Code—Reinforcing Steel.*

C. Precast/Prestressed Concrete Institute
PCI MNL 117—*Manual for Quality Control for Plants and Production of Architectural Precast Concrete Products.*

The following documents, while not a part of these specifications except for sections specifically referred to herein, describe recognized acceptable practices.

PCI MNL 116—*Manual for Quality Control for Plants and Production of Precast Prestressed Concrete Products.*
PCI MNL 119—*PCI Drafting Handbook—Precast and Prestressed Concrete.*
PCI MNL 120—*PCI Design Handbook—Precast and Prestressed Concrete.*
PCI MNL 127—*Recommended Practice for the Erection of Precast Concrete.*

NOTES TO SPECIFIERS

1.01.B.10   Glazing of precast concrete panels in plant. Delete when specified in this section.

1.01.B.11   Reglets for use with structural glazing gaskets. Delete when specified in this section.

1.01.C   Delete when furnished by precast concrete manufacturer. Add additional items as may be required for the particular project.

1.01.C.2   May include inserts/attachments for window or door frames, window washing equipment, etc.

1.01.D   Delete when testing agency is provided by precast concrete manufacturer or general contractor. Coordinate with Division 1, General Conditions.

1.02   The various standards should have a number affixed indicating the year of the latest issue (if such is available). The architect should verify and affix the number of the standard in effect at the time the project specifications are being prepared. List only those standards applicable to the specific project as different reference documents may contain conflicting statements.

In Canada:
National Building Code of Canada.
CAN 3-A23.3 Code for the Design of Concrete Structures for Buildings.
CAN 3-A23.4 Precast Concrete—Materials and Construction.
CSA A251 Qualification Code for Manufacturers of Architectural and Structural Precast Concrete.

1.02.A   This document is one of two current ACI Publications written in specification language. Other ACI documents should not be referenced in the contract documents. If items found in other ACI documents are desired, they should be phrased in mandatory language and incorporated into the contract documents.

1.03 Quality Assurance

A. Manufacturer qualifications:

The precast concrete manufacturing plant shall be certified by the Precast/Prestressed Concrete Institute, Plant Certification Program, prior to the start of production.

* * OR * *

The manufacturer shall, at their own expense, meet the following requirements:

1. Retain an independent testing or consulting firm approved by the architect/engineer and/or owner.

2. The basis of inspection shall be the Precast/ Prestressed Concrete Institute *Manual for Quality Control for Plants and Production of Architectural Precast Concrete Products*, MNL-117, and *Manual for Quality Control for Plants and Production of Precast and Prestressed Concrete Products*, MNL-116.

3. This firm shall inspect the precast concrete plant at two-week intervals during production and issue a report, certified by a registered engineer, verifying that materials, methods, products and quality control meet all the requirements of the specifications, drawings, and MNL-117 and/or MNL-116. If the report indicates to the contrary, the architect/engineer, at the precaster's expense, will inspect and, at architect's option, may reject any or all products produced during the period of non-compliance with the above requirements.

* * OR * *

Acceptable manufacturers:

1. _____

2. _____

3. _____

* * * * * *

B. Erector qualifications:
Regularly engaged for at least _____ years in erection of architectural precast concrete units similar to those required on this project.

C. Welder qualifications:
In accordance with AWS D1.1.

1.03.A   It is recommended that the architect establish procedures for approval of individual precast concrete manufacturers at least 10 days prior to bid date or identify approved bidders in the specification. The manufacture of architectural precast concrete requires a greater degree of craftsmanship than most other concrete products, and therefore requires some prequalification of the manufacturer. Manufacturers should have a minimum of 2 to 5 years of production experience in architectural precast concrete work of the quality and scope required on the project. Manufacturer should be able to show that plant has experienced personnel, physical facilities, established quality control procedures, and a management capability sufficient to produce the required units without causing delay to the project. Manufacturer should demonstrate adequate financial responsibility, including ability to post bond, if required. When requested by the architect, the manufacturer should submit written evidence of the above requirements.

1.03.B   Usually 2 to 5 years. Only competent workmen who are properly trained to handle and erect architectural precast concrete units should be employed. Specifications may leave the responsibility for erection of architectural precast concrete open for decision by the general contractor, or it may require that the erection be part of the precast concrete manufacturer's work, to be done by precaster's own forces or sub-contracted to specialized erection firm. The inclusion of erection in the precast concrete contract should be governed by local practices.

1.03.C   Qualified within the past year. Delete when welding is not required.

D. Testing:
In general compliance with testing provisions in MNL-117, *Manual for Quality Control for Plants and Production of Architectural Precast Concrete Products.*

E. Testing agency:

NOTES TO SPECIFIERS

1.03.E   Delete when provided by owner (or architect). The cost of this inspection should not be part of the precast concrete contract, but may be included as a lump sum allowance, determined by the architect. When an independent concrete laboratory is used to verify plant and product quality, it is recommended that it should be engaged by the owner or architect. The precaster should provide reasonable access to all relevant plant operations for the architect or a qualified testing laboratory, appointed by the architect, for the purpose of verifying the quality. The manufacturer should also supply materials required for independent testing, but should not include the cost of such testing.

1. Not less than _____ years experience in performing concrete tests of type specified in this section.

1.03.E.1   Usually 2 to 5 years.

2. Capable of performing testing in accordance with ASTM E 329 (CSA A283).

3. Inspected by Cement and Concrete Reference Laboratory of the National Institute of Standards and Technology.

F. Requirements of regulatory agencies:
Manufacture and installation of architectural precast concrete to meet requirements of _____ .

1.03.F   Local building code or other governing code relating to precast concrete. For projects in Canada, standards from the National Building Code of Canada should be listed in addition to or in place of the U.S. Standards.
CSA W47.1 Certification of Companies for Fusion Welding of Steel Structures.
CSA W59.1 General Specification for Welding of Steel Structures.
CSA W186 Welding of Reinforcing Bars in Reinforced Concrete Construction.
CSA A23.1 Concrete Materials and Methods of Concrete Construction.
CSA A23.2 Methods of Test for Concrete.

G. Allowable tolerances:

1. Manufacture wall panels so that dimensional tolerances shall be as follows:

1.03.G   These standard industry tolerances should only be modified to meet specific job conditions after careful review and consideration. Most manufacturers can meet closer tolerances, if required, but closer tolerances normally increase costs. Refer to Manual, Section 4.6.2. Some types of window and equipment frames require openings very accurately placed. When this is the case, the minimum practical tolerance should be defined with input from the precaster.

Overall length and width (measured
at neutral axis of ribbed members)
   10 ft or under .... ±⅛ in.
   10 to 20 ft .... +⅛ in., −³⁄₁₆ in.
   20 to 40 ft .... ±¼ in.
   Each additional 10 ft .... ±¹⁄₁₆ in. per 10 ft
Total thickness or flange thickness   −⅛ in., +¼ in.
Rib thickness .... ±⅛ in.
Rib to edge of flange .... ±⅛ in.
Distance between ribs .... ±⅛ in.
Angular variation of plane
of side mold .... ±¹⁄₃₂ in. per 3 in. of depth or
                   ±¹⁄₁₆ in., whichever is greater
Variation from square or designated skew
(difference in length of the two
diagonal measurements) .... ±⅛ in. per 6 ft of
               diagonal or ±½ in.,
               whichever is greater

Length and width of blockouts and
openings within one unit................................. ±¼ in.
Location and dimensions of blockouts hidden
from view and used for HVAC and utility
penetrations.. ±¾ in.
Dimensions of haunches .... ±¼ in.
Haunch bearing surface deviation
from specified plane .... ±⅛ in.
Difference in relative position of
adjacent haunch bearing surfaces
from specified relative position........................ ±¼ in.
Bowing.. L/360, max. 1 in.
Differential bowing between adjacent
panels of the same design .... ½ in.
Local smoothness .... ¼ in. in 10 ft
Warping................................. ¹⁄₁₆ in. per ft of distance
from nearest adjacent corner
Location of window opening within panel......... ±¼ in.
Position of plates.. ±1 in.
Tipping and flushness of plates .... ±¼ in.

Position tolerances. For cast-in items measured from
datum line location as shown on approved erection
drawings:
Weld plates .... ±1 in.
Inserts .... ±½ in.
Handling devices .... ±3 in.
Reinforcing steel and welded wire fabric......... ±¼ in.,
where position has structural implications
or affects concrete cover, otherwise ±½ in.
Tendons.. ±⅛ in.
Flashing reglets .... ±¼ in.
Flashing reglets at edge of panel.................... ±⅛ in.
Reglets for glazing gaskets............................ ±¹⁄₁₆ in.
Groove width for glazing gaskets................... ±¹⁄₁₆ in.
Electrical outlets, hose bibs, etc...................... ±½ in.
Haunches .... ±¼ in.

H. Job mockup:

1. After standard samples are accepted for color and texture, submit full-scale unit meeting design requirements.

2. Mockup to be standard of quality for architectural precast concrete work, when accepted by architect/engineer.

1.03.H.1 Full-scale samples or inspection of the first production unit are normally required, especially when a new design concept or new manufacturing process or other unusual circumstance indicates that proper evaluation cannot otherwise be made. It is difficult to assess appearance from small samples. It should be noted that larger samples require considerable time in their manufacture. They should not be specified unless considerable lead-time is allowed for the entire project. Delays in visiting plants for approvals may upset normal plant operations and the job schedule. These production samples are not to be confused with the pre-bid samples on which the architect based the initial selection of aggregates, color and texture. Slight variations are to be expected depending on the precaster and actual production materials. Refer to Manual, Section 3.2.

1.03.H.2 Use to determine range of acceptability with respect to color and texture variations, surface defects, and overall appearance. Refer to Manual, Section 3.2.4. Mockup should also serve as testing areas for remedial work. Precast concrete products which do not meet the color and texture range or the dimensional tolerances of this specification may be rejected at the option of the architect, if they cannot be satisfactorily corrected. It should also be specifically stated in the contract documents who the accepting authority will be—contractor, architect, engineer of

GUIDE SPECIFICATIONS

3. Incorporate mockup into work in location reviewed by architect/engineer after keeping unit in plant _____ for checking purposes.

I. Source quality control:

1. Quality control and inspection procedures to comply with applicable sections of MNL 117.

2. Water absorption test on unit shall be conducted in accordance with MNL 117.

1.04 Submittals

A. Samples:

1. Submit samples representative of finished exposed face showing typical range of color and texture prior to commencement of manufacture.

2. Sample size: approximately 12 in. × 12 in. and of appropriate thickness, representative of the proposed finished product.

B. Shop drawings:

General contractor shall expedite submittal with architect to conform with allotted shop drawing approval time shown on the precast concrete supplier's order acknowledgment.

1. Content:

a. Unit shapes (elevations and sections), and dimensions.

b. Finishes.

c. Reinforcement, joint, and connection details.

d. Lifting and erection inserts.

e. Location, dimensional tolerances, and details of anchorage devices that are embedded in or attached to structure or other construction.

NOTES TO SPECIFIERS

record, owner or jobsite inspector. One person must have final undisputed authority.

1.03.H.3   Delete when mockup is not to be included in work. State how long unit should be kept. Mockup is normally incorporated in the building, at least for production units.

1.03.I.2   Water absorption test is an early indication of weather staining (rather than durability). Verify the water absorption of the proposed face mix, which for average exposures and based upon normal weight concrete (150 lbs per cubic foot) should not exceed 5% to 6% by weight. As an improved weathering (staining) precaution, lower absorption between 3% to 4% (by weight) is feasible with some concrete mixes and consolidation methods. In order to establish comparable absorption figures for all materials, the current trend is to specify absorption percentages by volumes. The stated limits for absorption would, in volumetric terms, correspond to 12% to 14% for average exposures and 8% to 10% for special conditions. Refer to Manual, Section 3.6.

1.04.A   Number of samples and submittal procedures should be specified in Division 1. All approved samples should be initialed by architect. Pre-bid samples should be submitted a minimum of 10 days prior to bid date. Refer to Manual, Sections 3.2.2 and 3.2.3.

1.04.A.1   If the back face of a precast concrete unit is to be exposed, samples of the workmanship, color, and texture of the backup concrete should be shown as well as the facing. Refer to Manual, Section 3.5.17.

1.04.B   Specifications should state the number of copies required for approval or whether reproducibles are required. Current practice usually calls for two prints and one reproducible of shop drawing to be submitted for approval. When erection drawings contain all information sufficient for design approval, production drawings, except for shape drawings, need not be submitted for approval, except in special cases. However, record copies are frequently requested. Guidelines for the preparation of drawings are given in the PCI Drafting Handbook—Precast and Prestressed Concrete. Refer to Manual, Section 4.3.

1.04.B.1.e   Drawings normally prepared by precast concrete manufacturer and provided to general contractor for work by other trades.

    f. Other items cast into panels.

    g. Handling procedures, plans and/or elevations showing panel location, and sequence of erection for special conditions.

    h. Relationship to adjacent material.

  2. Show location of unit by same identification mark placed on panel.

C. Test reports:

Submit, on request, reports on materials, compressive strength tests on concrete and water absorption tests on units.

D. Design calculations:

Submit, on request, structural design calculations.

E. Design modifications:

  1. Submit design modifications necessary to meet performance requirements and field coordination.

  2. Variations in details or materials shall not adversely affect the appearance, durability or strength of units.

  3. Maintain general design concept without altering size of members, profiles and alignment.

1.05 Product Delivery, Storage, and Handling

A. Delivery and handling:

  1. Deliver all architectural precast concrete units to project site in such quantities and at such times to assure continuity of erection.

  2. Handle and transport units in a position consistent with their shape and design in order to avoid stresses which would cause cracking or damage.

  3. Lift or support units only at the points shown on the shop drawings.

  4. Place nonstaining resilient spacers of even thickness between each unit.

  5. Support units during shipment on non-staining shock-absorbing material.

  6. Do not place units directly on ground.

1.04.B.1.g If sequence of erection is critical to the structural stability of the structure, or for access to connections at certain locations, it should be noted on the contract plans and specified. Refer to Manual, Section 4.2.9.

1.04.B.1.h Details, dimensional tolerances and related information of other trades affecting precast concrete work should be furnished to precaster.

1.04.C The number and/or frequency of each type of test should be clearly stated in the specifications by listing the required testing or by reference to applicable standards such as PCI MNL 117. Schedule of required tests, number of copies of test reports, and how distributed are included in Testing Laboratory Services, Section _____ .

1.04.D Design team members of the precast concrete manufacturer should be under the direct supervision of a professional engineer experienced in design of precast concrete panel products and registered in the state in which the project is located. This team would prepare all design calculations for precast concrete products not defined or completed on the drawings and for steel reinforcement required for handling and erection loads. Refer to Manual, Section 4.1.4.

1.04.E.1 The precast concrete manufacturer should submit any changes to the general contractor along with any changes in contract dollar amounts. Changes should be approved or disapproved by the architect/engineer with fair consideration to advantages gained by all sub-contractors involved.

1.05.A Erector should coordinate arrival of precast concrete units and provide for possible storage and for erection in a safe manner within the agreed schedule and with due consideration for other trades. Handling procedures, including type and location of fastenings, should normally be left to the precaster, but the fastening devices should be located and identified on shop drawings.

GUIDE SPECIFICATIONS

B. Storage at jobsite:

1. Store and protect units to prevent contact with soil, staining, and physical damage.

2. Store units, unless otherwise specified, with non-staining, resilient supports located in same positions as when transported.

3. Store units on firm, level, and smooth surfaces to prevent cracking, distortion, warping or other physical damage.

4. Place stored units so that identification marks are discernible, and so that product can be inspected.

2. PRODUCTS

2.01 Materials

Unless otherwise stated within the specifications, all materials shall comply with the specifications, standards and codes given for each material covered in this section. The architect shall be furnished satisfactory certification reports that all materials incorporated in the architectural precast concrete products comply with the requirements herein specified.

A. Concrete:

1. Portland cement:
   a. ASTM C 150 (CSA A5), type _____ , _____ color.

   b. For exposed surfaces use same brand, type, and source of supply throughout.

2. Air-entraining agent: ASTM C 260.

3. Water reducing, retarding, accelerating, high range water reducing admixtures:

   ASTM C 494 (CAN 3-A266)

NOTES TO SPECIFIERS

1.05.B The ideal sequence of precast concrete erection is the unloading of units directly to their proper location on the structure without storing on the jobsite. If on-site storage is an absolute necessity to enable the erector to operate at the speed required to meet the established schedule, leaving the precast concrete units on the trailer eliminates extra handling or possible damage caused by improper on-site storage techniques.

2.01 Refer to Manual, Section 3.2.6.

2.01.A.1.a Type: [I(General use)], [III(High early strength)]. Color: (gray), (white), (buff). Gray should be used for non-exposed backup concrete. Finish requirements will determine color selected for face mix.

2.01.A.1.b To minimize color variation. Specify source of supply when color or shade is important. Cements should be selected to provide predictable strength and durability as well as proper color. Characteristics of certain special and colored cements should be investigated prior to use to be certain that they do not exhibit undesirable attributes of high slump loss, retrogression, plateau-strength or other aberrations under the variable environmental casting conditions. The precast concrete manufacturer should have the choice of using either normal or high early strength cement provided the color is acceptable and the specified ultimate strength is attained, and as long as the chosen type is used throughout the job, if part of the face mix.

2.01.A.2 Delete if air entrainment is not required. Air entrainment should be required to increase resistance to freezing and thawing where environmental conditions dictate. Air-entraining admixtures are preferred to air-entraining cements because their use allows adjustment in order to obtain the correct amount of air for the exposure condition.

2.01.A.3 Delete if water reducing, retarding, or accelerating admixtures are not required. These admixtures are not normally used unless a specific change is desired in the properties or workability of the mix. Calcium chloride, or admixtures containing significant amounts of calcium chloride,

GUIDE SPECIFICATIONS

4. Coloring agent: ASTM C 979

    a. Synthetic mineral oxide.

    b. Harmless to concrete set and strength.

    c. Stable at high temperature.

    d. Sunlight and alkali-fast.

5. Facing Aggregates:

    a. Provide fine and coarse aggregates for each type of exposed finish from a single source (pit or quarry) for entire job. They shall be clean, hard, strong, durable, and inert, free of staining or deleterious material.

    b. ASTM C 33 or C 330 (CSA A231).

    c. Material and color: _____ .

    d. Maximum size and gradation: _____ .

NOTES TO SPECIFIERS

should not be allowed. Use admixtures in strict compliance with manufacturer's directions. Use amounts as recommended by admixture manufacturer for climatic conditions prevailing at time of placing concrete. The selection of the particular admixture(s) should be left to the precast concrete manufacturer subject to approval of architect/engineer.

2.01.A.4 Investigate use of naturally colored fine aggregate in lieu of coloring agent. Delete if coloring agent is not required.
2.01A.4.b Consider effects upon concrete prior to final selection.

2.01.A.5.a Approve or select the size, color and quality of aggregate to be used. Base choice on visual inspection of concrete sample and on assessment of certified test reports. Use same type and source of supply to minimize color variation. Fine and coarse aggregates are not always from same source.

2.01.A.5.b Grading requirements (only) are generally waived or modified because of special requirements of gap-graded face mixes, the advantages in having backup mixes as consistent as possible with face mixes, and the requirements to produce panels with texture and finish matching approved samples. A specified gradation developed from the pre-bid sample may have to be attained by blending of pre-graded sizes. For exposed concrete in cold climates, the magnesium sulfate soundness loss should preferably be less than 5% unless at least five years historical experience under similar environmental use indicates proper potential durability. Aggregates should be free of silt and organic matter. Permissible tolerances for organic content of ASTM C 33, Standard Specifications for Concrete Aggregates, should be reduced by one-half with the further requirement that an organic test for coarse aggregate should exhibit a color no darker than one-third the standard reference color. Lightweight aggregates employed in exposed face mixes should conform to tolerances of ASTM C 330, Standard Specifications for Lightweight Aggregates for Structural Concrete, with the further proviso that their absorption, as defined in ASTM C 330, should not exceed 11%. It is not recommended that they be used in cold or humid climates (if exposed to the weather) unless their performance has been verified by tests or records of previous satisfactory usage in similar environments.

2.01.A.5.c Specify type of stone desired such as crushed marble, quartz, limestone, granite, or locally available gravel as well as color. Some lightweight aggregates, limestones, and marbles may not be acceptable as facing aggregates. Omit where sample is to be matched.

2.01.A.5.d State required sieve analysis. Omit where sample is to be matched.

GUIDE SPECIFICATIONS	NOTES TO SPECIFIERS

6. Backup Concrete Aggregates:
 	
  a. Provide fine and coarse aggregates from a single source (pit or quarry) for entire job. They shall be clean, hard, strong, durable, and inert, free of staining or deleterious material.
 	
  b. ASTM C 33 or C 330 (CSA A231).
	2.01.A.6 Delete when production requirements dictate that face mix be used throughout the panel.

7. Water: Free from deleterious matter that may interfere with the color, setting or strength of the concrete.	*2.01.A.7 Potable water is ordinarily acceptable.*
B. Reinforcing steel: 1. Materials:	*2.01.B.1 Grades of reinforcing steel are determined by the structural design of the precast concrete units. Panels are designed as crack free sections so benefit of higher grade steel is not utilized. Refer to Manual, Section 4.4.*

a. Bars: (CSA G30 series)
 	
   (1) Deformed steel:
    ASTM A 615, grade 60.
	2.01.B.1.a State plain, galvanized or epoxy coated. Use galvanizing or epoxy coating only where corrosive environment or severe exposure conditions justify extra cost. Availability of galvanized or epoxy coated bars should be verified. Refer to Manual, Sections 4.4.3 and 4.4.7.

(2) Weldable deformed steel: ASTM A 706.	*2.01.B.1.a(2) Availability should be checked. When not available establish weldability in accordance with AWS D1.4.*
(3) Galvanized reinforcing bars: ASTM A 767.	*2.01.B.1.a(3) Damage of the coating as a result of bending should be repaired with zinc-rich paint.*
(4) Epoxy coated reinforcing bars: ASTM A 775.	
b. Welded wire fabric: (CSA G30 series) (1) Welded steel: ASTM A 185.	*2.01.B.1.b Should be sheets, not rolls. State plain, galvanized or epoxy coated. Use galvanized or epoxy coated only where corrosive environment or exposure conditions justify extra cost. Refer to Manual, Section 4.4.2.*
(2) Welded deformed steel: ASTM A 497.	
(3) Epoxy coated welded wire fabric: ASTM A 884.	
c. Fabricated steel bar or rod mats: ASTM A 184.	
d. Prestressing strand: ASTM A 416 (CSA G279), grade _____ .	*2.01.B.1.d Occasionally used in long and/or thin panels to control cracking from handling or service load conditions. Grades 250 or 270. Refer to Manual, Section 4.4.4.*
C. Cast-in anchors: 1. Materials:	*2.01.C Loose attachment hardware usually specified under Miscellaneous Metals. Refer to Manual, Section 4.5.9.*
a. Structural steel: ASTM A 36.	*2.01.C.1.a For carbon steel connection assemblies.*
b. Stainless steel: ASTM A 666, type 304, grade _____ .	*2.01.C.1.b Stainless steel anchors for use only when resistance to staining merits extra cost. (A), (B).*
c. Carbon steel plate: ASTM A 283, grade _____ .	*2.01.C.1.c (A), (B), (C), (D).*
d. Malleable iron castings: ASTM A 47, grade _____ .	*2.01.C.1.d Usually specified by type and manufacturer. Grades 32510 or 35018.*
e. Carbon steel castings: ASTM A 27, grade 60-30.	*2.01.C.1.e For cast steel clamps.*

f. Bolts: ASTM A 307 or A 325.

g. Welded headed studs:
AWS D1.1, Chapter 4, Part F.

2. Finish:

a. Shop primer: FS TT-P-86, oil base paint, type I, or SSPC-Paint 14, or manufacturer's standard.

b. Galvanized: hot dip galvanized (ASTM A 153 or CSA G164), electroplated, or metallized.

c. Cadmium coating: ASTM A 165.

d. Zinc rich coating: MIL-P-21035, self curing, one component, sacrificial organic coating.

D. Receivers for flashing: 28 ga. formed _____ _____ , or polyvinyl chloride extrusions.

E. Sandwich panel insulation: _____ .

F. Grout:

1. Cement grout: Portland cement, sand, and water sufficient for placement and hydration.

2. Nonshrink grout: Premixed, packaged ferrous and non-ferrous aggregate shrink-resistant grout.

3. Epoxy-resin grout: Two-component mineral-filled epoxy-polysulfide, FS MMM-G-560 _____ , type _____ , grade C.

G. Bearing Pads:

1. Chloroprene (Neoprene): Conform to Division II, Section 25 of AASHTO Standard Specifications for Highway Bridges.

2. Random oriented fiber reinforced: Shall support a compressive stress of 3000 psi with no cracking, splitting or delaminating in the internal portions of the pad. One specimen shall be tested for each 200 pads used in the project.

3. Duck layer reinforced pad: Conform to Division II, Section 10.3.12 of AASHTO Standard Specifications for Highway Bridges, or Military Specification, MIL-C-882C.

4. Plastic: Multi-monomer plastic strips shall be non-leaching and support construction loads with no visible overall expansion.

2.01.C.1.f For low-carbon steel bolts, nuts and washers.

2.01.C.2 Galvanizing or other protection is required for corrosive environments and if humidity may affect hidden connections. Refer to Manual, Section 4.5.7.

2.01.C.2.a For exposed carbon steel anchors.

2.01.C.2.b For exposed carbon steel anchors where corrosive environment justifies the additional cost.

2.01.C.2.c Particularly appropriate for threaded fasteners.

2.01.C.2.d 95% pure zinc in dried film. For field spot painting.

2.01.D (stainless steel), (copper), (zinc). Coordinate with flashing specification to avoid dissimilar metals. Delete when included in Flashing and sheet metal section. Specify whether precaster or others furnish.

2.01.E Specify type of insulation such as foamed plastic (polystyrene and polyurethane), glasses (foamed glass and fiberglass), foamed or cellular lightweight concretes, or lightweight mineral aggregate concretes. Thickness of sandwich panel insulation governed by wall U-value requirements. Refer to Manual, Section 5.3.

2.01.F Indicate required strengths on contract drawings.

2.01.F.2 Grout permanently exposed to view should be non-oxidizing (non-ferrous).

2.01.F.3 Check with local suppliers to determine availability and types of epoxy-resin grouts.

2.01.G.1 AASHTO grade pads having a minimum durometer hardness of 50 and utilizing 100 percent chloroprene as the elastomer. Less expensive commercial grade pads are available, but are not recommended.

2.01.G.2 Standard guide specifications are not available for random-oriented, fiber-reinforced pads. Proof testing of a sample from each group of 200 pads is suggested. Normal service load stresses are 1500 psi, so the 3000 psi test load provides a factor of 2 over service stress. The shape factor for the test specimens should not be less than 2.

2.01.G.4 No standard guide specifications are available. Compression stress in use is not normally over a few hundred psi and proof testing is not considered necessary.

| GUIDE SPECIFICATIONS | NOTES TO SPECIFIERS |

GUIDE SPECIFICATIONS

5. Tetraflouroethylene (TFE) reinforced with glass fibers and applied to stainless or structural steel plates.

2.02 Mixes

A. Concrete properties:

1. Water-cement ratio: maximum 40 lbs of water to 100 lbs of cement.

2. Air entrainment: Amount produced by adding dosage of air entraining agent that will provide 19% ± 3% of entrained air in standard 1:4 sand mortar as tested according to ASTM C 185; or minimum 3%, maximum 6%.

3. Coloring agent: Not more than 10% of cement weight.

4. 28 day compressive strength: Minimum of 5000 psi when tested by 6 × 12 or 4 × 8 in. cylinders; or minimum 6250 psi when tested on 4 in. cubes.

B. Facing mix:

1. Minimum thickness of face mix after consolidation shall be at least one inch or a minimum of 1 1/2 times the maximum size of aggregate used; whichever is larger.

2. Water-cement and cement-aggregate ratios of face and backup mixes shall be similar.

C. Design mixes to achieve required strengths shall be prepared by independent testing facility or qualified personnel at precast concrete manufacturer's plant.

2.03 Fabrication

A. Manufacturing procedures shall be in general compliance with PCI MNL-117.

B. Finishes:

NOTES TO SPECIFIERS

2.01.G.5 ASTM D 2116 applies only to basic TFE resin molding and extrusion material in powder or pellet form. Physical and mechanical properties must be specified by naming manufacturer or other methods.

2.02 Refer to Manual, Section 3.2.6.

2.02.A The backup concrete and the surface finish concrete can be of one mix design, depending upon resultant finish, or the surface finish (facing mix) concrete can be separate from the backup concrete. Clearly indicate specific requirements or allow manufacturer's option.

2.02.A.1 Keep to a minimum consistent with strength and durability requirements and placement needs.

2.02.A.2 Gradation characteristics of most facing mix concrete will not allow use of a given percentage of air. PCI recommends a range of air entraining be stated in preference to specified percentage.

2.02.A.3 Amount used should not have any detrimental effects on concrete qualities. Delete if coloring agent is not required.

2.02.A.4 Vary strength to match requirements. In some special cases, strength below 5000 psi may be accepted. Strength requirements for facing mixes and backup mixes may differ. Also the strength at time of removal from the molds should be stated if critical to the engineering design of the units. The strength level of the concrete should be considered satisfactory if the average of each set of any three consecutive cylinder strength tests equals or exceeds the specified strength and no individual test falls below the specified value by more than 500 psi.

2.02.B Delete if separate face mix is not used.

2.02.B.1 Minimum thickness should be sufficient to prevent bleeding through of the backup mix and should be at least equal to specified minimum cover of reinforcement.

2.02.B.2 Similar behavior with respect to shrinkage is necessary in order to avoid undue bowing and warping.

2.02.C Proportion mixes by either laboratory trial batch or field experience methods using materials to be employed on the project for each type of concrete required. Tests will be necessary on all mixes including face, backup, and standard, which may be used in production of units. Water content should remain as constant as possible during manufacture.

2.03.B Finishing techniques used in individual plants may vary considerably from one part of the continent to another, and between individual plants. Many plants have developed specific techniques supported by skilled operators or special facilities. Refer to Manual, Section 3.5.

1. Exposed face to match approved sample or mockup panel.

2.03.B.1 Preferable to match sample rather than specify method of exposure. Refer to Manual, Section 3.2.2.

* * OR * *

1. Smooth finish:

   a. As cast using flat smooth non-porous molds.

2.03.B.1.a Difficult to obtain uniform finish. Refer to Manual, Section 3.5.2.

* * OR * *

1. Smooth finish:

   a. As cast using fluted, sculptured, board finish or textured form liners.

2.03.B.1a Many standard shapes of form liners are readily available. Refer to Manual, Section 3.5.4.

* * OR * *

1. Textured finish:

   a. Achieve finish on face surface of precast concrete units by form liners applied to inside of forms.

   b. Distress finish by breaking off portion of face of each flute.

2.03.B.1.b Delete if distressed finish is not desired. Refer to Manual, Section 3.5.8.

   c. Achieve uniformity of cleavage by alternately striking opposite sides of flute.

* * OR * *

1. Exposed aggregate finish:

   a. Apply even coat of retardant to face of mold.

   b. Remove units from molds after concrete hardens.

   c. Expose coarse aggregate by washing and brushing or lightly sandblasting away surface mortar.

   d. Expose aggregate to produce a _____ exposure.

2.03.B.1.d (light) (medium) (deep) Finishes obtained vary from light etch to heavy exposure, but must relate to the size of aggregates. Matrix can be removed to a maximum depth of one-third the average diameter of coarse aggregate but not more than one-half the diameter of smallest sized coarse aggregate. Refer to Manual, Section 3.5.3.

* * OR * *

1. Exposed aggregate finish:

   a. Immerse unit in tank of acid solution.

2.03.B.1.a Use reasonably acid resistant aggregate such as quartz or granite. Refer to Manual, Section 3.5.6.

* * OR * *

   a. Pressure spray with acid and hot water solution.

* * OR * *

   a. Treat surface of unit with brushes which have been immersed in acid solution.

GUIDE SPECIFICATIONS

    b. Protect hardware, connections and insulation from acid attack.

    c. Expose aggregate to produce a _____ exposure.

* * OR * *

1. Exposed aggregate finish:

    a. Use power or hand tools to remove mortar and fracture aggregates at the surface of units (bushhammer).

* * OR * *

1. Sandblasted finish:

    a. Sandblast away cement-sand matrix to produce a _____ exposure.

* * OR * *

1. Exposed aggregate finish:

    a. Hand place large facing aggregate, fieldstone, or cobblestones in sand bed over mold bottom.

    b. Produce mortar joints by keeping cast concrete 1/2 in. to 1 in. from face of unit.

* * OR * *

1. Honed or polished finish:

    a. Polish surface by continued mechanical abrasion with fine grit, followed by special treatment which includes filling of all surface holes and rubbing.

* * OR * *

1. Veneer-faced finish:

    a. Cast concrete over brick, tile, terra cotta or natural stone placed in the bottom of the mold.

    b. Connection of natural stone face material to concrete shall be by mechanical means.

* * * * * *

NOTES TO SPECIFIERS

2.03.B.1.c (light) (medium) (deep)

2.03.B.1.a Use with softer aggregates such as dolomite and marble. Refer to Manual, Section 3.5.7.

2.03.B.1 (light) (medium) (deep) Exposure of aggregate by sandblasting can vary from 1/16 in. or less to over 3/8 in. Remove matrix to a maximum depth of one-third the average diameter of coarse aggregate but not more than one-half the diameter of smallest sized coarse aggregate. Depth of sandblasting should be adjusted to suit the aggregate hardness and size. Refer to Manual, Section 3.5.5.

2.03.B.1.a Refer to Manual, Section 3.5.9.

2.03.B.1 Honing and polishing of concrete are techniques which require highly skilled personnel. Use with aggregates such as marble, onyx, and granite. Refer to Manual, Section 3.5.11.

2.03.B.1.a Full scale mockup units with natural stone in actual production sizes, along with casting and curing of the units under realistic production conditions are essential for each new or major application or configuration of the natural stones. Bowing should be carefully measured over several weeks in the normal storage area and the final details of stone sizes and fastening determined to suit the observed behavior. Refer to Manual, Sections 3.5.10 and 3.5.12.

2.03.B.1.b Provide a complete bondbreaker between the natural stone face material and the concrete. Ceramic tile, brick, and terra cotta are bonded to the concrete. Refer to Manual, Sections 3.5.10 and 3.5.12.

GUIDE SPECIFICATIONS

2. _____ back surfaces of precast concrete units after striking surfaces flush to form finish lines.

C. Cover:

1. Provide at least 3/4 in. cover for reinforcing steel.

2. Do not use metal chairs, with or without coating, in the finished face.

3. Provide embedded anchors, inserts, plates, angles and other cast-in items as indicated on shop drawings with sufficient anchorage and embedment for design requirements.

D. Molds:

1. Use rigid molds to maintain units within specified tolerances conforming to the shape, lines and dimensions shown on the approved shop drawings.

2. Construct molds to withstand vibration method selected.

E. Concreting:

1. Convey concrete from the mixer to place of final deposit by methods which will prevent separation, segregation or loss of material.

2. Consolidate all concrete in the mold by high frequency vibration, either internal or external or a combination of both, to eliminate unintentional cold joints, honeycomb and to minimize entrapped air on vertical surfaces.

F. Curing:

1. Precast concrete units shall be cured until the compressive strength is high enough to ensure that stripping does not have an effect on the performance or appearance of the final product.

NOTES TO SPECIFIERS

2.03.B.2 (Smooth float finish), (Smooth steel trowell), (Light broom), (Stippled finish). Use for exposed back surfaces of units.

2.03.C.1 Increase cover requirements when units are exposed to corrosive environment or severe exposure conditions. For exposed aggregate surfaces, the 3/4 in. cover should be from bottom of aggregate reveal to surface of steel. It is extremely important that all reinforcement and hardware items are accurately located in the mold and that they be maintained in this position while the concrete is placed and consolidated. Reinforcing cages must be sufficiently rigid to prevent dislocation during consolidation in order to maintain the required cover over the reinforcement.

2.03.C.2 If possible, reinforcing steel cages should be supported from the back of the panel, because spacers of any kind are likely to mar the finished surface of the panel. For smooth cast facing, stainless steel chairs may be permitted. The wires should be soft stainless steel and clippings should be completely removed from the mold.

2.03.D.2 Molds for architectural precast concrete should be built to provide proper appearance, dimensional control and tightness. They should be sufficiently rigid to withstand pressures developed by plastic concrete, as well as the forces caused by consolidation. A detailed check of the initial unit from the mold is further verification of the adequacy of the mold. Unless otherwise agreed in the contract documents, the molds are the property of the precast concrete manufacturer.

2.03.E.2 There are a number of ways to place, consolidate and finish concrete with varying degrees of hand and machine operation. The specific technique should be left to the precast concrete manufacturer since the final assessment of any procedure is the quality of the end product. The prime objective is to consolidate the concrete thoroughly, producing a dense, uniform product with fine surfaces, free of imperfections. Bonding between backup and face mix should be assured if backup concrete is cast before the face mix has attained its initial set.

2.03.F A wide variation exists in acceptable curing methods, ranging from no curing in some warm humid areas, to carefully controlled moisture-pressure-temperature-curing. Consult with local panel manufacturers to avoid unrealistic curing requirements.

2.03.F.1 Stripping strength should be set by the plant based on the characteristics of the product and plant facilities. It is the responsibility of the precaster to verify and document the fact that final design strength has been reached.

GUIDE SPECIFICATIONS

G. Panel identification:

 1. Mark each precast concrete panel to correspond to identification mark on shop drawings for panel location.

 2. Mark each precast concrete panel with date cast.

H. Acceptance:
 Architectural precast concrete units which do not meet the color and texture range or the dimensional tolerances may be rejected at the option of the architect, if they cannot be satisfactorily corrected.

2.04 Concrete Testing

A. Make one compression test at 28 days for each day's production of each type of concrete.

B. Specimens:

 1. Provide two test specimens for each compression test.

 2. Obtain concrete for specimens from actual production batch.

 3. 6 in. × 12 in. or 4 in. × 8 in. concrete test cylinder, ASTM C 31.

 * * OR * *

 3. _____ sized concrete cube,_____
 _____ .

 * * * * * *

 4. Cure specimens using the same methods used for the precast concrete units until the units are stripped, then moist cure specimens until test.

C. Keep quality control records available for the architect upon request for two years after final acceptance.

3. EXECUTION

3.01. Inspection

A. Before erecting architectural precast concrete, the general contractor shall verify that structure and anchorage inserts not within tolerances required to erect panels have been corrected.

B. Determine field conditions by actual measurements.

NOTES TO SPECIFIERS

2.03.H Refer to Manual, Section 3.5.20.

2.04.A This test should be only a part of an in-plant quality control program.

2.04.B.1 One test specimen may be used to check the stripping strength.

2.04.B.3 Specify size. Cube specimens are usually 4 in. units, but 2 in. or 6 in. units are sometimes required. Larger specimens give more accurate test results than smaller ones. Test results will be lower for 2 in. cubes. Source: (molded individually), (sawed from slab).

2.04.C These records should include mix designs, test reports, inspection reports, member identification numbers along with date cast, shipping records and erection reports.

3.01.B. Any discrepencies between design dimensions and field dimensions which could adversely affect installation in strict accordance with the contract documents should be brought to the general contractor's attention. If such conditions exist, installation should not proceed until they are corrected or until design requirements are modified. Beginning of installation can mean acceptance of existing conditions. Erector should set out joint locations prior to actual product installation. Refer to Manual, Section 4.1.4.

GUIDE SPECIFICATIONS

3.02 Erection

A. Clear, well-drained unloading areas and road access around and in the structure (where appropriate) shall be provided and maintained by the general contractor to a degree that the hauling and erection equipment for the architectural precast concrete products are able to operate under their own power.

B. General contractor shall erect adequate barricades, warning lights or signs to safeguard traffic in the immediate area of hoisting and handling operations.

C. Set precast concrete units level, plumb, square and true within the allowable tolerances. General contractor shall provide true, level bearing surfaces on all field placed concrete which are to receive precast concrete units. General contractor shall be responsible for providing lines, center and grades in sufficient detail to allow installation.

D. Provide temporary supports and bracing as required to maintain position, stability and alignment as units are being permanently connected.

E. Non-cumulative tolerances for location of precast concrete units shall be as follows:

Plan location from building grid datum .... ±½ in.
Plan location from centerline of steel .... ±½ in.

Top elevation from nominal top elevation
    Exposed individual panel .... ±¼ in.
    Nonexposed individual panel .... ±½ in.
    Exposed relative to adjacent panel .... ¼ in.
    Nonexposed relative to adjacent panel .... ½ in.

Support elevation from nominal elevation
    Maximum low .... ½ in.
    Maximum high .... ¼ in.

Maximum plumb variation over height of structure or 100 ft whichever is less .... 1 in.

Plumb in any 10 ft of element height .... ¼ in.

Maximum jog in alignment of matching edges ..... ¼ in.

Joint width (governs over joint taper) .... ±¼ in.

Joint taper maximum .... ⅜ in.
Joint taper over 10 ft length .... ¼ in.

Maximum jog in alignment of matching faces ...... ¼ in.

Differential bowing or camber, as erected, between adjacent members of the same design ...... ¼ in.

F. Set non-loadbearing units dry without mortar, attaining specified joint dimension with lead, steel, plastic or asbestos cement spacing shims.

G. Fasten precast concrete units in place by bolting or welding, or both, completing drypacked joints, grouting sleeves and pockets, and/or placing cast-in-place concrete joints as indicated on approved erection drawings.

NOTES TO SPECIFIERS

3.02.A   Sequence of erection of precast concrete units and access to various portions of the structure can be a dominant factor in determining the cost of the structure. The erector should have uninterrupted access to the structure during erection including personnel access to all floors. Power should be supplied for welding equipment and small tools. Arrangements should be made with the general contractor in advance to eliminate conflicts with other trades. General contractor should coordinate delivery and erection of precast concrete products with other jobsite operations. Refer to Manual, Section 4.2.9.

3.02.C   Controlled reference lines should be used because the characteristics of precast concrete make a surface elevation difficult to define. Where thickness is not of exact concern, lines used in erection should be controlled from exposed exterior precast concrete surfaces. Refer to Manual, Section 4.6.3.

3.02.E   It is recommended that the architectural precast concrete industry standard tolerances in Manual, Section 4.6.3 be included in the specifications or referenced. Final erection tolerances should be verified and agreed on when erection commences and—if different than those originally planned—stated in writing or noted on erection drawings. When precast concrete units require adjustment beyond design or tolerance criteria, precaster should discontinue affected work and advise architect. Matching precast concrete with other building elements at the construction site should involve the teamwork of different suppliers and architect/engineer to achieve an acceptable result in the form of optimum uniformity.

3.02.F   Shims should be near the back of the unit to prevent their causing spall on face of unit if shim is loaded. The selection of the width and depth of field-molded sealants, for the computed movement in a joint, should be based on the maximum allowable strain in the sealant. Refer to Manual, Sections 4.5.3 and 4.7.5.

3.02.G   The erector shall protect units from damage caused by field welding or cutting operations and provide noncombustible shields as necessary during these operations. Structural welds should be made in accordance with the erection drawings which should clearly specify type, extent, sequence and location of welds. If galvanizing is specified, all weld areas and exposed or accessible steel anchorage devices should be given a coat of liquid galvanizing (95% zinc-rich paint) immediately after cutting or welding (and slagging). Adjustments or changes in connections, which could involve additional stresses in the products or connections, should not be permitted without approval by the architect/engineer. Units should be erected in sequence indicated on the erection drawings. Refer to Manual, Section 4.5.8.

H. Temporary lifting and handling devices cast into the precast concrete units shall be completely removed or, if protectively treated, removed only where they interfere with the work of any other trade.

3.03 Repair

A. Repair exposed exterior surface to match color and texture of surrounding concrete and to minimize shrinkage.

3.03.A A certain amount of product repair is to be expected to patch minor spalls and chips. Repair is normally accomplished prior to final cleaning and caulking. It is recommended that the precaster execute all repairs or approve the methods proposed for such repairs by other qualified personnel. The precaster should be compensated for repairs of any damage caused by others. Repairs should be acceptable providing the structural adequacy of the product and the appearance is not impaired. Repairs can be made at any age of the concrete, provided suitable mixtures are used, proper bonding is attained, and reasonable curing is possible. A considerable trial period may be involved in determining a suitable mixture. The precast concrete manufacturer should develop appropriate repair mixtures and techniques during the production sample approval process. Refer to Manual, Sections 3.2.4 and 3.5.21.

B. Adhere large patch to hardened concrete with bonding agent.

3.03.B Bonding agent should not be used with small patches.

3.04 Cleaning

A. After installation and joint treatment: _____ shall clean soiled precast concrete surfaces with detergent and water, using fiber brush and sponge, and rinse thoroughly with clean water in accordance with precast concrete manufacturer's recommendation. Use cleaning materials or processes which will not change the character of exposed concrete finishes.

3.04.A State whether erector or precaster should do cleaning under the responsibility of general contractor. Refer to Manual, Section 3.6.14.

* * OR * *

A. Clean precast concrete panels with _____ .

* * * * * *

B. Use acid solution only to clean particularly stubborn stains after more conservative methods have been tried unsuccessfully.

3.04A (acid-free commercial cleaners), (steam cleaning), (water blasting), (sandblasting). Use sandblasting only for units with original sandblasted finish. Ensure that materials of other trades are protected when cleaning panels. If at all possible, cleaning of concrete should be done when the temperature and humidity allow rapid drying. Long drying periods increase the possibility of efflorescence and discoloration.

C. Use extreme care to prevent damage to precast concrete surfaces and to adjacent materials.

D. Rinse thoroughly with clean water immediately after using cleaner.

GUIDE SPECIFICATIONS	NOTES TO SPECIFIERS

3.05 Protection

A. All work and materials of other trades shall be adequately protected by the erector at all times.

3.05.A Particular care should be exercised to avoid damage to glass. Etching of glass may occur due to welding, cutting or other construction operations. Many glass suppliers recommend glass cleaning every two weeks during construction period following installation.

B. A fire extinguisher, of an approved type and in operating condition, shall be located within reach of all burning and welding operations at all times.

C. The erector shall be responsible for any chipping, spalling, cracking or other damage to the units after delivery to the jobsite unless damage is caused in site storage by others. After installation is completed, any further damage shall be the responsibility of the general contractor.

3.05.C After erection of any portion of precast concrete work to proper alignment and appearance, the general contractor should make provisions to protect all precast concrete from damage and staining.

3.06 Warranty

A. The precast concrete manufacturer shall guarantee the precast concrete products against defects in materials and workmanship for a period of one year after acceptance of the units by the owner.

3.06.A The owner or designated agent should be required to accept the precast concrete products within 30 days after completion of erection (not 30 days after building project is completed) to preclude inordinate delays that would extend the warranty period unfairly.

6.5 PAYMENTS

The following should be placed in the General Conditions section of the project specification and renumbered for that section:

A.1 Payments

A.1.1 Monthly progress payments equal to 90% of the in-plant value will be made to the precaster for all products fabricated and stocked in the precaster's plant prior to delivery. Progress payments shall not relieve the precaster from compliance with terms of the contract with the buyer.

A.1.1 Industry practice is to fabricate, in advance, products for each individual job in accordance with design requirements and dimensions for that job. Such products cannot normally be used on any other project. As monthly payment would be made for such products if they were fabricated on the site, monthly progress payments for such material fabricated off site and stored for delivery and erection as scheduled is justified.

A.1.2 Full payment for all products delivered and/or installed will be made within 35 days of completion of all work under the contract and acceptance by the architect.

METRIC CONVERSIONS

Conversion to International System of Units (SI)

To convert from	to	multiply by
Length		
inch (in.)	millimeter (mm)	25.4
inch (in.)	meter (m)	0.0254
foot (ft)	meter (m)	0.3048
yard (yd)	meter (m)	0.9144
Area		
square foot (sq ft)	square meter (sq m)	0.09290
square inch (sq in.)	square millimeter (sq mm)	645.2
square in (sq in.)	square meter (sq m)	0.0006452
square yard (sq yd)	square meter (sq m)	0.8361
Volume		
cubic inch (cu in.)	cubic meter (cu m)	0.00001639
cubic foot (cu ft)	cubic meter (cu m)	0.02832
cubic yard (cu yd)	cubic meter (cu m)	0.7646
gallon (gal) Can. liquid*	liter	4.546
gallon (gal) Can. liquid*	cubic meter (cu m)	0.004546
gallon (gal) U.S. liquid*	liter	3.785
gallon (gal) U.S. liquid*	cubic meter (cu m)	0.003785
Force		
pound (lb)	kilogram (kgf)	0.4536
pound (lb)	newton (N)	4.448
Pressure or Stress		
pound/square foot (psf)	kilopascal (kPa)**	0.04788
pound/square inch (psi)	kilopascal (kPa)**	6.895
pound/square inch (psi)	megapascal (MPa)**	0.006895
pound/square foot (psf)	kilogram/square meter (kgf/sq m)	4.882
Mass		
pound (avdp)	kilogram (kg)	0.4536
ton (short, 2000 lb)	kilogram (kg)	907.2
ton (short, 2000 lb)	tonne (t)	0.9072
Mass (weight) per Length		
pound/linear foot (plf)	kilogram/meter (kg/m)	1.488
pound/linear foot (plf)	newton/meter (N/m)	14.593
Mass per volume (density)		
pound/cubic foot (pcf)	kilogram/cubic meter (kg/cu m)	16.02
pound/cubic yard (pcy)	kilogram/cubic meter (kg/cu m)	0.5933
Bending Moment or Torque		
inch-pound (in.-lb)	newton-meter	0.1130
foot-pound (ft-lb)	newton-meter	1.356
Temperature		
degree Fahrenheit (deg. F)	degree Celsius (C)	$t_C = (t_F - 32)/1.8$
Energy		
British thermal unit (Btu)	joule (j)	1056
Other		
Coefficient of heat transfer (Btu/ft²/h/°F)	W/m²/°C	5.678
Modulus of elasticity (psi)	MPa	0.006895
Thermal conductivity (Btu-in./ft²/h/°F)	Wm/m²/°C	0.1442
Thermal expansion in./in./°F	mm/mm/°C	1.800
Area/length (in.²/ft)	mm²/m	2116.80
ounces/square foot	kilogram/square meter	0.305
ounces/cubic yard	kilogram/cubic meter (kg/cu m)	0.0371

*One U.S. gallon equals 0.8321 Canadian gallon.   **A pascal equals one newton/square meter

INDEX BY SUBJECTS

LIST OF ILLUSTRATED BUILDINGS
AND ARCHITECTURAL CREDITS